T5-CUL-924

Organic Polarography

Petr Zuman and Charles L. Perrin

INTERSCIENCE PUBLISHERS

a division of

JOHN WILEY & SONS, *New York • London • Sydney • Toronto*

The paper used in this book has pH of 6.5 or higher. It has been used because the best information now available indicates that this will contribute to its longevity.

Reprinted in full from

Advances in Analytical Chemistry and Instrumentation,
C. N. Reilley, Editor, Volume 2, pages 219–253.
Copyright © 1963, by John Wiley & Sons, Inc.

Progress in Physical Organic Chemistry,
S. G. Cohen, A. Streitwieser, and R. W. Taft, Editors,
Volume 3, pages 165–315.
Copyright © 1965, by John Wiley & Sons, Inc.

Progress in Physical Organic Chemistry,
A. Streitwieser and R. W. Taft, Editors,
Volume 5, pages 81–206.
Copyright © 1967, by John Wiley & Sons, Inc.

10 9 8 7 6 5 4 3 2 1
SBN 471 985902

Printed in the United States of America

Preface

Interest in organic polarography shows a steady increase, but, although organic chemists have become accustomed to routine utilization of I.R., N.M.R. and mass spectra and of gas chromatography, the use of polarography by the organic chemist remains rather the exception than the rule. Even though polarography cannot compete with sophisticated spectrometric techniques, there are some fields of organic chemistry, where the polarographic method can offer information that is accessible with difficulty—if at all—by other physical methods. Because the polarography of organic compounds reveals the dynamic properties of organic molecules, it is a useful tool in quantitative expressions of reactivities. Under appropriate conditions, organic electrode reactions follow a pattern similar to that of homogeneous organic reactions in solutions. The special feature of the electrochemical approach is that a single reagent—the electron—is involved in reactions with widely differing substrates and moreover that the energy of this reagent can be changed within certain limits. From this point of view the study of organic electrode processes can contribute to the study of organic mechanisms. Quantitative evaluation of substituent effects from polarographic data is simple and fast compared with isolation of analogous data from homogeneous kinetics measurements. And finally, because polarography enables continuous recording of concentrations of one or often more components of a reaction mixture, and because intermediates with half-times from 10 sec. to 10 min. can be identified and followed, polarography has proved to be a useful analytical tool in homogeneous kinetics. For all these applications it is essential to understand, at least in principle, the course of the organic electrode process involved.

One of the possible reasons for the limited application of polarography to the solution of problems or organic chemistry is that few organic chemists realize which problems can be solved by polarography and how the basic information can be obtained. In the absence of a modern textbook on this subject, we were therefore pleased by the suggestion made by our Publisher, that our contributions, published in *Progress in Physical Organic Chemistry* (S. G. Cohen, A. Streitwieser, and R. W. Taft, Eds.) and *Advances in Analytical Chemistry and Instrumentation* (C. N. Reilley, Ed.) should be reprinted. Two of these contributions deal with the techniques used in the collection of polarographic data and their interpretation; the third by C. L. Perrin gives a general picture of the type of organic electrode processes involved. Because of production techniques, it has been impossible to consider developments during the last couple of years. In the field of techniques important developments have been reported in the identification of intermediates and products by means of controlled potential electrolysis, rectangular voltage polarization (commutator methods), and studies of instantaneous current–time curves. Recently, frequent application has been made of techniques in which the potential is scanned for a given electrode surface (whereas d.c. polarography can be considered as a potentiostatic method), e.g., single sweep and triangular sweep (voltammetric) methods; these techniques can provide useful information, but it should be kept in mind that processes under potentiostatic and nonpotentiostatic conditions may differ.

In the field of individual organic electrode processes, a more detailed understanding of the individual steps has been achieved recently, and in some instances, schemes which were considered correct four years ago, are no longer valid.

Information about recent progress can be found in the biannual reviews by D. J. Pietrzyk in *Analytical Chemistry*.

We hope that even with the limitations imposed by rapid progress, this volume will contribute to a wider use of polarograph in organic chemistry.

P. Zuman

Contents

Some Techniques in Organic Polarography

PETR ZUMAN, *Polarographic Institute, Czechoslovak Academy of Science, Prague*

I. INTRODUCTION

According to the number of papers published, interest in organic polarography steadily increases. The problems discussed in most of these papers may be classified as follows:

(*a*) Description of new polarographically active groups, which undergo reduction or oxidation at the mercury dropping electrode, which form slightly soluble or complex compounds with mercury, or which produce catalytic electrode processes.

(*b*) Studies of environmental effects (solvent, supporting electrolyte, buffer composition, ionic strength, specific effects of cations and anions) on polarographic behavior.

(*c*) Studies of the correlation between structure, reactivity, and polarographic behavior of organic compounds.

(*d*) Studies of the mechanisms of electrode processes.

(*e*) Applications of polarographic curves to problems of pure and applied chemistry.

As these volumes are primarily concerned with analytical applications, only advances involving the last point (*e*) will be discussed here in some detail, although some of the problems belonging to other classes will be mentioned. Moreover, several of the problems under (*a*)–(*d*) have already been covered by recently published reviews (1–14).

The analytical applications of organic polarography can be divided into two large groups: direct and indirect methods.

The *direct methods* involve dissolution of the sample in an appropriate supporting electrolyte, recording of the current–voltage curve, and evaluation. These methods anticipate that the substance under study is soluble in the chosen supporting electrolyte and that it is electroactive under the conditions employed, i.e., that it gives a measurable wave on the current–voltage curve. These methods, of course, can be used only in those instances where the sample analyzed contains no interfering substances.

Considered as interfering are substances which are reduced at half-wave potentials differing only slightly from that of the substance to be analyzed. According to the slope of the waves involved and according to the relative concentrations of the particular components, from 0.1 to 0.2 v. is the smallest practical difference necessary when separate measurement of two adjacent waves is required.

Furthermore, interference occurs with those substances which are present in fivefold or greater excess than the substance to be determined and which are reduced at potentials more positive than the substance analyzed. On the other hand, an excess of a compound giving a reduction wave at more negative potentials (or an anodic wave at more positive potentials), even when present in several hundredfold excess, usually does not interfere in the analysis of a trace component reduced at more positive (or showing an anodic wave at more negative) potentials.

Elimination of interferences is sometimes possible with a proper choice of the supporting electrolyte. Changes in instrumentation (derivative and subtractive circuits, oscilloscopic methods, square-wave polarography, a.-c. polarography, etc.) help sometimes to resolve such coincidences and interferences. And finally, preliminary separations involving physical or chemical methods must sometimes be used in analysis of complex mixtures.

The fastest and simplest of these methods of resolution is the application of a suitable supporting electrolyte. It is therefore the method preferred whenever possible. In the past, the choice of supporting electrolyte has often been based on trial and error. A proper choice of the supporting electrolyte, which enables us to obtain well-developed waves and to eliminate the interferences, can nevertheless be made only if the polarographic behavior* is known and, if possible, also understood.

Indirect methods permit the determination of substances which do not exhibit a polarographic wave of practical applicability. Such polarographically inactive substances may sometimes be transformed by a chemical reaction into compounds that do exhibit useful polarographic waves in an appropriate supporting electrolyte. Alternatively, the substance to be determined may be allowed to react with a polarographically active reagent. The difference in the wave height of this reagent before and after the reaction is then proportional to the concentration of the substance to be analyzed.

To perfect an indirect method, a knowledge of the polarographic behavior of the electroactive substance is necessary just as for direct methods. Only when the polarographic behavior is known in sufficient detail is it possible to predict and to prevent difficulties which may arise for a given sample and for a chosen type of procedure.

Therefore, those aspects which need to be considered in a study of the polarographic behavior of a compound not previously investigated will be discussed first. As direct methods represent a routine treatment with possible difficulties experienced at times in the sample preparation (which is beyond the scope of this article), the interest is centered here on indirect methods, for these still offer numerous

* Polarographic behavior is a general expression for those changes of the form, heights, and half-wave potentials that occur with changes in pH, solvent, concentration of depolarizer and other components of the electrolyzed solution, with capillary constants, etc.

possibilities for application of the chemical knowledge and inventiveness of the chemist.

II. POLAROGRAPHIC BEHAVIOR

It is well understood that, because of the diversity of problems encountered in organic polarography, it is hardly possible to devise a general scheme of techniques which can be used in every case. In this section several techniques will be described which have been successfully applied to a number of systems.

1. Stock Solution and Supporting Electrolyte

The concentration range of depolarizer* recommended for both theoretical studies and practical applications is from 1 to 5×10^{-4} *M*. When necessary, this range can be extended to 1×10^{-5} to 1×10^{-3} *M*. At lower concentration, capacity current already presents difficulties; and at higher concentration, irregularities are sometimes observed. To prepare the solution for polarographic study in the above-mentioned concentration range, a small volume of a stock solution is usually added to an appropriate supporting electrolyte, which has been deaerated beforehand.† Solubility of the substance both in the stock solution and in the electrolyte mixture usually presents the first serious problem and one which may limit the polarographic study of a given organic substance.

Quantitative data concerning the solubility of less common organic substances are usually not available. Preliminary tests (e.g., using microscopic slides) provide us with rough but sometimes useful information, and the same may be said for remarks in papers or tables con-

* Depolarizer is a substance that alone or after a fast chemical reaction in the chosen supporting electrolyte undergoes electrolysis at the working electrode.

† The preliminary deaeration is important especially in cases where the depolarizer in a given supporting electrolyte is subjected to oxidation or where it undergoes another chemical reaction. After addition of a small volume of the stock solution, the inert gas is reintroduced for approximately 10–20 sec., followed by an immediate recording of the current–voltage curve. With this arrangement, a record of the polarographic curves can be obtained shortly after mixing and, moreover, the period of contact is well defined. As time changes occurring in a previously uninvestigated depolarizer, studied in a given supporting electrolyte, cannot be ruled out in advance, adoption of the above procedure as a general one is justified.

cerning solvents used for crystallization. As a result, the method of trial and error is usually employed. However, it should be stressed that, because of the great dilution used in organic polarography, many substances considered by organic chemists as completely insoluble in a given supporting electrolyte are sufficiently soluble for our purposes.

We usually try to prepare a $0.01M$ stock solution and, whenever possible, we prefer aqueous solution. This principle is substantiated by the fact that we have at least a limited knowledge of the electrochemistry of aqueous solutions. When the solubility in water is too low, mixtures of water with organic solvents are used. For lipophilic substances, non-aqueous solutions, melts (such as ammonium nitrate saturated with ammonia), or additions of detergents (15) are recommended.

When low solubility prevents the preparation of an aqueous $0.01M$ stock solution, we try to prepare a $1 \times 10^{-3}M$ stock solution, which is then mixed with an equal volume of the supporting electrolyte. When even this is insufficient, we try an organic solvent which is highly miscible with water, usually ethanol. Our aim is to prepare the stock solution (as well as the solution to be polarographed) with as little organic solvent as possible. A weighed amount of the substance is, therefore, dissolved in the organic solvent and water gradually added until turbidity occurs. 5 to 10% of excess of organic solvent is then used for the preparation of the stock solution; the same principle is followed in the preparation of the electrolyzed solution.

Whenever possible, we avoid application of heat in the preparation of the stock solution. When heating is necessary in order to accelerate the dissolution, it is necessary to check that the behavior of the "saturated" solution before and after heating is identical. Furthermore, it is important to control the stability of the stock solution over hours and over several days to ascertain the effect of diffuse daylight. When proved necessary, stock solutions should be prepared fresh. These factors are sometimes overlooked, with resulting confusion.

When low solubility or stability prevents the use of aqueous solutions or of mixtures of water with miscible organic solvents, non-aqueous solutions or solutions containing an excess of an organic solvent with a small percentage of water are used.

The most widely used solvents employed for this purpose are dioxane, dimethylformamide, acetonitrile, and Cellosolves. For aprotic substances, mixtures of methanol-benzene (1:1) are also recommended. Although application of these solvents involves some complications in the experimental techniques (e.g., an increase in the *iR* potential drop, definition of the reference electrode, liquid junction potentials, etc.), their use undoubtedly broadens the scope of organic polarography. Because solvation effects, which influence the behavior of depolarizers in aqueous solutions, may be absent in organic solvents and because some reactions, whose intermediates or products react in the presence of water, can be avoided in such solvents, it is sometimes claimed (16) that the polarographic behavior of organic substances in non-aqueous solvents is simpler than in water. Actually we can in some instances obtain valuable information, especially concerning electrode process proper,* when non-aqueous solutions are used. On the other hand, because of the limited solubility of electrolytes† under such conditions, and especially because of the difficulties involved in the definitions of acidity in non-aqueous solvents, we are less able to change over a wide range those parameters which help us to understand the course of the total electrode process. For this reason, and also because the present author is less acquainted with the work in non-aqueous solvents, the remaining discussion is limited to solutions in water and its mixtures with organic solvents.

An interesting concept based on the effect of detergents was introduced into polarography of slightly soluble substances by Proske (15). A detergent is added to an aqueous supporting electrolyte in order to increase either the solubility of the depolarizer or the solubility of a nonpolar solvent (like toluene) in water. The solubilized nonpolar solvent then helps to dissolve a slightly soluble depolarizer. The principal difficulty encountered with such techniques is caused by the substantial influence which surface-active substances, of which detergents are extreme examples, are able to exert on the entire polarographic behavior. Too little experience has been gained with such

* The total electrode process can be described as a sequence of preceding reactions (involving the transformations of the depolarizer into electroactive species), electrode process proper (a heterogeneous reaction involving among other steps the electron transfer), and succeeding reactions (involving transformation of the primary product of the electrode process proper).

† Lithium and tetraalkyl ammonium salts are usually more soluble and, therefore, more widely used in organic solvents. The latter salt often exhibits specific adsorption effects. It should be noted that several strong acids and bases are also soluble in these solvents, permitting at least the extreme points of the acidity scale to be reached.

"polarography in a gravy" for the technique to be generally recommended at present.

2. Limiting Currents

The stock solution having been prepared and the necessary content of the organic solvent estimated, preliminary examinations in some few typical supporting electrolytes are carried out. Usually we start with 0.1*M* sulfuric acid; 0.1*M* acetate buffer, pH 4.7; 0.1*M* phosphate buffer, pH 6.8; 0.05*M* borax (0.1*M* borate buffer), pH 9.3; and 0.1*M* NaOH. According to the character of the substance under study some other buffers are added to this line. For example, if α-hydroxyketones are studied, a Veronal buffer of pH 8.5, 0.1*M* trimethylamine with 0.1*M* trimethylammonium chloride, as well as 1*M* ammonia with 1*M* ammonium chloride are also used.

When the system under study is known or expected to be reduced at potentials more negative than about -1.9 v., 0.1*M* lithium hydroxide (for the range -1.8 to -2.0 v.) or 0.05*M* NR_4OH or a borate buffer prepared with tetraalkylammonium hydroxide are used. On the other hand, the five supporting electrolytes mentioned above allow us to detect even the most positive anodic waves available with the dropping mercury electrode. A prescribed volume of the stock solution is added to the deaerated supporting electrolytes, so as to produce a $2 \times 10^{-4}M$ solution. All supporting electrolytes are then inspected after the addition of the depolarizer to insure that no turbidity occurs. If turbidity is present, the concentration of the organic solvent is raised in all supporting electrolytes so that complete dissolution is assured in all solutions studied. We prefer to maintain the same concentration of organic solvents in all solutions to be compared.

The 2×10^{-4} molar solutions prepared in all five (or more) deaerated solutions are subjected to polarographic electrolysis. The resulting curves are then inspected to determine whether a cathodic or an anodic wave is present and, if so, in what pH region.*

* If a universal buffer is used first, a study of the whole pH region is usually carried out and the effect of the particular buffer components is not separated. Preliminary experiments of the kind described in the text show us immediately those pH regions where it is unnecessary to perform a more detailed study, i.e., where no wave exists at all. Since in some instances a reduction wave appears only in a narrow pH range, the curves in the above-mentioned buffers, where the intervals are 2 to 4 pH units, should be inspected carefully for ill-developed waves.

When a reduction wave is obtained, we check over 10- and 20-min. intervals to see whether this wave changes with time. Should the depolarizer undergo a chemical reaction in the given supporting electrolyte and the wave alter with time, such curves are then recorded always in a fresh solution. If the rate of reaction is rather fast, extrapolation to zero time (using three or four curves recorded at selected time intervals) is recommended.

After the time factor has been eliminated, the first pieces of information concerning the type of limiting current* are sought. In those supporting electrolytes, where polarographic waves were observed, curves are recorded at different concentrations of the depolarizer. A linear relationship between wave height and concentration of the depolarizer indicates that either a diffusion limiting current or a kinetic limiting current (limited by first-order kinetics)

* Four very important types of limiting currents are known in polarography:

(1) *Diffusion current* ($\bar{i}_d$), where the rate of transport of the depolarizer to the surface of the electrode is determined by its diffusion. The diffusion current follows the Ilkovič equation

$$\bar{i}_d = 0.627\, nFcD^{1/2}m^{2/3}t_1^{1/6}$$

(2) *Kinetic current* ($\bar{i}_k$), limited by the rate of a chemical reaction (whose formal rate constant is k_f and whose equilibrium constant is K) preceeding the electrode process proper. The equation

$$\bar{i}_k = 0.51 \cdot nFcD^{1/2}m^{2/3}t_1^{2/3}k_f^{1/2}K^{1/2}$$

is obeyed for fast reactions, i.e., those for which $(k_fK)^{1/2} < 0.05$ sec.$^{-1/2}$ Proton-transfer reactions, dehydration, and ring openings are the most familiar types of chemical reactions involved.

(3) *Catalytic current,* ($\bar{i}_c$) limited by catalytic effects of a polarographically inactive catalyst which causes a shift of the reduction potential of the polarographically active substance towards more positive potentials. Catalytic lowering of hydrogen overvoltage is the most common type of these currents.

(4) *Adsorption current,* ($\bar{i}_a$), caused by the adsorption of the oxidized or reduced form of the depolarizer at the surface of the electrode. The current, which is limited by the rate of the formation of the surface of the electrode covered by the adsorbed layer, follows the equation:

$$\bar{i}_a = 0.85\, nFzm^{2/3}t_1^{-1/3}$$

In the above equations n is the number of electrons transferred, F is the Faraday charge (96,484 coulombs), c is the concentration of the depolarizer, D is the diffusion coefficient, m is the outflow velocity of mercury, t_1 is drop time, and z is the number of moles of the substance adsorbed on 1 cm.^{2} of the surface of the electrode.

is involved. If the linear relationship is obtained even when the concentration is varied over a sufficiently broad range, adsorption and catalytic currents are unlikely to be involved.

On the other hand, when the height of the particular wave increases with increasing concentration of the depolarizer, but only in the concentration range below a certain limiting value, and when the height of the wave does not increase with further addition of the depolarizer above the limiting concentration, an adsorption or a catalytic current is involved.

When two or more waves are observed on polarographic curves, the effect of concentration of the depolarizer on the ratio of their heights is important. When the two waves correspond to two substances present in the sample, or to two forms present in a slowly established equilibrium, or when one electroactive form is formed in a fast first-order reaction from another which is reducible at more negative potentials, the ratio of the two waves remains constant and independent of the concentration of the depolarizer. On the other hand, a changing ratio of wave heights in well-buffered solutions shows that one of the waves is an adsorption or catalytic current or that kinetic reactions of higher order are involved.

Based on the results of these preliminary experiments, a more detailed pH dependence using universal buffers (usually Britton-Robinson or McIlvaine) is studied and the entire pH range where a polarographic wave appears is covered. The number of waves and the change of their heights with pH value are followed. Limiting diffusion-controlled current and limiting adsorption currents are usually pH independent. Kinetic and catalytic currents are usually pH dependent when the rate of the corresponding chemical reaction is pH dependent. Kinetic currents corresponding to a proton-transfer reaction decrease with increasing pH value and have the familiar S-shape form of a dissociation curve. Because the dehydration reactions are acid and base catalyzed, a bell-shaped pH dependence is often observed for this type of kinetic current. Catalytic currents corresponding to hydrogen evolution usually decrease sharply with increasing pH values.

When several waves are observed in the pH interval studied, the effect of concentration of the depolarizer, mentioned above, is followed for each of these waves. The study of the pH dependence helps us to locate those conditions where the additional criteria discussed below can be best applied for determining the type of the electrode process involved.

The effect of mercury pressure and capillary characteristics is studied next. In a very simple procedure, the height of the wave is compared with the height of the mercury column, h, the latter being measured between the level of mercury in the reservoir and the tip of the capillary (h_{exp}) and then corrected for the interfacial tension at the surface of the growing drops (h_{back}):

$$h = h_{\mathrm{exp}} - h_{\mathrm{back}}$$

where

$$h_{\mathrm{back}} = \frac{3.1}{m^{1/3}t^{1/3}}$$

This dependence is measured at pH values where the limiting wave height has reached its maximum value and at such pH values where the limiting wave height is only about 10% of this maximum value.

When the limiting current is proportional to the square root of mercury column ($\bar{i} \sim h^{1/2}$), a limiting diffusion current (or a catalytic current) is involved.*

A limiting current that is independent of mercury pressure ($\bar{i} \sim h^0$) is a limiting kinetic current (or catalytic current). The current is independent of h only when it corresponds to about 10% of the diffusion limiting current or less.

If the limiting current is a linear function of the height of mercury column ($\bar{i} \sim h^1$), this is taken as a proof that a limiting adsorption current is involved.†

Finally, a limiting current which increases with decreasing height of the mercury column is usually a catalytic or an autocatalytic current.

These tests are simple, but not always convincing. Sometimes it is

* All wave heights should be carefully corrected for capacity currents. It is recommended that a plot be constructed of $\bar{i} - h^{1/2}$ dependence as a small additive error arising from an improper measurement of the wave height is easily detected graphically, whereas, in numerical computation, in which the constant value of the product $\bar{i} \cdot h^{1/2}$ is proved, such additive errors can be misinterpreted. The axes should always include the zero values both of current $\bar{i}$ and of $h^{1/2}$ in order that a substantial intercept of the linear $\bar{i}$–$h^{1/2}$ plot on one of the axes may be detected.

† Such a dependence is observed only in the concentration range where the wave shows no further increase with increasing concentration of the depolarizer.

difficult to distinguish unambiguously which plot is best fitted by the experimental points. Several other tests were, therefore, devised. As the different types of limiting currents differ widely in their dependence on the drop time, t_1, the effect of t_1 is studied under conditions where mercury outflow velocity, m, is kept constant.*

A simple way to study the drop-time relationship is to measure the limiting currents and drop time with the capillary in the normal, vertical position ($\bar{i}'$, t_1') and with the same capillary in a horizontal position ($\bar{i}''$,t_1''). If the tip of the capillary is the same distance below the mercury level in the reservoir in both cases, the outflow velocity, m, remains unchanged and the change of the position of the capillary results only in a change of the drop time. For diffusion and kinetic currents the following relations then hold true:

$$\frac{\bar{i}_d'}{\bar{i}_d''} = \left(\frac{t_1'}{t_1''}\right)^{1/6} \qquad \frac{\bar{i}_k'}{\bar{i}_k''} = \left(\frac{t_1'}{t_1''}\right)^{2/3}$$

The effect of the drop time on the limiting current can be followed over a wider range when the drop time is artificially regulated, i.e., when the mercury drop is detached at preset time intervals using a mechanical device monitored usually by a relay system. The measured mean limiting current is then plotted graphically as a function of the drop time ($\bar{i}$ vs. t_1).

For diffusion currents, an approximately $^1/_6$-order parabola is obtained, and for kinetic currents a $^2/_3$-order parabola; for catalytic currents, different shapes of these plots are obtained. Most typical are the plots obtained for adsorption currents, which often show one or more maxima.

The same patterns are found when the changes of the instantaneous current with time are observed during the life of one drop (i–t curves). These curves are recorded preferably on the first drop (after the particular potential was applied), using an oscilloscope or a string galvanometer to follow the fast changes of the current. These curves are of the most fundamental importance in any attempt toward the understanding of electrode processes.

Again the i–t curves of diffusion-controlled currents show a $^1/_6$ parabola, currents limited by the rate of a chemical reaction follow a $^2/_3$ parabola, and adsorption currents give i–t curves with increasing

* Upon changing the height of the mercury column h, both t_1 and m are changed.

and decreasing portions. Autocatalytic currents show a steadily increasing i–t curve, sometimes with an induction period.

Most of the currents depend on $m^{2/3}$. A pronounced effect of the outflow velocity was observed for the maxima of second kind (round maxima on the limiting current), which usually increases substantially more than diffusion and adsorption current with the increase of mercury pressure, h.

An additional test is the comparison of the relative wave heights. Whereas adsorption currents at concentrations of $2 \times 10^{-4}M$ depolarizer are often comparable with diffusion currents obtained in equimolar solutions,* kinetic currents are sometimes (e.g., for formaldehyde, glucose) substantially smaller than hypothetical diffusion currents. Conversely, certain catalytic currents can be several hundred times as high as the corresponding diffusion currents. Thus, when no polarographic wave is observed in a solution of a substance which was believed to be reducible at the dropping mercury electrode, a substantially more concentrated solution should be polarographed (e.g., $0.1M$ solutions when solubility permits). Sometimes kinetic currents are observed in such solutions.

Certain kinetic currents can be detected from the effect of temperature. Diffusion currents increase with increasing temperature by about 1.8% deg.$^{-1}$. Limiting currents showing a more pronounced increase (of about 5–20% deg.$^{-1}$) are kinetic currents. But not all kinetic currents show this dependence. For catalytic and adsorption currents, the effect of temperature is not conclusive.†

Further evidence can be gained from the study of the effect of buffer composition and concentration on limiting currents. Diffusion and adsorption currents are usually little influenced by changes of buffer concentration or kind. Substantial changes are observed for such kinetic currents for which the chemical reaction involved is subjected to a general catalysis (17–22) and for several catalytic hydrogen evolutions (23,24). Usually the limiting current increases with in-

* Because of the independence of adsorption currents (and of the increase of diffusion currents) obtained with $1 \times 10^{-3}M$ and more concentrated solutions, adsorption currents are usually small compared with diffusion currents.

† Rate of the formation of adsorption layer and the adsorption equilibrium can be influenced in opposite directions with increasing temperature. Certain types of adsorption currents exhibit a smaller dependence on temperature than do diffusion currents.

creasing buffer concentration (at constant ionic strength) and depends on the pK_a value of the acid buffer component. Some kinetic currents corresponding to a proton transfer [e.g., of tropylium (25)] are independent of the buffer effects. In some cases (22,26), the increase of the limiting current is believed to involve not only the chemical reaction preceding the electrode reaction proper, but also the latter heterogeneous reaction.

Cations and anions of strong electrolytes present in the supporting electrolyte usually exert little influence on the normal limiting currents. An increase of ionic strength normally causes a small decrease in the limiting current due to the increase of viscosity influencing the diffusion coefficient, D. Pronounced effects were observed recently for calcium ions in the reduction of terephthalic acid (27). On the other hand, the decrease of current, observed sometimes on limiting currents of anions (28) and of some other compounds, e.g., some halogen derivatives, are usually very sensitive to the kind and charge of cations. In such cases the effect of neutral salts can be used to decide whether the electroactive species bears a positive or a negative charge.

Some large cations, like tetrabutylammonium ion, can act both as a charged particle and as a surface-active substance. The presence of a surface-active substance often distorts the shape of a polarographic wave.

Addition of certain surface-active substances (like the strongly adsorbable camphor) causes a decrease of the limiting current. Slower electrode processes are substantially retarded by a surface layer of an adsorbed substance. This effect can be applied for the detection of the fast step in the overall electrode process. For example, the addition of camphor and similar substances to mononitro compounds causes, at higher pH values, a decrease in the first four- or six-electron step to a value corresponding to a one-electron process (29). Hence, it is assumed that the uptake of the first electron is more mobile than the subsequent electrode processes. Kinetic currents are often particularly sensitive to the addition of a surface-active compound. With adsorption currents, concurrent adsorption of depolarizer and surface-active compound sometimes takes place, with resulting changes in the adsorption waves obtained.

Limiting currents are also affected by the solvent used (30). These effects are little understood so far. For diffusion-controlled currents the effect of the change in solvation, which in turn affects the values of diffusion coefficients, is usually considered. Currents limited by the rate of a chemical reaction are also altered because of changes both in the rate and in the equilibrium of the chemical reaction involved. Adsorption currents sometimes disappear after addition of an organic solvent to an aqueous solution; this is perhaps caused by the increased solubility of the adsorbate and the concurrent adsorption of the organic solvent on the mercury drop.

3. Half-Wave Potentials

Simultaneously with the study of changes in the limiting currents, the effect of the individual parameters on the half-wave potentials is also observed.

The effect of pH is discussed first. Most organic electrode processes are pH dependent. A pH-independent half-wave potential is observed for reversible redox systems* which do not involve proton transfer and for irreversible systems, in which no proton transfer occurs before the slow electron-transfer step. For most systems which exhibit pH dependence, the half-wave potentials are shifted towards more negative values with increasing pH values. The $E_{1/2}$–pH plot can usually be approximated by linear portions. The slope of these portions for a reversible system depends on the number of electrons (n_e) and on the number of protons (m_{H}) transferred during the electrode process. At 25°C. the slope $dE_{1/2}/d\mathrm{pH}$ is given† by $0.059 \times m_{\mathrm{H}}/n_e$. The intercepts of the linear parts correspond to dissociation constants of the reduced and oxidized form, respectively. If the two forms, corresponding to two adjacent linear portions, differ in the dissociation of the reduced form, the slope decreases at higher pH values. Dissociation of the oxidized form results in an increase of the $dE_{1/2}/d\mathrm{pH}$ slope.

For irreversible systems a very common pattern for the $E_{1/2}$–pH plot is: a pH-independent portion at low pH values, a pH region where the half-wave potential is shifted to more negative values (which can usually be approximated by one or more linear sections), and another pH-independent region at high pH values. Frequently, one of these regions is outside the available acidity range. For sys-

* The meaning of the term reversibility as used in polarography is explained in Part 5 of this section.

† For a system where dissociation occurs only for the reduced form.

tems where proton transfer precedes the slow electrode process, such behavior can be interpreted as follows: In acid solutions the electrolyzed substance is present in the bulk of the solution in a protonated form which is reducible. In the pH range where the half-wave potential is pH dependent, the substance is present in the bulk of the solution in the conjugate base form, but only reduction of the protonized acid form occurs. The protonation reaction to form the reducible species occurs in the vicinity of the electrode surface. When the proton-transfer reaction is not sufficiently fast, as at higher pH values, the reduction of the unprotonized form takes place in a pH-independent step. When the half-wave potentials of the protonated and unprotonated form at higher pH values differ sufficiently, two separate waves are obtained.

In addition to this quite common scheme, several other schemes explaining the pH dependence of the half-wave potentials have been described, involving, for example, proton-transfer in several steps, either preceding the slow electron transfer process (or involved in such a slow process) or following the first reversible electron transfer.

Shifts of half-wave potential with a change in the concentration of the depolarizer are usually not pronounced.* For reversible systems the half-wave potentials should be practically unaffected by a change in the depolarizer concentration unless the oxidized or reduced form dimerizes. The half-wave potentials of many irreversible systems are dependent on the concentration of the depolarizer, but the relationships have been little understood so far; as a result, observations of this type are presently of little diagnostic value. Particularly marked are the shifts of half-wave potentials with concentration change for systems involving a formation of mercury compounds and/or adsorption.

Also the effects of a change in the drop time† on half-wave potentials are usually not pronounced. Reversible systems are practically uninfluenced, shifts are sometimes observed for irreversible systems. The effect of drop time has been used for diagnostic purposes for systems where the wave form corresponds to that predicted by theory

* Correction for the changes in iR potential drop, caused by the change in the limiting current with increased concentration, should not be neglected.

† Either changed by a change in mercury head or by a mechanical regulation of the drop time.

for reversible systems (cf. Part 5 of this Section), but where the half-wave potentials are shifted when compared with the thermodynamic equilibrium potentials obtained with potentiometric methods.

A typical example of this type is the behavior of ascorbic acid where a shift of half-wave potentials was observed (31). It was assumed that reversible electron-exchange reaction is followed by a slower chemical reaction, viz., hydration of the dehydroascorbic acid formed.* Another example are reversible systems with a subsequent step involving slow dimerization (33).

The effects of temperature, buffer composition, and capacity on the half-wave potentials have been studied very little so far. On the other hand, some interest has been paid to the effects of neutral salts in the supporting electrolyte, particularly with respect to specific effects of cations and anions (34). In general, the half-wave potentials of negatively charged particles are shifted with increasing concentration of cations, those of positively charged particles with increasing concentration of anions. This should enable us to distinguish the charge of the electrolyzed particle. However, in several examples, shifts were observed even for electrolysis of particles assumed to be uncharged. Either a charged intermediate is involved in such reactions, or the dipole, formed perhaps under the influence of the electrical field, behaves in a manner similar to that of a particle bearing a unit charge.

Pronounced changes of half-wave potentials resulting from the use of different solvents are little understood and hence of little value in the explanation of electrode processes. Changes in iR potential drop, in liquid junction potential, and the choice and potential of the reference electrode present difficulties. Hence, comparison with an internal standard (e.g., naphthoquinone or potassium ion) is recommended.

4. Determination of the Number of Electrons Transferred

The height of the polarographic wave can be measured in microamperes and the number of the electrons transferred (n) can be calculated using the Ilkovič equation (cf. Sec. II.2). A knowledge of

* It has been, nevertheless, shown recently (32) that the reduction wave of dehydroascorbic acid does not correspond to that predicted by this simple theory. The actual mechanism appears to be more complicated.

the value of the diffusion coefficient, D, is, of course, necessary for this computation. These values have rarely been measured for organic substances under conditions of polarographic electrolysis and hence this method for determination of the value of n has limited application. Moreover, the extended form of Ilkovič equation (35) should be employed for determination of precise values of n.

The principle of wave-height measurement is applied in a method where comparison is made with a standard substance for which the number of electrons, n, is known precisely under conditions used. In this method the wave height of the substance under study is compared with the wave height of the standard substance, and this measurement is made using equimolar solutions in the same supporting electrolyte. The standard substance must be a well-defined chemical entity and its molecular shape and molecular weight should not be substantially different from that of the substance studied. It is also advantageous for both the studied and the standard substance to possess some identical structural features. It is advisable to use more than one standard compound simultaneously. As standard substances quinones, nitro compounds, and some carbonyl compounds are suitable. We observe whether the wave heights are approximately equal, or whether the wave of the studied species is approximately half or twice as high as the standard wave and then deduce whether the number of electrons consumed is identical, half, or twice of that for standard substance.

In such a comparison it is assumed that the diffusion coefficients of both the studied and the standard substances are practically identical. This condition is best fulfilled when the studied and standard grouping are both attached on the same molecular frame. If the reduction potentials of these groupings are sufficiently separated, two waves—that of the studied grouping and that of the standard grouping—are observed on the polarographic curve. From the ratio of wave heights of the studied and standard grouping, the value of n for the studied grouping can be computed.

An example of this type is the reduction of ω-aminoacetophenone and similar α-aminoarylketones (36). Of the two waves observed at pH about 8, the more negative one is identical in height and in half-wave potential to that of the parent arylketone (e.g., acetophenone). Under these conditions, acetophenone is reduced with consumption of two electrons. As the more positive wave is practically equal in height to the more negative wave of the CO group, it can be deduced that in the more positive wave the $C–N^+$ bond is reduced with consumption of two electrons.*

A widely used technique for the determination of n is based on the determination of the slope of the wave (cf. Sec. II.5). The application of this method is restricted to reversible systems. Any attempt to use the slope of the wave for the determination of the total number of electrons, n, transferred in an irreversible electrode process is incorrect. This misleading practice is widely employed, but a wide acceptance of a wrong method does not diminish the failure of such an approach.

Coulometry and its modifications for small volumes and small currents (37) are also used for the determination of the number, n. The quantity of electricity necessary to reduce a distinct amount of the studied substance is measured at the potential corresponding to limiting current. This potential is usually controlled by a potentiostat. The cathodic and anodic compartments should be separated in order to prevent subsequent electrolytic changes in the electrolysis product. Under such conditions when electrolysis is carried out with a mercury pool instead of with a dropping mercury electrode, side reactions at the constant surface of the electrode can take place that do not occur at the dropping electrode. Coulometric methods are hence usually not very accurate and are best suited for systems where $n = 1$ or 2. For higher values of n the decision is sometimes difficult. Polarographic control of the course of the constant potential electrolysis is recommended.

The most unambiguous method for determination of the value of n is the identification of the product resulting from electrolysis at the dropping mercury electrode. Once it has been determined which grouping is electrolyzed and which is formed, it is usually simple to estimate the number of electrons consumed in this transformation.

Application of constant potential electrolysis, used in the coulometric methods, is most useful also in the determination of the electrolysis product. Either a stationary mercury electrode or a dropping mercury electrode is used as the working electrode. Greater quantities of the electrolysis product can be prepared when the mer-

* By introduction of appropriate groupings it is possible to determine the value of n for previously unstudied compounds.

cury pool electrode is used, but the conditions may differ from those obtained with the dropping electrode. The difference, resulting from side reactions mentioned above, is mainly caused by a prolonged period of contact of the electrode and the product. With the dropping mercury electrode, smaller quantities of product are obtained, but more confidence can be placed in the nature of the product because it is formed under conditions identical to that in polarography.

The potential—usually corresponding to the limiting current—is kept constant either by use of a potentiostat or by manual settings. When the dropping electrode is employed and the limiting current is extended over a range greater than about 0.3 v., the latter method has proved sufficient. The recording of a polarographic wave each 3–6 hours and careful deoxygenation is recommended.

Products formed at the mercury pool electrode can usually be identified using even the methods of classical organic chemistry. When the electrolysis is carried out with the dropping electrode, methods applicable for identification in dilute solutions, such as paper chromatography, spectrophotometry, determination of dissociation constants, or polarography, must be applied.

In those cases in which the product of the electrode process is itself electroactive in the same solution where the electrolysis was performed, the product can be identified using oscillographic polarography, single-sweep methods, periodical change of the applied voltage, stripping methods (preferably using the hanging mercury drop electrode), and finally from the course of polarographic curves.

In oscillographic polarography (38) performed in solutions containing the substance under study, some new incisions can be observed on the dE/dt vs. $f(E)$ curves in addition to the incisions corresponding to the electrolysis of the original bond. These "new" incisions, which correspond to an effect of the product of the primary electrode process, are sometimes absent on the first curve recorded after beginning of the polarization. To check the identity of a product, an appropriate amount of the assumed electrolysis product is added to the solution containing the studied substance. When the "new" incision increases and when coincidence of depolarization potentials has been proved in at least three or four different supporting electrolytes, the identity of the electrolysis product can be taken as proved.

In a similar manner the identity of the product of a primary electrode process can be proved using the additional peaks, observed in the patterns of the i–E curves obtained with single-sweep methods (39).

Both of the above-mentioned methods use oscilloscopic recording, and the electrolysis is performed only for a short period (usually 10^{-1}–10^{-3} sec.). Because of the short time span, these methods sometimes indicate the presence of intermediates too short lived to be detected by other methods. The specific conditions under which the electrolysis is carried out and especially the cyclization of the polarization process may result in different products of the electrolytic reaction being obtained from those found in classical polarography.

More similar to the conditions in classical polarography are experimental conditions used in commutator and in stripping methods.

The commutator method (40,41) is based on polarization with a periodically changed rectangular voltage. In the most widely used modification, the voltage tapped off the potentiometer of a polarograph increases regularly and the fixed voltage on an auxiliary potentiometer is chosen to be in the region of the limiting current of the studied wave. The polarizing voltage applied to the cell is periodically switched between the regularly increasing voltage source and the auxiliary one having a constant value. If the product formed at the potential of the limiting current is capable of exhibiting a polarographic wave, this wave can be observed on the curve recorded with the commutator method. The identity of the product is then proved from the behavior and half-wave potentials of this wave in the usual way.

The stripping methods are most conveniently carried out using the hanging mercury drop electrode (42,43). This type of electrode is polarized at a potential corresponding to the limiting current for the species from which the electrolysis product is being produced. After polarization for some 10–60 sec., the electrode is then polarized by a continuously changing voltage which scans either to more negative or to more positive potentials. A peak current attributed to electrolysis of the product is then obtained. The product is identified by comparing its peak with that of an authentic substance.

Using the four methods described above, the electrolysis product can be examined only in solutions in which it was formed. In other cases, where the product of the electrode process is electro-inactive under conditions under which it is formed, but polarographically active under other conditions, it is necessary to use controlled potential electrolysis. The electrolysis can be then carried out, e.g., in acid solutions, and the product identified in alkaline media.

When an organic compound is reduced in several steps—either because of stepwise reduction of one group or because of successive reduction of several groupings—it is sometimes possible to identify the product of the electrolysis directly from the course of the polarographic wave itself. The product formed during polarographic electrolysis at the first, most positive step can be identified by comparison of the more negative wave (or waves) with waves of the assumed product, these latter experiments being performed in the same supporting electrolyte. Again, comparison in several supporting electrolytes is necessary. The aforementioned reduction of ω-piperidinoacetophenone (36) can be quoted here as an example. This substance is reduced in two waves; the behavior and half-wave potentials of the more negative wave was identified as the wave of acetophenone. Thus, the more positive wave is to be attributed to the reduction of the $C\text{–}N^+$ bond.

5. Reversibility of Electrode Processes

For the description of the rate of establishment of the oxidation–reduction equilibria at the surface of the dropping electrode, the term *reversibility* is often used. In reversible systems, the equilibria are established rapidly, whereas in irreversible systems there is a slow step in the total electrode process and hence the electrode reaction proceeds practically in one way only. The meaning of the words "rapid" and "slow" depends on the electrochemical method used in the study of the system. In classical polarography we describe as reversible those systems for which the half-wave potentials of the oxidized and of the reduced form are practically identical and the wave of which possesses a slope predicted by the Nernst theory. Both these conditions must be fulfilled.

When possible, the polarographic behavior of both oxidized and reduced forms should always be compared. The identity of half-wave potentials of these two forms is the most simple and convincing proof of reversibility.

When only specimens of the oxidized form are available, the reduced form can sometimes be prepared by a chemical reaction within the polarographic vessel. The familiar reducing agents, such as hydrazine, hydroxylamine, sulfite, or hydrosulfite, usually react with the studied system but undergo certain side reactions in addition to reduction. Platinum or palladium sols, saturated with hydrogen, result in catalytically increased anodic waves and are, therefore, not often useful. The most successful reduction procedure seems to be an addition of platinum asbestos to the solution of the oxidized form and subsequent introduction of hydrogen. Half-wave potentials of the oxidized form after deaeration with nitrogen and of the reduced form after introduction of hydrogen are compared.

When one of these forms—either oxidized or reduced—is not sufficiently stable, auxiliary methods must be used for the determination of reversibility, such as oscillographic polarography, single-sweep methods, commutator and stripping methods.

In oscillographic polarography (38) using applied current, the incisions on the dE/dt vs. $f(E)$ curves are most frequently compared. Identical potentials for the incisions of the oxidized and of the reduced form (in the cathodic and anodic branch) are taken as a proof of reversibility. Similarly interpreted are identical peak potentials for the oxidized and reduced forms in single-sweep methods (39). Because of the short time periods used in these oscillographic methods, the meaning of the term "reversibility" in oscillographic and in classical polarography need not be identical. Systems described as reversible in oscillographic methods may be irreversible in classical polarography and vice versa.

Conditions more similar to those present in classical polarography are operative in the following two auxiliary methods, commutator and stripping methods. In these methods, the electrolysis product is electrolytically prepared at the surface of the dropping mercury electrode under polarographic conditions. In the commutator method, using polarization with periodically changed rectangular voltage (40,41) (cf. p. 238), the half-wave potential of the normal polarographic i-E curve is compared with the half-wave potential of the electrolysis product obtained with the commutator. Identity of half-wave potentials is one proof of reversibility.

In the stripping method, a constant surface electrode [a hanging mercury drop electrode (42,43) or a mercury pool electrode] is polarized continuously first from positive to negative potentials. The voltage scanning is then interrupted and the polarization is continued for some time at a potential corresponding to the limiting current of the studied wave. (This period for "preparation" of the electrolysis product can be very short and in fact the reversal of the direction of the polarization can follow "immediately" after arrival at the negative potential corresponding to the limiting current.) Then the direction of voltage scanning is reversed and the electrode is polarized from negative to more positive potentials. If an anodic peak appears at the same potential as that of the cathodic peak in the first run, the system is described as reversible.

The second condition to be fulfilled by a wave of a reversible system is the shape of the polarographic wave. For a diffusion-controlled reversible system, not involving semiquinone formation or dimerization, the polarographic wave is described by an equation (44):

$$E = E_{1/2} + \frac{RT}{nF} \ln \frac{i_d - i}{i}$$

To prove the reversibility of a system possessing above-mentioned characteristics, the diffusion limiting current, i_d, is measured. Then the current, i, in the rising part of the polarographic wave is measured at several potentials, E_r (against an arbitrarily chosen zero, e.g., the half-wave potential $E_{1/2}$). For each point in the E_r–i_r curve, the value of $\ln (i_d - i_r)/i_r$ is computed and a graph is constructed showing the dependence of $\ln (i_d - i_r)/i_r$ on E_r. For a reversible system, this graph is linear with a slope of $2.3\ RT/nF$ (i.e., $0.059/n$ at 20°C.). For $\ln (i_d - i_r)/i_r = 0$, the value* of $E_r \equiv E_{1/2}$.

For reversible systems involving semiquinones, dimerization, formation of insoluble and complex compounds with mercury, as well as for reversible electrode systems accompanied by an antecedent chemical reaction, other equations are applicable and have been derived. These equations are verified in a similar manner using the appropriate logarithmic terms (so-called *logarithmic analysis*).

Irreversible systems and systems having complicated mechanisms yield either a linear plot with a slope other than $2.3\ RT/nF$, a curved plot, or a plot consisting of several linear parts in the logarithmic analysis.

* This method is sometimes suggested for accurate measurements of half-wave potentials. In our opinion a direct measurement from polarographic curves is sufficient.

For irreversible systems giving a linear plot, the slope of the curves equals to 2.3 $RT/\alpha nF$. The value of the transfer coefficient, α (or better, of the product αn), can be determined in this way. For such cases of totally irreversible waves, it is possible (45) to compute from the polarographic wave the values of the formal rate constants of the electrode process.

Sometimes the slope of the logarithmic analysis plot for an irreversible process yields value of n equal to the thermodynamic value, but for a number of electrons, n_1, which is smaller than the value of n determined from the limiting diffusion current. Such cases are sometimes interpreted as involving a reversible step with a smaller number of electrons, n_1, than that totally exchanged.

6. Mechanism of the Electrode Process

The study of the role of the various parameters on polarographic behavior, as mentioned in preceding sections, usually enables us to draw some conclusions concerning the mechanism of the electrode process.* Only when such conclusions concerning a mechanism are drawn and their significance is understood, is it possible to predict disturbing effects and limitations of organic polarography for analytical purposes.

Our ideas concerning the mechanism of an electrode process are sometimes supported by a study of structural effects, usually their influence on half-wave potentials. The changes in half-wave potentials exhibited by changes in the structure of the electroactive species can be discussed in two ways: either qualitatively, based on knowledge

* Two explanations are usually given to the term "mechanism of the electrode process": The first involves description of all stable particles, starting from the substance added to the solution proceeding to all products of the electrode process, including the determination of the electroactive species and of all stable intermediates. The second involves, in addition to this, information concerning the position of the electroactive species relative to the electrode and the structure of the activated complex of the potential determining reaction. The first meaning is based on experience obtained by methods described in preceding paragraphs. It is especially this meaning which is discussed in this paragraph. The second meaning of the word mechanism is nowadays based mainly on speculation! Analogy with homogeneous reaction kinetics, some structural effects, and complicated and not fully understood effects of temperature are sometimes used to support these views.

of structural effects in general organic chemistry, or quantitatively, using quantum chemical structural indices or Hammett and Taft empirical substituent constants, σ (10,13). In the last case the half-wave potentials are plotted as a function of the substituent constants, σ. The positive or negative slope of this dependence may be interpreted as due to nucleophilic or electrophilic reaction in the potential determining step.

III. INDIRECT METHODS

After the polarographic behavior of the electroactive substance has been established, using the criteria given in the preceding section, the polarographic waves can be used either in *direct* or in *indirect* methods. The latter will be discussed in some detail in this section. The indirect methods will be first classified according to the origin of the wave which is measured: (*1*) This wave arises from an electroactive substance which was prepared by a chemical reaction from the electroinactive substance to be determined; (*2*) it is a wave of an electroactive reagent which reacts with the electro-inactive analyzed substance. This reaction must proceed stoichiometrically and, if possible, fast. Subclassification of the indirect methods may be made with respect to the type of the chemical reaction involved.

A special type of indirect determination is the polarometric (amperometric) titration, where the equivalence point, and hence the amount of the substance, is determined from a plot of the dependence of the wave height on the volume of the titrant added. Because these methods have been reviewed elsewhere (46,47), we shall restrict ourselves to cases where wave heights—but not equivalent volumes—are measured.

1. Formation of an Electroactive Substance

The possibilities of performing chemical reactions to transform the analyzed substance into a compound showing a measurable polarographic wave are relatively little exploited. A few examples, outlined in the following paragraphs, should focus the attention of the analytical chemist using polarography on the possibility of applying such types of chemical reactions, which can offer a high yield of the electroactive compound.

A. NITRATION

Nitration is one of the most popular reactions for indirect methods, permitting the transformation of the analyzed substance into a nitro compound. Nitric acid, either pure or with potassium nitrate or sulfuric acid, and in some examples nitrous acid [morphine (48–50), estrone (51)] is used as the nitration mixture.

This method has been used for hydrocarbons, especially aromatic [e.g., benzene (52) and its homologs (53–55)], phenols (56–58), aniline derivatives (59,60), and other compounds bearing a phenyl ring [e.g., tyrosine, tryptophane and phenylalanine (61–68), phenylethylbarbituric acid (69), etc.].

The composition of the reaction mixture, the temperature, and the time necessary to carry out the nitration depend on the substance analyzed. A systematic study is required in each particular case in order to obtain a single nitration product (if possible) in a quantitative reaction. Polarographic curves are recorded in slightly acidic or in slightly alkaline solution (pH 4–10). In strongly acid solutions, current maxima and the hydroxylamine wave prevent accurate measurements. In more alkaline solutions the nitro compounds formed undergo often further chemical reactions.

B. NITROSATION

Nitroso compounds usually undergo easy polarographic reduction and their waves can be used for indirect determinations. The nitroso group is usually introduced in an acid solution to which sodium nitrite has been added.

With this method, secondary amines can be determined in the presence of primary and tertiary amines (70–72) after transformation into *N*-nitroso compound. Among the phenolic compounds, formation of nitroso derivatives has been claimed for resorcinol (73) and for some flavones (74).

The reaction with nitrous acid is usually quenched by addition of sodium hydroxide or ammonia. In this way the excess nitrous acid is converted into nitrite ion which does not interfere. The nitrosation reaction is usually not quantitative and hence experimental conditions and especially the reaction period should be carefully controlled.

C. CONDENSATION

For analytical purposes in organic polarography, the condensations which are most important are those which result in the formation of an azomethine (>C=N—) bond. These condensation reactions transform the carbonyl group into a product which is reduced in well-developed waves and at more positive potentials than the parent species. These methods were found especially useful for the determination of compounds bearing isolated (nonconjugated) keto or aldehyde groups. According to the reactive amine, aldimines, ketimines, oximes, hydrazones, and semicarbazones are formed and determined. Reaction conditions depend on the amine used. For reactions with ammonia and primary amines, a pH above the $(pK_a)_{amine}$ is preferable. Semicarbazide and hydroxylamine react in slightly acidic media, whereas 2,4-dinitrophenyl hydrazine reacts in 2*N* HCl. Girard's reagents (which are betainylhydrazines) condense in glacial acetic acid or in methanolic solutions. The reaction time and temperature depend both on the amine used and on the carbonyl compound studied. Both the reaction rates and the equilibrium conditions are of importance, and it can generally be concluded that a systematic quantitative study of kinetics and equilibria is of great importance for the proper analytical use of those reactions. Also the suitability of these and perhaps other reagents can be judged only on the basis of rate and equilibrium constants.

The polarographic curve is recorded either in the reaction mixture containing the excess of the reagent and at the pH value at which the condensation was carried out (e.g., for ketimines, aldimines, oximes, and semicarbazones), or in another buffer (e.g., betainylhydrazones). In the latter case, hydrolysis rates and equilibria must be considered.

These methods have been found useful in the determination of saturated ketones (75–82), ketoacids (75,83), aldehydes (75,78,84,85), and oxidation products of ascorbic acid (75,86).

Differences in the rate and equilibrium constants enable us to analyze some mixtures, e.g., of ketones and aldehydes using transformation into semicarbazones (78), of pyridoxal in the presence of pyridoxal-5-phosphate (85), or of cyclopentanone and cyclohexanone derivatives (80,81).

Condensation of diamines results sometimes in cyclic compounds. Thus, several oxidation products of enediols react with *o*-phenylenediamine. The waves of the cyclic products formed are sufficiently separated one from another to allow complete analysis of some mixtures (87). Similarly, an excess of hexamethylenediamine is suitable for determination of cyclohexanone and cyclopentanone in mixtures containing aldehydes (88).

The formation of aldimines can be also used in the opposite way, i.e., in the determination of primary amines. Piperonal was recommended as condensing reagent. The method was developed for the determination of traces of methylamine in commercial samples of dimethylamine (72).

Another reagent for condensation with amines is carbon disulfide. In alkaline solutions dithiocarbamates are formed which exhibit anodic waves corresponding to mercury salt formation. This reaction was used for determination of amino acids (89–91) in one procedure and of carbon disulfide in another (92). In the determination of amino acids the reaction is performed in a buffer of pH 9.0 to 9.3, either in an aqueous solution saturated with carbon disulfide (4–5 hours), or in water–acetone mixed solvent with an excess of carbon disulfide (20–30 min.). In the determination of carbon disulfide, the analyzed gas is introduced into an ethanolic solution containing 1% diethylamine. The reaction takes place practically instantly.

D. ADDITION

Methods based on the addition of bromine to an unsaturated bond are used for determination of number of these bonds and of olefins. α,β-dibromides are formed in these reactions; these compounds yield reduction waves corresponding to the reductive splitting of the bromine atoms, i.e., elimination. The addition reaction is performed in methanolic solutions (93) in the presence of sodium bromide. The excess of bromine is removed by ammonia and the polarographic curves are recorded using lithium chloride as the supporting electrolyte. The reaction time depends on the structure of the olefin in question.

E. OXIDATION

Among the oxidation reagents, periodate is the most widely used in organic polarography. This is probably due to its selectivity toward α-diols, α-aminoalcohols, α-hydroxyacids, etc., and due to the fact that the reaction products are subject to few side reactions under the mild conditions used. The oxidation of polyalcohols is advantageously performed at pH 4 to 7 (in the absence of daylight), that of aminoalcohols at somewhat higher pH values. The quantity of the aldehydes formed in these reactions is determined either after separation by distillation, or after removal of the excess of periodic acid by addition of an excess of arsenite in acid media. When the resulting solution is made alkaline, the wave of formaldehyde can be recorded unaffected by the presence of either trivalent or pentavalent arsenic.

Glycols (94), glycerol (95,96), and other polyalcohols (96), as well as serine (97), were determined using these methods. Mixtures of ethyleneglycol and 1,2-propyleneglycol can be analyzed (94) using the waves of formaldehyde and acetaldehyde formed. After hydrolysis this method can be used for determination of ethylene and propylene chlorohydrine (98).

Permanganate is used in the oxidation of lactic acid, the wave of the acetaldehyde formed being measured (99). Ferric ions are sufficiently powerful to oxidize acetoin to diacetyl (100) and iodate reacts with adrenaline and noradrenaline to form iodoadrenochrome and iodonoradrenochrome (101).

Oxidation of simple alcohols, especially of methanol and ethanol, has been carried out using several different oxidizing agents and methods (102–105), but none of the suggested methods has found wide acceptance so far. This seems to be caused by the poor reproducibility of the oxidation, volatility of both the alcohols analyzed and the aldehydes formed, reactivity and polymerization of the aldehydes, and the kinetic character of the formaldehyde wave.

F. COMPLEX FORMATION

An organic complex-forming compound can be determined using a wave of a complex formed either with a metal ion added to the solution or with a slightly soluble salt. Only strong complex-forming agents (usually chelating) can be determined by the first technique. Sometimes even structurally related substances show different complex stability and half-wave potentials under given experimental conditions (pH value and type of the metal added). These methods may thus be made selective and allow analyses of certain mixtures of complex-forming reagents. If the waves are measured at several pH values and for several added metal ions, the reliability and selectivity of these methods are increased.

Propylenediamine (106) (in solutions of cupric ions), 1,2-diaminocyclohexane in the presence of hexamethylenediamine and aminomethylcyclopentylamine (107) (in solutions containing nickel ions), as well as traces of nitrilotriacetic acid in ethylenediaminetetraacetic acid (108) (after reaction with cadmium ions) can be determined in this way.

In the second group of methods based on complex formation, the complexing reagent to be determined is added to a suspension of a slightly soluble salt of a heavy metal. After the equilibrium between the solid phase and solution is established, the concentration of the complex-bound metal in the liquid phase is measured polarographically. These methods are often not sufficiently selective, and moreover the wave height is influenced by the complex stability constants and diffusion coefficients of the particular compounds. For analysis of mixtures, preliminary separation is frequently necessary.

A typical example of this type of method is the determination of amino acids (109–111), based on reaction of these compounds with an excess of insoluble copper phosphate at controlled pH. The waves of the soluble copper–amino acid complexes formed are measured. Preliminary separation (e.g., by chromatography) is necessary for the analysis of a mixture of amino acids.

2. Concentration Change of the Reagent

A number of polarographically active reagents—both inorganic and organic—react with a wide variety of organic compounds. Such reactions are often universal for a group of substances or for a particular functional grouping and thus the measurement of the wave height of the reagent before and after reaction possesses the character of a functional group analysis rather than of an analysis of a specific compound. This is demonstrated here by a few examples.

A. CONDENSATION

Whereas free sulfurous acid exhibits a polarographic reduction wave in acid solutions, the reaction product of sulfite with carbonyl compounds are polarographically inactive in the same potential range. Thus, the decrease of the wave height of sulfurous acid can be taken as a measure of the content of carbonyl groups present. This method has been recommended for determination of acetone (112) and for several other ketones and aldehydes (113).

Polarographically reducible phthalaldehyde reacts with amines to form a cyclic compound that exhibits no wave in the potential range studied. The decrease of the phthalaldehyde wave is, therefore, a function of the concentration of the amine added. The differences in the reaction rates and in reaction ratio of phthalaldehyde:amine prevent application of this method in complex mixtures; however, in mixtures of simple primary amines the content of $—NH_2$ groups could be detected. Applications for ammonia (114) and amino acids (115) have been described.

B. ADDITION

Reaction of maleic anhydride with unsaturated bonds also possesses analytical applicability. Either the decrease of the wave of maleic anhydride or, more frequently, the change in the concentration of maleic acid is measured after hydrolysis. Butadiene in gaseous mixtures (116) and styrene and vinylacetate (117) were determined in this way.

C. OXIDATION

The Malaprade reaction can be also followed by measuring the decrease of the first wave of periodic acid. The first polarographic wave, corresponding to the electroreduction of periodate to iodate, rises directly from the potential of dissolution of mercury. The method allows the determination of the total amount of α-dihydroxy, α-hydroxyamino, or α-diamino groupings in the presence of simple alcohols and amines. The optimum reaction conditions are the same as given for methods based on measurement of the aldehydes formed (cf. p. 246).

Glycols (96), glycerol (96), sugars (96,118), and serine (119) were determined in this way. The method was also used for the study of mechanism of the Malaprade reaction (96,120) and for the analysis of mixtures of epimers (96).

Oxidation with dichromate in nitric acid media has been used in the determination of ethanol (121); the decrease of the wave of dichromate was followed. Even the chromic oxide formation in glacial acetic acid may be followed (122) in this way.

D. FORMATION OF SLIGHTLY SOLUBLE AND COMPLEX COMPOUNDS

If the analyzed substance forms with a metal ion a slightly soluble or complex compound, the substance can be determined from the decrease of the wave of the metal ion. Thus, the decrease in the height of the wave of copper ions can be used in the determination of Cardiazole (123) (pentamethylene tetrazole) which forms insoluble copper compounds and of reducing sugars (124) after reaction with Fehling's solution.

References

1. Elving, P. J., in Mitchell, Jr., Ed., *Organic Analysis*, Vol. II, Interscience, New York, 1954, p. 195.
2. Elving, P. J., *Ric. Sci. Suppl.*, **30,** *Contr. Teor. Sper. Polarograf.*, **V,** 205 (1960).
3. Elving, P. J., in P. Zuman and I. M. Kolthoff, Eds., *Progress in Polarography*, Vol. II, Interscience, New York–London, 1962, p. 625.
4. Koršunov, I. A., *Zavodsk. Lab.*, **24,** 543 (1958).
5. Nürnberg, H. W., *Angew. Chem.*, **72,** 433 (1960).
6. Tirouflet, J., *Commun. 84e Congr. Soc. Savantes*, Dijon, 1959, p. 215.
7. Tirouflet, J., in I. Longmuir, Ed., *Advances in Polarography*, Vol. II, Pergamon Press, Oxford, 1960, p. 740.
8. Tirouflet, J., and M. Person, *Commun. 84e Congr. Soc. Savantes*, Dijon, 1959, p. 209.
9. Wawzonek, S., *Anal. Chem.*, **21,** 61 (1949); **22,** 30 (1950); **24,** 65 (1954); **28,** 638 (1956); **30,** 661 (1958); **32,** 1445 (1960); **34,** 182R (1962).
10. Zuman, P., *Ric. Sci. Suppl.*, **30,** *Contrib. Teor. Sper. Polarograf.*, **V,** 229 (1960).
11. Zuman, P., *Chem. Listy*, **54,** 1244 (1960); *J. Polarog. Soc.*, **7,** 66 (1961).
12. Zuman, P., *Chem. Listy*, **55,** 261 (1961); *Z. Chem.*, in the press.
13. Zuman, P., in P. Zuman and I. M. Kolthoff, Eds., *Progress in Polarography*, Vol. I, Interscience, New York–London, 1962, p. 319.
14. Zuman, P., and S. Wawzonek in P. Zuman and I. M. Kolthoff, Eds., *Progress in Polarography*, Vol. I, Interscience, New York–London 1962, p. 303.
15. Proske, G., *Anal. Chem.*, **24,** 1834 (1952).
16. Given, P. H., and M. E. Peover, in I. S. Longmuir, Ed., *Advances in Polarography*, Vol III, Pergamon Press, Oxford, 1960, p. 948.
17. Hanuš, V., *Proc. Intern. Polarog. Congr. Prague, 1st Congr., 1951*, Part I, p. 811.
18. Hans, W., and K. H. Henke, *Z. Elektrochem.*, **57,** 595 (1953).
19. Wiesner, K., M. Wheatley, and J. M. Los, *J. Am. Chem. Soc.*, **76,** 4858 (1954).

20. Green, J. H., and A. Walkley, *Australian J. Chem.*, **8**, 51 (1955).
21. Kemula, W., Z. R. Grabowski, and E. T. Bartel, *Roczniki Chem.*, **33**, 1125 (1959).
22. Majranovskij, S. G., and L. I. Liščeta, *Collection Czech. Chem. Commun.*, **25**, 3025 (1960).
23. Knobloch, E., *Collection Czech. Chem. Commun.*, **25**, 3330 (1960).
24. Zuman, P., and M. Kuik, *Collection Czech. Chem. Commun.*, **24**, 3861 (1959).
25. Zuman, P., J. Chodkowski, and F. Šantavý, *Collection Czech. Chem. Commun.*, **26**, 380 (1961).
26. Volková, V., *Nature*, **185**, 743 (1960).
27. Krupička, J., Private communication (February 1962).
28. Frumkin, A. N., and N. Nikolaeva-Fedorovich, in P. Zuman and I. M. Kolthoff, Eds.; *Progress in Polarography*, Vol. I, p. 223. Interscience, New York–London, 1962 (gives earlier references).
29. Kastening, B., and L. Holleck, *Z. Elektrochem.*, **64**, 823 (1960) and references given herein.
30. Schwabe, K., in P. Zuman and I. M. Kolthoff, Eds., *Progress in Polarography*, Vol. I, Interscience, New York–London, 1962, p. 333.
31. Kern, D. M. H., *J. Am. Chem. Soc.*, **76**, 1011 (1954).
32. Ono, S., M. Takagi, and T. Wasa, *Bull. Chem. Soc. Japan*, **31**, 356 (1958).
33. Hanuš, V., *Chem. Zvesti*, **8**, 702 (1954).
34. Kůta, J., *Acta Chim. Acad. Sci. Hung.*, **9**, 119 (1956) and references given.
35. Koutecký, J., and M. v. Stackelberg in P. Zuman and I. M. Kolthoff, Eds., *Progress in Polarography*, Vol. I, Interscience, New York–London, 1962, p. 21.
36. Zuman, P., and V. Horák, *Collection Czech. Chem. Commun.*, **26**, 176 (1961).
37. Meites, L., in P. Zuman and I. M. Kolthoff, Eds., *Progress in Polarography*, Vol. II, Interscience, New York–London, 1962, p. 515.
38. Kalvoda, R., in P. Zuman and I. M. Kolthoff, Eds., *Progress in Polarography*, Vol. II, Interscience, New York–London, 1962, p. 449.
39. Vogel, J., in P. Zuman and I. M. Kolthoff, Eds., *Progress in Polarography*, Vol. II, Interscience, New York–London, 1962, p. 429.
40. Kalousek, M., *Collection Czech. Chem. Commun.*, **13**, 105 (1948).
41. Rálek, M., and L. Novák, *Collection Czech. Chem. Commun.*, **21**, 248 (1956).
42. Kemula, W., *Advances in Polarography* (*Proc. IInd Intern. Congr. Polarog., Cambridge, 1959*), Vol. I, Pergamon Press, London, 1960, p. 135.
43. Říha, J., in P. Zuman and I. M. Kolthoff, Eds., *Progress in Polarography*, Vol. II, Interscience, New York–London, 1962, p. 383.
44. Heyrovský, J., and D. Ilkovič, *Collection Czech. Chem. Commun.*, **7**, 198 (1935).
45. Delahay, P., *New Instrumental Methods in Electrochemistry*, Interscience, New York, 1954, p. 72f.
46. Kolthoff, I. M., *Anal. Chim. Acta*, **2**, 606 (1948).
47. Doležal, J., and J. Zýka, *Polarometrické titrace* (Polarometric Titrations), SNTL, Prague, 1961.
48. Baggesgaard-Rasmussen, H., C. Hahn, and K. Ilver, *Dansk. Tidskr. Farm.*, **19**, 41 (1945).

49. Baggesgaard-Rasmussen, H., *Bull. Federation Intern. Pharm.*, **21**, 233 (1947)
50. Lund, H., *Acta Chim. Scand.*, **12**, 1444 (1958).
51. Gry, O., *Dansk. Tidsskr. Farm.*, **23**, 139 (1949).
52. Škramovský, S., and J. Teisinger, *Casopis Lekaru Ceskych*, **82**, 621 (1943)..
53. Srbová, J.: *Pracovni Lekar.*, **4**, 47 (1952); see also *Anal. Chem.*, **24**, 917 (1952).
54. Landry, A. S., *Anal. Chem.*, **21**, 674 (1949).
55. Šedivec, V., *Collection Czech. Chem. Commun.*, **23**, 57 (1958); *Chem. Prumysl*, **8**, 180 (1958).
56. Roubal, J., and J. Zdražil, *Pracovni Lekar.*, **2**, 187 (1950); **3**, 148 (1951).
57. Roubal, J., J. Zdražil, *Proc. Intern. Polarog. Congr., Prague, 1st Congr., 1951*, Pt. I, 724 (1951), Pt. III, 542 (1952).
58. Bergerová-Fišerová, V., S. Škramovský, and J. Menšíková, *Pracovni Lekar.*, **6**, 229 (1955).
59. Matsumoto, K., *J. Pharm. Soc. Japan*, **73**, 1375 (1953); *Leybold Polarograph. Ber.*, **2**, 87 (1954).
60. Novotný, B., *Cesk. Farm.*, **3**, 302 (1954); *Chem. Tech. (Berlin)*, **6**, 662 (1954).
61. Monnier, D., and Y. Rusconi, *Helv. Chim. Acta*, **34**, 1297 (1951).
62. Rusconi, Y., D. Monnier, and P. E. Wenger, *Helv. Chim. Acta*, **34**, 1943 (1951).
63. Monnier, D., and Y. Rusconi, *Anal. Chim. Acta*, **7**, 567 (1952).
64. Monnier, D., Z. Besso, *Helv. Chim. Acta*, **35**, 777 (1952).
65. Besso, Z., D. Monnier, and P. E. Wenger, *Anal. Chim. Acta*, **7**, 286 (1952).
66. Monnier, D., and R. Guerne, *Anal. Chim. Acta*, **19**, 90 (1958).
67. Wenger, P. E., D. Monnier, and J. Vogel, *Microchim. Acta*, **3–4**, 406 (1957).
68. Monnier, D., J. Vogel, and P. E. Wenger, *Anal. Chim. Acta*, **22**, 369 (1960).
69. Kalvoda, R., *Cesk. Farm.*, **3**, 300 (1954).
70. Smales, A. A., and H. N. Wilson, *J. Soc. Chem. Ind.*, **67**, 210 (1948).
71. English, F. L., *Anal. Chem.*, **23**, 344 (1951).
72. Leclercq, M., *Mem. Poudres*, **35**, 365 (1953).
73. Davídek, J., and O. Manoušek, *Cesk. Farm.*, **7**, 399 (1958).
74. Davídek, J., and O. Manoušek, *Cesk. Farm.*, **7**, 73 (1958).
75. Zuman, P., *Collection Czech. Chem. Commun.*, **15**, 839 (1950).
76. Zuman, P., and M. Březina, *Chem. Listy*, **46**, 599 (1952).
77. Březina, M., and P. Zuman, *Chem. Listy*, **47**, 975 (1953).
78. Souchay, P., and M. Graizon, *Chim. Anal.*, **36**, 85 (1954).
79. Coulson, D. M., *Anal. Chim. Acta*, **19**, 284 (1958).
80. Wolfe, J. K., E. B. Hershberg, and L. F. Fieser, *J. Biol. Chem.*, **136**, 653 (1940).
81. Prelog, V., and O. Häfliger, *Helv. Chim. Acta*, **32**, 2088 (1949).
82. Březina, M., J. Volke, and V. Volková, *Collection Czech. Chem. Commun.*, **19**, 894 (1954).
83. Neish, W. J. P., *Rec. Trav. Chim.*, **72**, 105, 1098 (1953).
84. Zuman, P., and F. Šantavý, *Chem. Listy*, **46**, 393 (1952); *Collection Czech. Chem. Commun.*, **18**, 28 (1953).
85. Manoušek, O., and P. Zuman, *J. Electroanal. Chem.*, **1**, 324 (1960).
86. Zuman, P., *Chem. Listy*, **46**, 521 (1952).

87. Wasa, T., M. Takagi and S. Ono, *Bull. Chem. Soc. Japan*, **34**, 518 (1961).
88. Hall, M. E., *Anal. Chem.*, **31**, 2007 (1959).
89. Zahradník, R., and L. Jenšovský, *Chem. Listy*, **48**, 11 (1954).
90. Zahradník, R., *Collection Czech. Chem. Commun.*, **21**, 447 (1956).
91. Zahradník, R., V. Mansfeld, and B. Souček, *Cesk. Farm.*, **4**, 119 (1955); *Pharmazie*, **10**, 364 (1955).
92. Zuman, P., R. Zumanová, and B. Souček, *Collection Czech. Chem. Commun.*, **18**, 632 (1953).
93. Rjabov, A. V., and G. D. Panova, *Dokl. Akad. Nauk SSSR*, **99**, 547 (1954).
94. Warshowsky, B., and P. J. Elving, *Ind. Eng. Chem., Anal. Ed.*, **18**, 253 (1946).
95. Elving, P. J., B. Warshowsky, E. Shoemaker, and J. Margolit, *Anal. Chem.*, **20**, 25 (1948).
96. Zuman, P., and J. Krupička, *Collection Czech. Chem. Commun.*, **23**, 598 (1958).
97. Boyd, M. J., and K. Bambach, *Ind. Eng. Chem., Anal. Ed.*, **15**, 314 (1943).
98. Cannon, W. A., *Anal. Chem.*, **22**, 928 (1950).
99. Dirscherl, W., and M. V. Bergmeyer, *Biochem. Z.*, **320**, 46 (1949).
100. Greenberg, L. A., *J. Biol. Chem.*, **147**, 11 (1943).
101. Henderson, J., and A. S. Freedberg, *Anal. Chem.*, **27**, 1064 (1955).
102. Zapletálek, A., *Collection Czech. Chem. Commun.*, **11**, 28 (1939).
103. Šťastný, J., *Chem. Obzor*, **19**, 119 (1944).
104. Schmidt, O., and R. Manz, *Klin. Wochschr.*, **33**, 857 (1955).
105. Kubis, J., *Casopis Lekaru Ceskych*, **1959**, 852; *Pracovni Lekar.*, **9**, 465.
106. Horton, A. D., P. F. Thomason, and M. T. Kelley, *Anal. Chem.*, **27**, 269 (1955).
107. Hall, M. E., *Anal. Chem.*, **31**, 1219 (1959).
108. Daniel, R. L., and R. B. LeBlanc, *Anal. Chem.*, **31**, 1221 (1959).
109. Jones, T. S. G., *Biochem. J.*, **42**, LIX (1948).
110. Martin, A. J. P., and R. Mittelmann, *Biochem. J.*, **43**, 353 (1948).
111. Blaedel, W. J., and J. W. Todd, *Anal. Chem.*, **32**, 1018 (1960).
112. Novák, J. V. A., cf., J. Heyrovský, *Polarographie*, Springer, Vienna, 1941, p. 368.
113. Strnad, F., *Chem. Listy*, **13**, 16 (1949).
114. Norton, D. R., and C. K. Mann, *Anal. Chem.*, **26**, 1180 (1954).
115. Norton, D. R., and N. H. Furman, *Anal. Chem.*, **26**, 1116 (1954).
116. Warshowsky, B., P. J. Elving, and J. Mandel, *Anal. Chem.*, **19**, 161 (1947).
117. Whitnack, G. C., *Anal. Chem.*, **20**, 658 (1948).
118. Takiura, K., and K. Koizumi, *J. Pharm. Soc. Japan*, **78**, 961 (1958).
119. Ladik, J., and I. Székács, *Nature*, **184**, 188 (1959).
120. Zuman, P., J. Sicher, J. Krupička, and M. Svoboda, *Nature*, **178**, 1407 (1956); *Collection Czech. Chem. Commun.*, **23**, 1237 (1958).
121. Monnier, D., and W. F. Rüedi, *Helv. Chim. Acta*, **38**, 402 (1955).
122. Krupička, J., and J. Kadlec, *Collection Czech. Chem. Commun.*, **24**, 1783 (1959).
123. Parrák, V., *Cesk. Farm.*, **3**, 42 (1954).
124. Dlezek, J., *Proc. Intern. Polarog. Congr. Prague, 1st Congr., 1951*, Pt. I, p. 740.

Mechanisms of Organic Polarography

By Charles L. Perrin
University of California, San Diego

CONTENTS

SYMBOLS

A	area of electrode
a_{e^-}	the activity of the electrons in the electrode
α	transfer coefficient
D_i	diffusion coefficient of species i
E	potential of the electrode
E^0	standard potential of the half-cell, sometimes formal potential
$E_{1/2}$	half-wave potential
F	the Faraday: 96,500 coulombs
h	height of mercury column
i	cathodic current, averaged over the life of the drop
i_d	average limiting current, under diffusion control
i_k	average limiting current, under kinetic control
i_l	average limiting current, at very negative E
$k_{b,h}$	heterogeneous pseudo-first-order rate constant, for backward reaction
$k_{b,h}^0$	$k_{b,h}$ at $E = 0$
$k_{f,h}$	heterogeneous pseudo-first-order rate constant, for forward reaction
$k_{f,h}^0$	$k_{f,h}$ at $E = 0$
k_h	heterogeneous pseudo-first-order rate constant, at $E = E^0$
m	rate of mercury flow
n	total number of electrons in electrode reaction, per mole of substrate
n_a	number of electrons involved in potential-determining step
[O]	bulk concentration of oxidized species
$[O]_s$	concentration of oxidized species at electrode surface
pK'	pH at which two polarographic waves are of equal height
R	gas constant
[R]	bulk concentration of reduced species
$[R]_s$	concentration of reduced species at electrode surface
T	absolute temperature
t	drop time

I. Introduction

A. HISTORICAL

Since Heyrovský's discovery of polarography in 1922, Shikata's pioneer study of the polarography of an organic compound, and Müller and Bamberger's recognition of the significance of the half-wave potentials of organic compounds, a vast literature of organic polarography has developed. However, it is only in the past fifteen years that there has been appreciable application of the methodology of physical organic chemistry to the elucidation of the mechanisms of electrode reactions of organic compounds. In recent years, amazing progress has been made, and it is the aim of this review to describe what is now known about the mechanisms involved.

Numerous reference books are available on the principles, technique, and results of polarography. The standard reference is the second edition of Kolthoff and Lingane's *Polarography* (1), in two volumes. Others include Zuman's *Organic Polarographic Analysis* (2), the English translation of Březina and Zuman's *Polarography in Medicine, Biochemistry, and Pharmacy* (3) (results and applications), Meites' *Polarographic Techniques* (4) (technique), Delahay's *New Instrumental Methods in Electrochemistry* (5) (principles), Charlot, Badoz-Lambling, and Trémillon's *Electrochemical Reactions* (6) (principles), Allen's *Organic Electrode Processes* (7) (applications, technique), and Müller's chapter, *Polarography*, in the series, *Physical Methods of Organic Chemistry* (8) (principles, technique). Also, Wawzonek's extensive biennial bibliographies of organic polarography (9) are excellent compendia of research in the field.

Several previous reviews have covered mechanisms of organic polarography. Among them are articles by Laitinen (10), Page (11), Tanford and Wawzonek (12), Gardner and Lyons (13), Elving (14), Nürnberg (15), Elving and Pullman (16), Zuman and Wawzonek (17), and Elving (18).

B. OBJECTIVES

This review is intended for physical organic chemists. The treatment assumes that the reader is familiar with the methodology and results of mechanistic organic chemistry, but does present the principles of electrochemistry as needed. Insofar as is possible, the description of electrode reaction mechanisms is in the terminology usually used in discussing mechanisms of homogeneous reactions.

The electron is to be considered as just another reagent on the organic chemist's shelf. It is a very versatile reagent, capable of cleaving single bonds and adding to double bonds and other unsaturated and aromatic functional groups, and the products of cleavage and addition can undergo a multitude of subsequent steps. This review is concerned with the mechanisms of the reactions of organic compounds with electrons, much as other reviews have covered mechanisms of the reactions of organic compounds with solvent, or electrophiles, or free radicals, or photons, or any of the other reagents available to organic chemists.

The current state of our understanding of the mechanisms of organic electrode processes is not so highly developed as our understanding of mechanisms of homogeneous processes. Since an electrochemical reaction is a heterogeneous one, proceeding at an electrode surface, a thorough description of its mechanism must await a thorough understanding of the structure of the layer of solvent in the neighborhood of the electrode. As a result, most of the descriptions of electrode mechanisms have succeeded merely in listing the products of the electrode reaction and the intermediates formed during the reaction. Only in a few cases has there been significant progress toward describing the structure of the transition states involved. Therefore, this review is in large part limited to cataloguing the intermediates and products of electrode reactions, without much attention to the structure of the transition state and the alignment of the substrate relative to the electrode surface.

The first large portion of this review is devoted to an exposition of the principles of electrochemistry, polarography, and electrode kinetics, followed by a discussion of the methodology involved in drawing mechanistic conclusions from polarographic experiments. The remainder is devoted to a review of the mechanisms proposed to explain various organic electrode reactions. This section is devoted almost exclusively to mechanisms in classical polarography, and references to other techniques, such as high-frequency polarography and chronopotentiometry, are made when necessary, but without elaboration. Since most organic electrode processes have been investigated at the dropping mercury electrode functioning as a cathode, this review is devoted mostly to reduction processes. Anodic processes are mentioned occasionally; the principles outlined in the first section are still applicable, with suitable changes of sign. In accord with my own interests, I am emphasizing the usual organic functional groups—aromatic systems, carbonyl, nitro, and halide—and I pay less attention to heterocyclics and sulfur compounds.

This review is optimistic in tone, concerned with cataloging some reactions for which mechanisms are established, but ignoring older hypotheses replaced by newer ones. It is not the purpose of this article to imply that all reactions are understood; there are many whose mechanisms are still not understood, or are subjects of controversy, but such reactions are omitted in order to emphasize the progress that has been made. Also, I hope that the outline of principles and methodology will enable and encourage physical organic chemists to enter the field of mechanistic organic polarography.

II. Theory

A. ELECTROCHEMISTRY

The following discussion is adapted largely from Delahay (5) and from Müller (8).

1. The Nernst Equation

For a general oxidation reduction equilibrium

$$aA + bB + ne^- = yY + zZ$$

maintained between an electrode and the surrounding solution, we may write an equilibrium constant

$$K = \frac{a_Y^y a_Z^z}{a_A^a a_B^b a_{e^-}^n}$$

where a_{e^-} is the activity (fugacity) of the electron, or its "escaping tendency." The conventional standard state for electron activity is the normal hydrogen electrode, in which (from which) the electron has unit escaping tendency. Nevertheless, it is common in polarographic practice to choose the saturated calomel electrode as the standard state. For definiteness in this article, the polarographic standard state has been chosen, although this review is primarily concerned only with differences of potential among closely related compounds. Activities of chemical species refer to activities at the electrode surface, but activities are to be approximated by concentrations in this article. Generally, A and Y are oxidized and reduced forms, respectively, of an organic molecule, and B and Z are protons and water, respectively. For the case of a solid or solvent, the activity is constant, and may be set equal to unity.

The standard free energy change, ΔF^0, associated with reducing a moles of A to y moles of Y is related to the equilibrium constant and to E^0, the standard potential of the half-cell, by

$$\Delta F^0 = -RT \ln K = -nFE^0$$

According to the accepted convention, the standard potential is positive if A + B, in their standard states, are spontaneously reduced to Y + Z, in their standard states, when coupled with the reference electrode. To make clear the signs of electrode potentials, I state

$$Na^+ + e^- = Na; \; E_{sce}^0 = -2.47 \text{ v.}, \; E_{nhe}^0 = -2.71 \text{ v.}$$

The E^0 is negative because sodium ion does not oxidize either hydrogen or mercury spontaneously. In order for an electrode to be in equilibrium with sodium metal (unit activity) and sodium ion ($1M$), the potential of that electrode must be $E = -2.47$ v., relative to the sce. Notice that in order to achieve equilibrium with such a reducing system, the potential must be quite negative. Then

$$E^0 = \frac{RT}{nF} \ln K = \frac{RT}{nF} \ln \frac{a_Y{}^y a_Z{}^z}{a_A{}^a a_B{}^b} - \frac{RT}{F} \ln a_{e^-} \tag{1}$$

Consider a solution initially in equilibrium with an electrode. If a_{e^-} is increased in this electrode, $a_Y{}^y a_Z{}^z / a_A{}^a a_B{}^b$ must be increased in order to maintain the equilibrium, i.e., some A must be reduced to Y at the electrode surface, with the accompanying passage of a cathodic current. The quantity, $-(RT/F) \ln a_{e^-}$, is represented by E, the potential of the working electrode. Increasing a_{e^-} is equivalent to decreasing E, or making the electrode more reducing. Rearranging eq. (1) leads to the Nernst equation for the potential of an electrode in equilibrium with the solution:

$$E_{eq} = E^0 - \frac{RT}{nF} \ln \frac{a_Y{}^y a_Z{}^z}{a_A{}^a a_B{}^b} \tag{2}$$

2. *Current–Potential Plots*

Since most polarographic reactions are reduction processes, it is customary to plot cathodic current in the positive direction on the ordinate. Also, the abscissa represents $-E$ increasing (E becoming more reducing) from left to right (Fig. 1). Thus, points at the right correspond to potentials with greater electron activity. Notice that it is potential which is plotted, rather than applied voltage, which is the observed quantity. The former is obtained by correcting the latter for the IR drop across the cell.

3. *Electrode Kinetics*

Previously it was implied that for a given value of the potential, equilibrium could be maintained at the electrode surface by the passage of an appropriate current. We must now consider the kinetic aspects of this approach to equilibrium.

Let the general unimolecular oxidation–reduction reaction be abbreviated by

$$\mathrm{O} + ne^- = \mathrm{R}$$

If the reduction of O and the oxidation of R are first-order processes, the *net* rate of reaction, in moles per second per square centimeter of electrode surface, is given by

$$-\frac{dN_{\mathrm{O}}}{dt} = \frac{dN_{\mathrm{R}}}{dt} = k_{f,h}\,[\mathrm{O}]_s - k_{b,h}\,[\mathrm{R}]_s \tag{3}$$

where $[\mathrm{O}]_s$ and $[\mathrm{R}]_s$ are concentrations, in moles/cc., at the electrode surface. These rate constants represent rate constants of heterogeneous processes and are expressed in cm./sec. Furthermore, they depend upon the applied potential.

If the forward reaction is an nth order reaction in the electrons, we might expect $k_{f,h}$ to be proportional to $a_{e^-}{}^n$ and $k_{b,h}$ to be independent of a_{e^-}. However, on increasing a_{e^-} (decreasing E), the rate of the reverse reaction will decrease, as a result of the electric field at the electrode, which hinders the transfer of electrons from R to the electrode. The transfer coefficient, α $(0 < \alpha < 1)$, represents the fraction of an increase in $-E$ which favors the reduction, and $(1 - \alpha)$ then represents the fraction which hinders the oxidation.* Then

$$\begin{aligned} k_{f,h} &= k_{f,h}{}^0 a_{e^-}{}^{\alpha n} = k_{f,h}{}^0 e^{-\alpha nFE/RT} \\ k_{b,h} &= k_{b,h}{}^0 / a_{e^-}{}^{(1-\alpha)n} = k_{b,h}{}^0 e^{(1-\alpha)nFE/RT} \end{aligned} \tag{4}$$

where k^0's are rate constants when a_{e^-} is unity, or $E = 0$ (potential of the standard state reference electrode). The cathodic current, in amperes, is related to the rate of reaction by multiplying the latter by the area of the electrode, A, times the charge involved per mole of O reduced:

$$i = nFA\{k_{f,h}{}^0\,[\mathrm{O}]_s\,e^{-\alpha nFE/RT} - k_{b,h}{}^0\,[\mathrm{R}]_s\,e^{(1-\alpha)nFE/RT}\} \tag{5}$$

* The similarity of the transfer coefficient to the Brönsted α of general acid catalysis may help to clarify the former to physical organic chemists. A change in E represents a change in the equilibrium $\mathrm{O} + ne^- \underset{k_{b,h}}{\overset{k_{f,h}}{\rightleftharpoons}} \mathrm{R}$, just as a change in $\mathrm{p}K_a$ represents a change in the equilibrium $\mathrm{B} + \mathrm{HA} \underset{k_b}{\overset{k_f}{\rightleftharpoons}} \mathrm{BH}^+ + \mathrm{A}^-$. Both α's reflect the fraction of the change in equilibrium, determined by k_f/k_b, which appears in the rate of the forward reaction. For most electrode processes, α has been found to lie in the range 0.3 to 0.7.

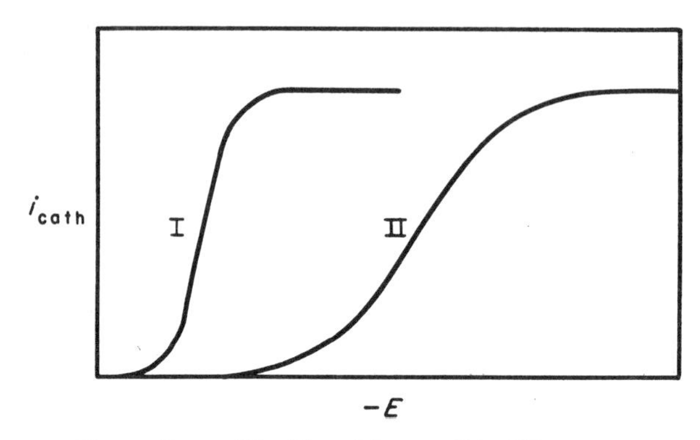

Fig. 1. Reversible (I) and irreversible (II) waves.

If concentrations are expressed in the polarographically more convenient units, millimoles per liter, the current is expressed in microamperes. Equation (5) may be simplified by referring potentials to the standard potential,

$$i = nFA\{k_{f,h}{}^0 e^{-\alpha nFE^0/RT}[\mathrm{O}]_s e^{-\alpha nF(E-E^0)/RT} - k_{b,h}{}^0 e^{(1-\alpha)nFE^0/RT}[\mathrm{R}]_s e^{(1-\alpha)nF(E-E^0)/RT}\}$$

at which there is no net current whenever $[\mathrm{O}]_s = [\mathrm{R}]_s$:

$$k_{f,h}{}^0 e^{-\alpha nFE^0/RT} = k_{b,h}{}^0 e^{(1-\alpha)nFE^0/RT} = k_h$$

or

$$i = nFAk_h\{[\mathrm{O}]_s e^{-\alpha nF(E-E^0)/RT} - [\mathrm{R}]_s e^{(1-\alpha)nF(E-E^0)/RT}\} \quad (6)$$

Thus the kinetics of the electrode process are parameterized by E^0 and the single rate constant, k_h, rather than by the two rate constants, $k_{f,h}{}^0$ and $k_{b,h}{}^0$. The quantity, $nFAk_h$, represents the exchange current per mole, i_0, at the equilibrium potential, where the anodic and cathodic currents exactly balance.

In many polarographic situations, the solution contains only O, in which case the second term within the braces vanishes, and the current is a simple exponential function of the potential. The overvoltage, η, is defined as the potential which must be added to the equilibrium potential in order to produce a requisite amount of current; for a cathodic process, it is a negative quantity, and is obviously dependent upon k_h, α, and the current required.

For the situation in which an intermediate is formed, according to the mechanism

$$O + n_a e^- = I \xrightarrow{(n - n_a)e^-} R$$

it is generally possible to consider one of the two steps as rate-limiting (or potential-determining, in this context). For example, if the first step, involving n_a electrons, is rate-limiting, then the intermediate is almost immediately transformed to R, with the consumption of $(n - n_a)$ more electrons. The kinetics of the total reaction are solely those of the first step. In the reverse reaction, R is in rapid equilibrium with $I + (n - n_a)e^-$, and the rate-limiting step is still the transfer of n_a electrons. The expression for the net current is similar to that above [eq. (6)], except n_a replaces n in the exponents (but not in the factor $nFAk_h$), $[I]_s$ replaces $[R]_s$, and E^0 is the standard potential for $O + n_a e^- = I$.

4. *Electrolysis at High Electron Activity*

A high electron activity corresponds to very negative E ($E \ll E^0$), so that the second term in braces is negligible compared to the first. Also, the cathodic current is very large, so that O is rapidly being converted to R at the electrode surface. (Of course, it is possible that the reaction is so slow that the solvent or supporting electrolyte is more readily reduced, without ever achieving reduction of the substrate. In such a case, no reduction of the substrate would be observed in the accessible potential range.) If there were no transfer of material between interface region and the bulk of the solution, the electrolysis would continue only until essentially all the O at the surface has been converted to R. If the solution is well-stirred, the large current would continue as long as O can be brought to the electrode surface and the resulting R carried away. The intermediate case is considered in the next section.

B. POLAROGRAPHIC LIMITING CURRENTS

1. *Diffusion Current*

As E is made increasingly more negative, the cathodic current increases exponentially. This current represents reduction of O to R, and must eventually become limited by the rate at which O is supplied to the electrode surface. Once this situation is attained, a further increase of electron activity cannot lead to any greater current, so that a limiting current, independent of potential, is observed. At potentials corresponding to this plateau, O is reduced to R as rapidly as O arrives at the electrode surface, and a concentration gradient is set up. Under polarographic conditions, only diffusion, under the influence of this concentration gradient, serves to replenish $[O]_s$ and maintain the limiting current.

In polarography, the dropping mercury electrode (dme) is, for convenience, the most common working electrode. The problem of a current limited by the rate of diffusion to an expanding spherical electrode was solved by Ilkovič. The limiting current, averaged over the lifetime of the drop, is given by

$$i_d = 607nm^{2/3} t^{1/6} D_O^{1/2} [O] \tag{7}$$

where current is in microamperes, m is the rate of flow of mercury, in milligrams per second, t is the drop time, in seconds, D_O is the diffusion coefficient of species O, in square centimeters per second (Diffusion coefficients of most organic molecules are of the order of 10^{-5} cm.^{2}/sec.), and concentrations are bulk concentrations, in mmoles/l.

The proportionality between current and concentration is the basis for the analytical applications of the diffusion current. It should also be noted that i_d is proportional to n, the total number of electrons transferred. Diffusion coefficients of organic molecules generally increase by about 2–4% per degree, so that i_d increases by about 1–2% per degree. Finally, i_d is proportional to $m^{2/3}t^{1/6} = m^{1/2}(mt)^{1/6}$. The quantity, $(mt)^{2/3}$, is proportional to the area of the mercury drop, independent of the height of the mercury column, h. But the rate of mercury flow is proportional to the pressure head in the mercury column, so $m^{1/2}$ is proportional to $h^{1/2}$. Therefore, i_d is proportional to $h^{1/2}$ for a diffusion-controlled current.

2. Kinetic Current

Consider the reaction

$$A \underset{k_b}{\overset{k_f}{\rightleftharpoons}} O \xrightarrow{ne^-} R$$

in which A is reducible only at a potential more negative than that at which O is reducible, and must form O by chemical reaction in order for reduction to proceed at the less negative potential. If the rate of the chemical reaction is sufficiently high, O will be regenerated at the electrode surface as rapidly as it is consumed by electrolysis, until all the A is consumed. In such a case, the current is limited only by the rate of diffusion of A + O. If the rate of the chemical reaction is very low, O can be replenished at the electrode surface only by the diffusion of more O to the electrode surface. In this case the limiting current is a diffusion current whose height is proportional to the actual concentration of O. For intermediate rates, the limiting current will be intermediate between proportionality to [O] + [A] and proportionality to [O], depending upon the values of the rate constants.

The problem of a current, i_k, limited by both a chemical reaction and diffusion to an expanding spherical electrode was solved by Koutecký (19), subject to the simplifications $D_A = D_O$ and $[A] \gg [O]$, i.e., $k_f \ll k_b$. (If [O] is appreciable relative to [A], the limiting current, i_l, is the sum of the diffusion current of O, i_d, and a kinetic current, i_k, determined by the rate of conversion of A to O. Then the equation below is still valid for $i_k = i_l - i_d$.) Koutecký's result cannot be expressed in closed form, but is approximated by the equation

$$\frac{i_k}{i_d} = \frac{0.87\beta}{1 + 0.87\beta} \tag{8}$$

where i_d is the diffusion current to be expected if A were reducible at the same potential as O, or if A were transformed into O very rapidly, and $\beta^2 = k_f^2 t/k_b = k_f K_e t$. For large values of β ($\beta > 10$), i_k is practically equal to i_d. Since drop times are generally around 4 sec., $\beta > 10$ corresponds to $k_f K_e > 25$ sec.$^{-1}$, and the chemical reaction is so fast that the current is almost entirely diffusion-controlled. For small values of β ($\beta < 0.1$, $k_f K_e < 2.5 \times 10^{-3}$ sec.$^{-1}$), the current is limited almost entirely by the rate of the chemical reaction, and i_k is proportional to β, viz.,

$$i_k = 0.87\beta i_d = 528nm^{2/3}t^{2/3}D_A^{1/2}k_f^{1/2}K_e^{1/2}[A] \tag{9}$$

This treatment has been used to evaluate rate constants of many rapid reactions from the magnitude of the kinetic currents. Although such data are of obvious interest to physical organic chemists, they will not be reviewed in an article on polarographic mechanisms. The interested reader is referred to Brdička's review article (20).

Notice that for this latter case of a slow chemical reaction, the kinetic current is still proportional to n and [A], but the quantity

$m^{2/3}t^{2/3}$ is proportional to the area of the electrode, but independent of h, the height of the mercury reservoir. For intermediate rates ($2.5 \times 10^{-3} < k_f K_e < 25$ sec.$^{-1}$), the current is limited partly by diffusion and partly by chemical kinetics, so that the dependence of limiting current on h is between zero- and half-order. Also, the temperature coefficient of k_f is considerably greater than that of D, so that i_k often increases by about 10% per degree. An apparent exception to this rule occurs with pyruvic acid, whose first wave shows a temperature coefficient of only 1.9% per degree (21). However, the reaction of pyruvate anion with protons is diffusion-controlled and would have a small activation energy.

A second-order reaction,

$$\mathrm{A} + \mathrm{B} \underset{k_b'}{\overset{k_f'}{\rightleftharpoons}} \mathrm{O} \xrightarrow{ne^-} \mathrm{R}$$

can usually be run with a large excess of B (usually protons or water), so that pseudo-first-order kinetics are maintained. The second-order rate constant, k_f', is converted to the first-order rate constant, $k_f'[\mathrm{B}]$, and this latter is to be interpreted as the k_f of the above discussion.

3. Catalytic Current

Consider the reaction scheme

$$\mathrm{O} + ne^- \rightarrow \mathrm{R}$$

$$\mathrm{R} + \mathrm{Z} \underset{k_b}{\overset{k_f}{\rightleftharpoons}} \mathrm{O}$$

in which Z is not reducible at potentials at which O is reduced. The reduction of Z must require a large overvoltage, since the equilibrium potential of Z is such that it is capable of oxidizing R. The electrolytic reduction of O catalyzes the reduction of Z, since O is regenerated by the chemical reaction. This repeated electrolytic reduction of O thus leads to an increased current.

The problem of a current, i_c, limited by both diffusion and the rate of regeneration of O was solved for the case of the dme by Koutecký. The results are not expressible in closed form, but they have been tabulated by Delahay (5), in terms of the parameter, $\rho = (k_f + k_b)[\mathrm{Z}]t$. For small values of ρ ($\rho < 0.2$), the regeneration of O is too slow to produce an appreciable increase in current, so that the limiting current is merely a diffusion current of O. For large values of ρ ($\rho > 10$), the limiting current, i_c, is given by

$$i_c = 0.812\rho^{1/2} i_d = 0.812(k_f + k_b)^{1/2}[\mathrm{Z}]^{1/2}t^{1/2}i_d$$

and the current is proportional to $m^{2/3}t^{2/3}$, but independent of h, although this is not always observed. Also, a large temperature dependence of i_c is expected.

4. Adsorption Current

Many organic compounds exhibit an anomalous prewave at potentials less negative than the main wave, which is observed at high concentrations. The sum of the two waves is proportional to the concentration, as required by the Ilkovič equation [eq. (7)] for a diffusion-controlled wave. However, the first is proportional to concentration only at low concentrations; as the concentration increases, the prewave reaches a limiting height. This prewave is generally attributed to facile reduction of an adsorbed layer. At low concentrations, the available sites on the mercury surface are not completely filled, so an increase in concentration can lead to an increase in adsorption current. But once the available sites are filled, no increase in adsorption current can arise from further increase in concentration, and only the normal wave increases in height. If adsorption is sufficiently rapid, the adsorption current is proportional to the rate of growth of the mercury surface, or to $m^{2/3}t^{-1/3} = m(mt)^{-1/3}$. Therefore, an adsorption current is directly proportional to h. Generally, adsorption currents seem to be nearly independent of temperature.

C. WAVE SHAPES AND HALF-WAVE POTENTIALS

1. Reversible Waves

Reversible waves are those for which the electrode reaction is sufficiently rapid, compared to diffusion, that equilibrium is always maintained at the electrode surface. The shape of the current–potential curve for the reversible reduction of one soluble species to another is given by the equation

$$E = E^0 - \frac{RT}{nF} \ln \frac{D_\mathrm{O}^{1/2}}{D_\mathrm{R}^{1/2}} + \frac{RT}{nF} \ln \frac{i_d - i}{i} \tag{10}$$

where E^0 is the formal potential (the standard potential, with suitable adjustment for concentrations of protons, etc., which balance the redox equation), i is the current for any given potential E, and i_d is the diffusion current, given by the Ilkovič equation (see Fig. 1). For the case in which both oxidized and reduced forms are present, there are both cathodic ($i_- > 0$) and anodic ($i_+ < 0$) limiting currents, and the general equation is

$$E = E^0 - \frac{RT}{nF} \ln \frac{D_O^{1/2}}{D_R^{1/2}} + \frac{RT}{nF} \ln \frac{i_- - i}{i - i_+} \tag{11}$$

When $i = i_d/2$ [or $(i_- + i_+)/2$, in the general case], the last term equals zero. The potential at which the current equals half the limiting current is known as the "half-wave potential," $E_{1/2}$, and is seen to equal $E^0 - (RT/nF) \ln (D_O^{1/2}/D_R^{1/2})$.

By comparing the shape of this polarographic wave with the curve obtained from a potentiometric titration, Müller (8) has drawn the analogy that the polarogram represents "an electron titration of the interface."

Since the diffusion coefficients of oxidized and reduced forms do not differ widely, $E_{1/2}$ is quite close to E^0. (Even a factor of two for the ratio of diffusion coefficients leads to a difference between $E_{1/2}$ and E^0 of only 0.009 v.) Since the limiting current represents the reduction of O as rapidly as it diffuses to the surface, a current of half that value represents the reduction of exactly half the O molecules which diffuse to the surface. If D_O equaled D_R, $[O]_s$ would equal $[R]_s$, and $E_{1/2}$ would equal E^0. When $D_O \neq D_R$, the different rates of diffusion upset the equality of $[O]_s$ and $[R]_s$ for the condition $i = i_d/2$, and $E_{1/2}$ no longer quite equals E^0.

It is important to note that $E_{1/2}$ is independent of both [O] and [R]. However, since E^0 above is a formal potential, $E_{1/2}$ will depend upon the concentrations of other components of the redox reaction, usually protons. For the reaction $O + mH^+ + ne^- = R$

$$E_{1/2} = E^0 - \frac{RT}{nF} \ln \frac{D_O^{1/2}}{D_R^{1/2}} + m \frac{RT}{nF} \ln a_{H^+} = E_{1/2}^0 - 0.059 \frac{m}{n} \text{pH} \tag{12}$$

where $E_{1/2}^0$ is the half-wave potential at pH = 0. This equation assumes that the solution is well-buffered, so that the electrode reaction does not appreciably consume or produce protons and thereby change the pH at the interface. Müller (22) and Kolthoff and Lingane (23) have provided a thorough discussion of reversible oxidations and reductions in unbuffered or poorly buffered solutions. An interesting corollary is the observation that carbon dioxide–bicarbonate buffers react too slowly to provide adequate buffering action, but addition of the enzyme, carbonic anhydrase, speeds the buffering action (24).

Half-wave potentials of reversible processes are nearly independent of the temperature.

2. Irreversible Waves

The shape of the current–potential curve for a reduction limited by both slow electron transfer and diffusion to an expanding spherical electrode was solved by Koutecký (19). The result cannot be expressed in closed form, but is approximated by

$$\frac{i}{i_d} = \frac{0.87\lambda}{1 + 0.87\lambda} = \frac{0.87k_{f,h}\sqrt{t/D_O}}{1 + 0.87k_{f,h}\sqrt{t/D_O}} \tag{13}$$

where $\lambda = k_{f,h}t^{1/2}D_O^{-1/2}$, and i_d is the limiting current at very negative potentials, where the electron activity is so high and the electrode reaction so rapid that the current is limited only by diffusion.

A qualitative explanation of the shapes of irreversible waves may be expressed as follows: At the less cathodic potentials, the rate of reduction is very low, and equilibrium is not established at the electrode surface, so that the observed current is less than the current for a reversible reduction. At these potentials i/i_d is directly proportional to $k_{f,h}$. As the electron activity is increased, the rate of reduction increases, since $k_{f,h}$ increases exponentially with a decrease in E [eq. (4)]. But the current rise is not as steep as that for a reversible wave, since only the fraction αn of a decrease in E facilitates the reduction. Eventually, at sufficiently cathodic potentials, the second term in the denominator of eq. (13) becomes much larger than unity, and i approaches i_d. Then the rate of reduction is so great that every O molecule which diffuses to the electrode is immediately reduced, and the current is limited only by diffusion (see Fig. 1).

So far we have been vague in referring to "fast" and "slow" electrode reactions, to contrast reversible and irreversible reactions, without specifying what these terms mean, in a quantitative sense. Delahay (25) has derived the more precise requirement that $k_h/D_O^{1/2}$ must be greater than 5 sec.$^{-1/2}$ in order for the reduction to be considered reversible. For an organic molecule with $D = 10^{-5}$ cm.^{2}/sec, this is equivalent to the requirement that k_h be at least 1.5×10^{-2} cm./sec., or an exchange current of at least 40 μamp. (for a two-electron reduction of a species present in 1mM concentration, and mt = 4 mg.).

It is important to recognize the specialized usage of the terms "reversible" and "irreversible" in electrochemistry. These terms reflect *rates* of electrode processes, and the irreversibility of an electrode reaction means only that the reaction does not proceed sufficiently rapidly to maintain equilibrium at the electrode surface. Of course, an electrode reaction involving chemically irreversible steps will be electrochemically irreversible, since if a forward reaction is irreversible, the backward reaction certainly cannot proceed sufficiently rapidly. But it is also possible for an electrochemical reaction to be irreversible solely because the electron transfer is slow; a simple inorganic example of this is the (electrochemically) irreversible reduction of aqua-cobalt ion, $Co(OH_2)_6^{++}$.

From Koutecký's exact solution, it follows that when $i = i_d/2$, the half-wave potential of an irreversible reduction is given by

$$E_{1/2} = \frac{RT}{\alpha nF} \ln \frac{k_{f,h}{}^0 t^{1/2}}{0.76 D_O{}^{1/2}} = E^0 + \frac{RT}{\alpha nF} \ln \frac{k_h t^{1/2}}{0.76 D_O{}^{1/2}} \tag{14}$$

which shows that smaller values of k_h shift $E_{1/2}$ to values increasingly more negative than E^0. Although $E_{1/2}$ depends on all the quantities in the above equation, the significance of $E_{1/2}$ in irreversible reductions is clear. When $i = i_d/2$, the rate of the electrode reaction is sufficient to reduce exactly half the molecules that diffuse to the electrode, and $E_{1/2}$ represents the electron activity required to produce that specific rate.

Notice especially that for irreversible waves, the half-wave potential [eq. (14)] depends upon the kinetic parameters of the electrochemical process. But for reversible waves the kinetic parameters are such that equilibrium is always maintained at the electrode surface. Although the position of equilibrium (measured by E^0) depends upon the relative magnitudes of $k_{f,h}{}^0$ and $k_{b,h}{}^0$, it does not depend upon the kinetics of the electrochemical process. Therefore, the kinetic parameters have disappeared from eq. (12).

Since the half-wave potential of an irreversible process depends upon the value of a rate constant, such half-wave potentials depend upon temperature.

The treatment for an anodic wave is similar, and leads to eq. (14), except with the sign of the second logarithmic term reversed, and α replaced by $(1 - \alpha)$. A small rate constant shifts the anodic $E_{1/2}$ to values more positive than E^0. Thus, the anodic and cathodic half-wave potentials for an irreversible oxidation–reduction are not equal, but are shifted by nearly equal amounts to higher and lower values than E^0. The $E_{1/2}$ of a reversible reaction is almost identical with its E^0, and independent of [O] or [R], but the rate of electrochemical reaction for an irreversible reaction is so low at $E = E^0$ that the potential must be shifted to more or less cathodic values in order to produce an appreciable cathodic or anodic current. Thus, a solution containing both O and R shows two separated waves, an anodic wave of height proportional to [R] and a cathodic wave of height proportional to [O].

From eq. (13) for an irreversible wave, it follows that

$$\ln \frac{i}{i_d - i} = \ln 0.87\lambda = \ln \frac{0.87 k_{f,h} t^{1/2}}{D_O^{1/2}} = \ln \frac{0.87 k_{f,h}^0 t^{1/2}}{D_O^{1/2}} - \frac{\alpha n F}{RT} E \tag{15}$$

[The discrepancy between the constants of eq. (14) and (15) arises from the use of the approximate solution, eq. (13), in deriving eq. (15).] Thus it is clear that a plot of $(RT/F) \ln [i/(i_d - i)]$ against $-E$ is linear with slope αn, as has been observed experimentally for many years. Such plots will henceforth be referred to as "log plots." As defined, the "log plot slope" equals αn, but it has been common practice to plot $-E$ against $\log [i/(i_d - i)]$ to obtain a slope equal to $0.059/\alpha n$, and these plots are also referred to as log plots, with slopes expressed in volts.

Many irreversible waves show nearly linear log plots, but with slopes unequal to n, and often corresponding to the transfer of an apparently non-integral number of electrons. For a reversible wave, according to eq. (10), the log plot slope should correspond to an integral n. A slope corresponding to a non-integral value of n has generally been taken as evidence for the irreversibility of an electrode process, although the converse is not necessarily true, as will be seen in the following examples.

3. *Pseudo-Reversible Waves*

There are numerous reactions for which log plots indicate the transfer of an integral number of electrons, but which are known on other grounds to be irreversible. Most of these reactions involve reversible electron transfer, followed by irreversible chemical reaction. We will mention only a few mechanisms which can lead to an apparently reversible wave.

Kivalo (26) has treated the following reaction scheme by a steady-state method

$$\mathrm{O} + ne^- \underset{k_{b,h}}{\overset{k_{f,h}}{\rightleftharpoons}} \mathrm{R} \underset{k_b}{\overset{k_f}{\rightleftharpoons}} \mathrm{P}$$

Subject to the assumption that diffusion of O, R, and P to and from the electrode surface is slow compared to the rates of electron transfer and chemical reaction, four limiting cases may be distinguished:

(a) The electron transfer is much faster than the reversible chemical reaction: Then the log plot slope equals the number of electrons transferred, even though the product of the electrode reaction reacts further. The half-wave potential is no longer approximately equal to the standard potential for the process $\mathrm{O} + ne^- = \mathrm{R}$, but is shifted to more positive potentials. This shift is due to the depletion of R at the electrode surface, which drives the reaction to the right.

(b) The rapid electron transfer is slower than the reversible chemical reaction: Then the characteristics of the polarographic wave are the same as in case (a), even though the electron transfer is relatively slow.

(c) The chemical reaction is irreversible ($k_b = 0$), but is slower than the electron transfer. The polarographic wave still appears reversible: However, the half-wave potential is shifted to more positive values to an extent dependent upon the magnitude of k_f, since $k_{\mathrm{obs}} = k_{f,h}k_f/(k_{b,h} + k_f) \approx k_{f,h}k_f/k_{b,h}$.

(d) The chemical reaction is irreversible ($k_b = 0$) and is faster than the electrode process: Only in this case is the polarographic wave irreversible, with log plot slope $= \alpha n$. The rate-limiting step is the reduction, and R is transformed into P immediately upon being formed. In this case the half-wave potential does not depend upon k_f, but only upon the kinetic parameters of the electron transfer, since $k_{obs} = k_{f,h}k_f/(k_{b,h} + k_f) \approx k_{f,h}$.

If the reaction scheme is second-order, according to

$$O + ne^- = R \qquad R + A = P + B$$

the above conclusions are still valid if pseudo-first-order conditions can be achieved, i.e., if A and B are present in large excess. The most common situation will be that in which A is H^+ or H_2O.

4. *Electrode Reactions Involving a Change in the Number of Molecules*

Consider the electrode reaction

$$O + \frac{n}{2} e^- \underset{k_{b,h}}{\overset{k_{f,h}}{\rightleftharpoons}} R \qquad 2R \underset{k_b}{\overset{k_f}{\rightleftharpoons}} R_2$$

with $k_f[O] > k_b$, so that the dimer, R_2, is predominant. If all rate constants are sufficiently large, the reaction will be reversible and the Nernst equation, relating [O], $[R_2]$, and E^0 for the overall process, will hold. The analog of eq. (10) for a purely cathodic wave is (27)

$$E = E^0 - \frac{RT}{nF} \ln \frac{d_O}{d_R^{1/2}} + \frac{RT}{nF} \ln \frac{(i_d - i)^2}{i}$$

where $d_i = 607nm^{2/3}t^{1/6}D_i^{1/2}$. When $i = i_d/2$

$$E_{1/2} = E^0 - \frac{RT}{nF} \ln \frac{d_O}{d_R^{1/2}} + \frac{RT}{nF} \ln \frac{i_d}{2}$$

$$= E^0 - \frac{RT}{nF} \ln \frac{D_O^{1/2}}{D_R^{1/2}} + \frac{RT}{nF} \ln \frac{[O]}{2} \qquad (17)$$

Notice that the half-wave potential depends upon [O]. An increase in [O] shifts $E_{1/2}$ to less cathodic potentials because the overall reduction is second order in O.

Similar equations hold for the case in which the oxidized form is dimeric. However, most organic polarography, except for that of sulfide–disulfide systems, is concerned with the above stoichiometry.

For the case of a rapid and reversible dimerization, but a slow electron transfer, the cathodic current–potential curve has the form

$$E = E^0 - \frac{RT}{\alpha nF} \ln \frac{d_O}{nFAk_h} + \frac{RT}{\alpha nF} \ln \frac{(i_d - i)^2}{i}$$
$$E_{1/2} = E^0 - \frac{RT}{\alpha nF} \ln \frac{D_O^{1/2}}{k_h t^{1/2}} + \frac{RT}{\alpha nF} \ln \frac{[O]}{2} \tag{18}$$

The half-wave potential depends upon k_h, but also upon concentration, as expected for a second-order reaction.

The case of a rapid, irreversible dimerization ($k_b = 0$), following a rapid, reversible electron transfer, was solved by Koutecký and Hanuš (28). The current–potential curve may be approximated by

$$E = E^0 + \frac{RT}{3nF} \ln \frac{k_f[O]t}{1.51} + \frac{RT}{3nF} \ln \frac{i_d(i_d - i)^2}{i^3}$$
$$E_{1/2} = E^0 + \frac{RT}{3nF} \ln \frac{k_f[O]t}{0.75} \tag{19}$$

Here the half-wave potential depends upon both k_f and [O].

III. Interpretation of Polarograms

A. DETERMINATION OF n

1. Product Analysis

Certainly one of the most important steps in establishing the mechanism of the process giving rise to a polarographic wave is the determination of the product of the electrode reaction. The most unambiguous method is to carry out the reduction (or oxidation) on a preparative scale and identify the products. To guarantee that the preparative electrolysis produces the same product as is formed in the polarographic wave, it is essential to carry out the electrolysis at a potential controlled near the beginning of the polarographic limiting current. Otherwise it is possible that the isolated product is one which arises from further reduction of the polarographic product at more cathodic potentials.

One of the difficulties with preparative electrolysis is that in polarography only minute amounts of product are formed. Polarographic currents are measured in microamperes, and recording a polarographic wave requires only about ten minutes. The electrolysis amounts to only of the order of 10^{-9} moles of product formed. Obviously it is necessary to prolong greatly the time of electrolysis or to increase the current, by increasing the concentration of reactant and the size of the electrode, and by stirring the solution to increase the rate of mass transfer to the electrode. The first alternative is clearly closer to polarographic conditions, but the latter may be quite different. An increase in [O] will facilitate an electrode reaction involving dimerization of an intermediate [eq. (19)] at the expense of a competing unimolecular process, such as further reduction, which is not affected by concentration. Stirring the solution may sweep a metastable intermediate away from the electrode before it undergoes further reduction. Also, a polarographic product may react further during the longer time scale of preparative electrolysis, but not during the life of the mercury drop. However, if due consideration is given to the possible pitfalls, controlled potential electrolysis (cpe) is an important first step in elucidating the mechanism of an electrode reaction.

Meites (29) has provided an excellent review of the principles, technique, and applications of cpe.

2. Coulometry

An alternative to identifying the product of an electrode process is to determine the number of electrons involved. It is usually an easy matter to guess the nature of the product if its oxidation state relative to that of the reactant is known. Controlled potential coulometry (cpc) involves measuring the amount of electricity required to effect complete electrolysis of the polarographically active material. The charge q is related to the current, to the volume of solution V, and to n by the equation

$$q = \int i dt = nF[\mathrm{O}]_{\mathrm{initial}} V$$

Macroscale cpc is subject to the same pitfalls as is cpe, but convenient procedures for microscale cpc have been developed, and an integral number of electrons is a reliable indication of the course of an electrode reaction.

Meites (29) has provided an excellent review of the principles, technique, and applications of cpc.

3. Estimation of n from the Magnitude of i_d

The Ilkovič equation [eq. (7)] provides another means of evaluating n. Since m and t for a given capillary and column height may be easily determined from separate experiments, an estimate of n can be obtained directly from the limiting current of the polarographic wave, if D_O can be determined. For small values of n, only a crude estimate of D_O is necessary to distinguish among possible integral values of n. It is especially easy to distinguish between one- and two- electron processes on the basis of a very approximate value for D_O, but diffusion coefficients are not known with sufficient accuracy to distinguish between two many-electron processes.

The diffusion coefficient of the polarographically active substance itself may be determined, or may be approximated by the diffusion coefficient of a structurally similar reference molecule. Several methods are available for the determination of diffusion coefficients (30,31):

(a) Direct determination of diffusion coefficients, with a diffusion cell.

(b) Evaluation of diffusion coefficients from ionic mobilities, as determined from equivalent conductances, λ_i:

$$D_i = (RT/n_iF^2)\lambda_i$$

(c) Estimation of diffusion coefficients from the Stokes–Einstein equation, where η = viscosity of solvent, and r = radius of particle: $D = kT/6\pi\eta r$. This equation also serves to relate diffusion coefficients in solutions of different viscosities.

(d) Estimation of diffusion coefficients from comparison of polarographic diffusion currents: This method involves determining D of a reference compound from its limiting current, for which n has been established by some other method. Then D of the species under investigation is approximated by D of the reference compound. Thus nitrobenzene has been shown to undergo four-electron reduction, since D for nitrobenzene can be approximated by D of benzoate anion, known from limiting conductance data; from a comparison of limiting currents, it can further be concluded that nitrobenzene and m-dinitrobenzene are both reduced with the same number of electrons (32). Common reference compounds include nitrobenzene, which gives a four-electron wave in neutral solution, benzoquinone, which gives a two-electron wave, and benzophenone, which gives two one-electron waves in acid. A variant of this method is to use the height of one well-understood wave of the compound being investigated in order to estimate n for another wave. For example, phenacyl bromide shows two waves of equal height, with the second at the same half-wave potential as the two-electron wave of acetophenone. These results are consistent with a two-electron reduction of phenacyl bromide to bromide ion plus acetophenone, which is reduced further at more negative potentials.

The primary advantage of using diffusion coefficients to estimate n is the simplicity of the method. It eliminates the necessity of carrying out a time-consuming cpe, requiring either product analysis or current measurement. Often values for the diffusion coefficients of substrate or related molecules are available in the literature. If not, it is easy to apply method (d), to determine n by comparison of the observed limiting current with the limiting current of a suitable reference compound. A further advantage of this method is that n is evaluated from the limiting current under polarographic conditions, rather than from data accumulated under different conditions, with a longer time scale.

B. MULTIPLE STEP POLAROGRAMS

Often a polarogram will exhibit more than one wave, and a mechanistic interpretation should explain each separate wave. However, a multiplicity of waves can be due to any of several causes, and it is not always easy to decide which is responsible. Two very important causes of wave doubling are successive reductions, via an intermediate, and separate reduction of two species.

1. *Successive Reductions Steps*

$$\mathrm{O} + n_1e^- \rightarrow \mathrm{I} \xrightarrow{n_2e^-} \mathrm{R}$$

If I, the product of the first polarographic wave, is itself reducible at more cathodic potentials, another wave should be observed at such a potential. At the potential of the first plateau, the electron activity is high enough to reduce the substrate, O, only to some intermediate reduction level, I. But at more cathodic potentials, the electron activity is high enough to reduce I also, as soon as it is formed from O. Of course, in order to show two separate waves, the second step must be slower than the first; otherwise I would be immediately reduced to R at any potential capable of reducing O to I. In short, if it is proposed that a reduction shows two waves because it proceeds via a stable intermediate, this mechanism can be tested by ascertaining whether I is indeed reduced at a potential equal to the second $E_{1/2}$ of O.

A corollary of this discussion is that if it is proposed in a mechanism that the substrate, O, is reduced to some product I, and if I is known to be reducible at a potential E_{I}, then the polarogram of O should exhibit a second wave at E_{I}. If I is known to be reducible at less cathodic potentials than O, then I cannot be the product of the first wave of O. Thus the product of the first (four-electron) reduction wave of nitrobenzene cannot be azobenzene, since azobenzene is reduced at less negative potentials. The product must be *N*-phenylhydroxylamine, since in acid this shows a reduction wave at the same $E_{1/2}$ as the second wave of nitrobenzene in acid (32). Nitrosobenzene is a likely intermediate, but it is reduced more readily than is nitrobenzene, and, therefore, is reduced immediately on being formed. (At pH 0, the half-wave potentials of these four compounds, nitrobenzene, azobenzene, phenylhydroxylamine, and nitrosobenzene, are -0.16, $+0.07$, -0.61, and $+0.2$ v., respectively.) Obviously, it is also necessary that the product be stable under the reaction conditions if it is indeed the product.

In the case of a polarogram exhibiting more than one wave, it is not necessary that the intermediate formed by electrolysis on the first plateau be a stable one. Often a metastable intermediate radical is formed in a reversible, one-electron process. This radical then undergoes irreversible dimerization [according to eq. (19)], or further reduction, or disproportionation. The competition between the first possibility, consuming only one electron, and the second, consuming a total of two, will be governed by both concentration and electrode potential. At very negative potentials the electron activity is so high that the radical is reduced more rapidly than it dimerizes or disproportionates. Another possibility is that I is less easily reducible than O, but undergoes a rapid rearrangement to a species more readily reducible than O. If the rearrangement is slow, two waves will be observed, with the second corresponding to the direct reduction of I. If the rearrangement is rapid, only one wave will be observed, since the rearrangement product of I is immediately reduced at any potential capable of reducing O.

Nor is it justifiable to infer that no intermediate is formed if a polarogram exhibits only a single wave. An intermediate may be formed, but may be reduced even more easily than the substrate. Then, on the diffusion plateau, O is immediately reduced to I, but I may be even more rapidly reduced to R. Nitrosobenzene is presumably such an intermediate.

2. Polarography of Equilibrium Mixtures

The substrate, O, may exist in the reaction mixture in two forms, O_1 and O_2, in equilibrium. If the two forms are reduced at different potentials, and if the interconversion is slow, there will be separate waves for each, even though they are reduced to the same product. The ratio of the two wave heights is equal to the ratio of the concentrations of the two species in solution, and both waves have the characteristics of a diffusion-limited wave (i_d proportional to $h^{1/2}$). If the rate of interconversion is high, only one wave will be observed, since as soon as electrolysis depletes the interface of the more readily reduced form, it is immediately regenerated from the other. For intermediate rates of interconversion, the limiting current of the more readily reduced form consists of a diffusion current ($i_d \propto h^{1/2}$) proportional to its equilibrium concentration, plus a kinetic current (i_k independent of h), determined by the rate at which it can be replenished by chemical reaction [eq. (8)]. The total current, on the second plateau, corresponds to diffusion-limited reduction of both O_1 and O_2, but the ratio of the two separate wave heights is not equal to the ratio of the concentrations. The first wave is augmented by the kinetic current, and the second wave is diminished by the same amount. The most common such case is that of an acid and its conjugate base, which are reducible at different potentials, as a result of the different substitutent effects of a COOH group and a CO_2^- group. Although there are equal concentrations of the two forms at pH = pK_a, the pH at which the two waves are of equal height will differ from pK_a. Since the acid is generally more readily reduced than its conjugate base, this latter pH, usually designated by pK', is greater than pK_a, by an amount dependent upon the magnitude of $k_f'[H^+]$.

In general, it is difficult to attribute a multiple polarographic wave unambiguously to either stepwise reduction or separate reduction of two species in slow equilibrium. That a wave is kinetic is not difficult to show, from its independence of h, and the fact that the ratio of wave heights differs from the concentration ratio. But if both waves are diffusion-controlled, it is necessary to show that cpe on the two plateaus leads to two different products (implying successive reduction) or the same product (implying separate reduction of two species, but possibly disproportionation of the intermediate).

3. Adsorption Prewaves

An adsorption wave at less cathodic potentials than the main wave is easy to spot by its unusual dependence on concentration and on h.

C. KINETIC ORDERS

A considerable number of mechanistic inferences in organic chemistry are made on the basis of kinetic orders. The transition state of a homogeneous reaction, nth order in a given reactant, is considered to contain n molecules of that reactant. Although much less attention has been focussed on describing the structures of transition states in heterogeneous electrode processes, it is still possible to carry over the notion of kinetic order to such reactions. It is necessary to realize that an increase in the concentration of a reactant may affect both the half-wave potential and the limiting current, but it is the effect on the half-wave potential which is kinetically significant.

In most of the previous discussion, we have assumed that the reduction and oxidation reactions are first-order in oxidant and reductant, respectively. On the basis of such an assumption, we have stated that for both reversible and irreversible reactions, $E_{1/2}$ is independent of concentration, if neither product nor reactant is dimeric. On the other hand, if the electrode reactions are first-order, but complicated by a dimerization of the reduced form, the half-wave potential depends logarithmically on [O].

We have seen that it is not possible to consider an n-electron reduction as nth order in electrons, since the reverse reaction, an n-electron oxidation, is impeded by an increase in the escaping tendency of electrons from the electrode. This effect is taken into account by the introduction of the transfer coefficient, α, whereby the reduction may be viewed as being of order αn in electrons. By analogy with the Brönsted α, it is reasonable to consider the transition state as being α of the distance along from reactant to product, with αn electrons having been transferred at the transition state. A consideration of the structure of the electrical double layer surrounding the electrode (33) shows that α is not constant, but must decrease as the potential becomes more cathodic beyond the electrocapillary maximum. This behavior is consistent with Hammond's Principle (34), applied to electrode reactions. At very cathodic potentials, the electron activity in the electrode is very high, and the reduction is very exoergonic, so the transition state is close to reactants, corresponding to a smaller value of α.

So far we have touched only briefly on kinetic orders of coreactants in polarographic reduction. By far the most common such coreactant is hydrogen ion, and the remainder of this section is devoted to a consideration of the role of protons in facilitating reductions. This discussion is adapted from Elving's excellent review (18).

For reversible waves, eq. (12) describes the dependence of $E_{1/2}$ on pH. A plot of $-E_{1/2}$ vs. pH should have a slope equal to $0.059m/n$ volts, where m is the difference in the number of protons between oxidized and reduced forms. This slope need not be constant over the entire pH range, since m will increase or decrease by one on increasing the pH beyond the pK_a of oxidized or reduced forms, respectively. Thus the plot will consist of a set of straight lines, with kinks at the pK_a's. Analytic forms of such curves are described by Kolthoff and Lingane (1) and by Elving (18); they are similar to potentiometric curves. Analogous behavior is to be expected if the reduced form is dimeric, if E^0 in eq. (17) is interpreted as a formal potential, dependent upon pH.

For a one-step irreversible reaction involving m protons [eq. (12)], eq. (14) is still valid, and a change in pH can affect both α and $k_{f,h}{}^0$. If the effect of pH on α is neglected (by no means justifiable—see Ref. 18), the effect of pH is that on $k_{f,h}{}^0$. Thus, if the rate constant at the zero of potential is $k_{f,h}{}^0[\mathrm{H}^+]^m$, differentiation with respect to pH leads to

$$\frac{dE_{1/2}}{d\mathrm{pH}} = -\frac{0.059}{\alpha n}\frac{d \log k_{f,h}{}^0}{d \log [\mathrm{H}^+]} = -0.059\,\frac{m}{\alpha n} \tag{20}$$

Since this derivation assumes only that m protons are included in the transition state, it does not distinguish between addition of the protons prior to electron transfer and addition of the protons simultaneous with electron transfer. The first possibility is equivalent to a rapid preequilibrium of substrate with its m-fold protonated form (slow preequilibrium can be distinguished by the kinetic nature of the wave), and the second is equivalent to protonation induced by an increased basicity of the transition state, due to the negative charge of αn electrons it has acquired.

Again, m need not be constant over the entire pH range, but the $E_{1/2}$–pH plot will consist of straight lines with kinks at the pK_a's of the substrate. At low pH values the substrate may be completely converted to a conjugate acid, so that one fewer proton is required to convert substrate to transition state. There will also be kinks at what might be called the pK_a's of the transition state. This phenomenon may be considered as arising from two simultaneous reactions

$$\mathrm{O} + ne^- + m\mathrm{H}^+ \underset{}{\overset{K_m{}^{\ddagger}}{\rightleftharpoons}} \mathrm{OH}_m{}^{-(n-m)\ddagger}$$

$$\mathrm{O} + ne^- + (m-1)\,\mathrm{H}^+ \underset{}{\overset{K_{m-1}{}^{\ddagger}}{\rightleftharpoons}} \mathrm{OH}_{m-1}{}^{-(n-m+1)\ddagger}$$

At higher pH, the second reaction, with its lower hydrogen-ion requirements, may become faster than the first, so that m will decrease by one. An example of this situation occurs in the reduction of α-bromoalkanoic acids. In acid solution, reduction of the acid is a pH-independent process. Slightly above the pK_a of the acid, where anion predominates, the reduction is pH-dependent, and the transition state contains one proton more than the substrate anion. Near pH 8 the reduction again becomes pH-independent, since it is then easier to reduce the anion directly by a process not requiring a proton.

We have already considered the possibility that the prior equilibrium of the substrate with its protonated form is not rapid compared to the electrode reaction. If so, the two forms will be reduced at different potentials, with a possible kinetic enhancement of the first wave if the protonation is not too slow. The protonation is first-order in $[\mathrm{H}^+]$, and, therefore, the height of the kinetic wave depends upon pH. The cases of slow proton transfer are usually those in which the site of protonation is different from the reaction site, but many cases involve both rapid proton transfer at the reaction site and slower proton transfer at a second site. Such a process shows two waves, with the first having a kinetic component to the current. Not only does the ratio of the two wave heights depend upon pH, but also the half-wave potentials of the two waves are pH-dependent. An example of this behavior is the reduction of the pyruvic acid–pyruvate anion system, in which protonation of the anion leads to a kinetic enhancement of the first wave, but for both waves, the $E_{1/2}$'s for reduction of the carbonyl group are pH-dependent.

The cases involving protonation of the primary reduction product subsequent to a reversible electron transfer exhibit different behavior, depending upon the rate and reversibility of the proton transfer. If the proton transfer is reversible, the half-wave potential will depend upon pH according to the Nernst equation and eq. (12), just as though the protons were added prior to electron transfer. If the proton transfer is irreversible, Kivalo's treatment, extended to pseudo-first-order protonation, is applicable. If irreversible protonation is faster than reversible electron transfer, the polarographic wave will be irreversible, with $E_{1/2}$ independent of the rate of proton transfer. If irreversible protonation is slower than the reversible electron transfer, the wave will appear reversible, but $E_{1/2}$ will depend upon the rate of proton transfer. This does not necessarily mean that $E_{1/2}$ depends upon pH, since measurements are often restricted to alkaline solution, because in acid, reduction of protons may precede reduction of substrate. In the alkaline solutions the most important proton donor may be solvent, rather than lyonium ion (depending upon the Brönsted α), and the rate of proton removal from solvent molecules is independent of pH. A likely example of this situation is the reduction of aromatic hydrocarbons, although the mechanism is extended by addition of yet another electron and another proton.

Proton transfer subsequent to an irreversible electron transfer can have no effect on $E_{1/2}$, since the potential-determining step is electron transfer, and rates of subsequent steps are immaterial.

Finally, a warning about the structure of the electrical double layer must be injected. The electric field at the interface acts so as to increase the concentrations of cations, and to decrease the concentrations of anions, at the electrode surface. Thus the pH in the vicinity of the electrode may be a few units less than that in the bulk of the solution. However, this augmentation of acidity depends upon the ionic strength of the medium, and even upon specific ion effects, so that measurements with different buffer systems may not be consistent with each other. It is probably this factor which has prevented distinguishing between preprotonation and protonation simultaneous with electron transfer. The half-wave potential for a process of the first type would depend only on pH (specific acid catalysis), whereas the half-wave potential for a process of the second type would depend upon concentrations of all acids (general acid catalysis). But the structure of the double layer is too sensitive to the composition of the medium to permit the distinction to be made experimentally.

IV. Examples of Reversible Systems

A. CRITERIA OF REVERSIBILITY

For a reversible unimolecular reaction, the half-wave potentials of oxidized and reduced forms are identical and independent of concentration. Furthermore, a mixture of the two should show a single combined cathodic–anodic wave, again with the same $E_{1/2}$. And this $E_{1/2}$ should agree with the potentiometrically determined standard potential. By contrast, the cathodic and anodic half-wave potentials of irreversible reactions are unequal, but bracket E^0. If the stoichiometry of the reaction is not one-to-one, the half-wave potential depends upon concentration, but for a reversible reduction to dimeric product, $E_{1/2}$ must be related to E^0 by eq. (17).

Often it is not possible to determine the anodic half-wave potential of the product, and other, less unequivocal, tests for reversibility must be applied. An electrode reaction involving bond-breaking or bond-making (except O—H and N—H bonds), as judged by the nature of the product, is generally irreversible. Another criterion is that $E_{1/2}$ for a reversible process is nearly independent of temperature. Also, for a reversible process, the relation of $E_{1/2}$ to pH must follow eq. (12).

A common criterion for reversibility is the slope of the log plot. For a reversible reaction, this slope equals the number of electrons transferred; for an irreversible reaction, αn is necessarily less than n, and often non-integral.

Irreversible reactions involving some reversible steps exhibit some manifestations of reversibility. The log plot slope may indicate reversible one-electron reduction, but the anodic wave of the product does not occur at the same $E_{1/2}$. Such a result indicates that the reversible electron transfer is followed by an irreversible chemical reaction, slower than the electron transfer. An example of this situation is the two-electron reduction of aromatic hydrocarbons, whose log plot slopes indicate the potential-determining steps to be a reversible one-electron reduction, followed by protonation. Subsequent addition of another electron and another proton does not affect the half-wave potential. The $E_{1/2}$ is independent of pH, as would be expected if water is the most important proton donor, but the result is also consistent with irreversible two-electron transfer, with α exactly 0.5.

Several high-frequency methods have been applied to ascertain whether an electrode process involves reversible electron transfer. These require rapid reversal of the applied potential so as to scan the anodic wave of the primary reduction product. Under such conditions, an irreversible reduction with reversible electron transfer may appear reversible if anodic scanning can be carried out before the primary reduction product reacts further. On the other hand, a reaction which is reversible under polarographic conditions may be irreversible under such rapid-scan conditions if the electrode process is not sufficiently fast to proceed to equilibrium at the higher frequencies involved. Equation (13) and the discussion following it are still applicable to high-frequency methods if t is interpreted as the period of scanning, rather than as the drop time.

A different approach is to modify the reaction system so as to eliminate irreversible steps subsequent to a reversible electron transfer. The use of aprotic solvents has been especially helpful in achieving this aim. A large number of organic compounds add an electron reversibly, but the reduction product undergoes irreversible protonation. Polarography in aprotic solvents, such as acetonitrile and dimethylformamide (DMF), permits an investigation of the electron transfer process alone, and a study of the properties of the primary reduction product.

The most common examples of reversible organic systems are quinones–hydroquinones, azoaromatics–hydrazoaromatics, and arylnitroso–arylhydroxylamine. These will be discussed in turn, along with related systems which are irreversible. Systems which become reversible in aprotic media will be discussed in the section on irreversible processes.

B. QUINONES AND DIKETONES

1. Quinones and Hydroquinones

In general, quinones and hydroquinones yield reversible polarographic waves, with the half-wave potential nearly equal to the standard potential. Because of the simplicity of the reaction, this is the system of choice for testing new techniques in polarography. When high accuracy has not been necessary, polarography has been a more convenient method of determining standard potentials of a series of related quinones than has potentiometry. A polarogram is easy to record, requires only small amounts of material, and can be applied to systems in which one of the forms is labile over the longer periods of time required for potentiometric titration. For example, polarography has been used for the determination of the effect of the size of *meta*-bridged rings on reducibility of quinones (35).

Also, polarography provides a simple method of investigating the stability of semiquinones. If the semiquinone is very stable toward disproportionation, the polarogram will exhibit two one-electron waves, corresponding to stepwise reduction or oxidation. As the stability decreases, the two waves move toward each other, and finally merge into one two-electron wave. Müller (36) has derived an equation for the shape of the polarographic wave which takes into account the possibility of semiquinone formation, with disproportionation constant, K_d = [semiquinone]^2/[quinone][hydroquinone]. He has also studied the effect of pH on the stability of semiquinones. By way of warning, it should be mentioned that there is an unusual situation when $K_d = 4$, such that the two waves just merge to give a two-electron wave, with a log plot slope corresponding to reversible one-electron reduction. But this special situation should not be mistaken for an irreversible two-electron reduction with $\alpha = 0.5$.

In aprotic solvents, anthraquinone (**1**) (37) and benzoquinone (38,39) show two reversible waves, each corresponding to the addition of one electron and no protons. The semiquinone anion (**2**) is stable, and so is the hydroquinone dianion (**3**), which reacts with ethyl bro-

1 + e^- $\underset{\text{DMF}}{\rightleftharpoons}$ **2** $\underset{\text{higher } a_{e^-}}{\overset{e^-}{\rightleftharpoons}}$ **3** $\xrightarrow{\text{EtBr}}$ **4**

O, O (**1**); $O^{\ominus}$, O· (**2**); $O^{\ominus}$, $O^{\ominus}$ (**3**); OEt, OEt (**4**)

mide to form the diethyl ether (**4**) and with acetic anhydride to form the diacetate. Lithium ion shifts both waves to more positive potentials (40). One-electron cpe of anthraquinone (**1**) (41), phenanthrenequinone (**5**), and acenaphthenequinone (**6**) (42) in DMF give

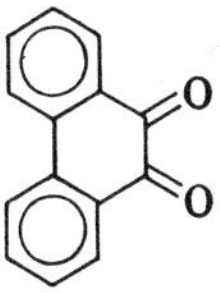

5 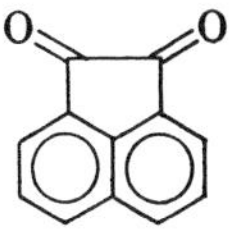 **6**

the semiquinone anions, whose ESR spectra have been observed.

Addition of proton donors, such as phenol or benzoic acid, shifts the second wave to more positive $E_{1/2}$, until it merges with the first. In the presence of proton donors, the semiquinones are not stable to disproportionation, and two-electron reduction, in one step, is observed, as in protic solvents. Even *ortho*-substituted stilbenequinones (**7**, R = Me, *t*-Bu) show the same dependence of wave shape upon concentration of proton donor, indicating that the steric effects on adding protons to the semiquinone are of limited importance (43).

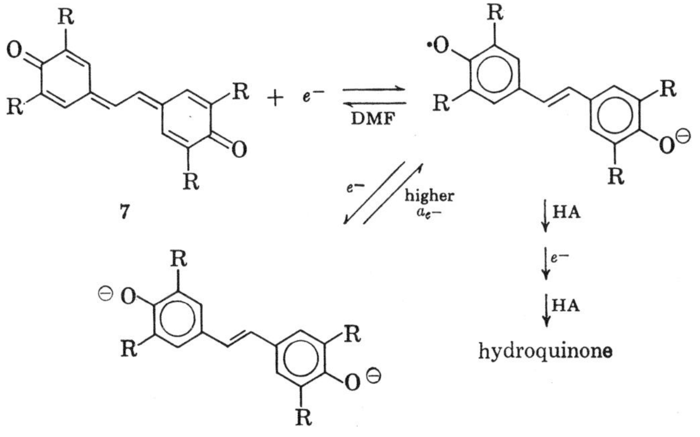

7 + e^- ⇌ (DMF); e^-, higher a_{e^-}; HA, e^-, HA → hydroquinone

Only a few quinones are reduced irreversibly. 1-Hydroxyanthraquinone (**8**) undergoes a reversible two-electron reduction, but the initial product tautomerizes (44). This behavior is similar to that of α-diketones, discussed later.

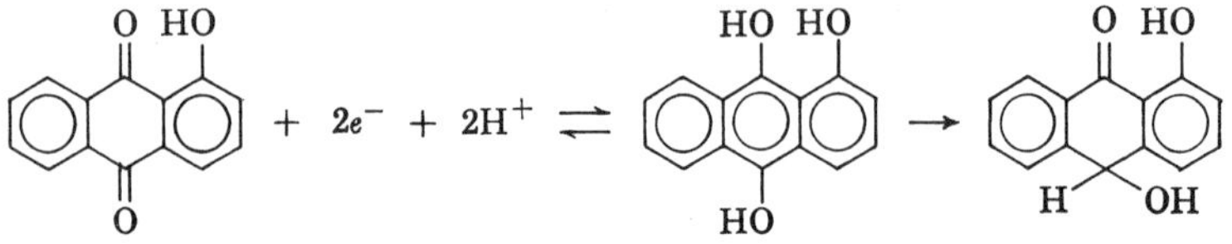

$$\mathbf{8} + 2e^- + 2H^+ \rightleftharpoons \text{(hydroquinone)} \rightarrow \text{(anthrone derivative)}$$

Anthraquinone-1,5-disulfonate (**9**) is reduced reversibly at pH > 11,

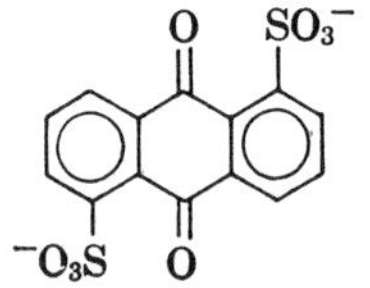

9

but at pH < 11 the anodic and cathodic half-wave potentials are unequal (45). The author has suggested that the irreversibility is due to hydrogen-bonding of the protons of the hydroquinone with the sulfonate substituent, whereby the proton loss from the hydroquinone is presumably slow.

Chromans (e.g. **10**) represent an interesting example of a hydroquinone derivative which is oxidized reversibly, but whose oxidation product (**11**) rearranges to another quinone (**12**) which is reduced at a different $E_{1/2}$ (46). This mechanism corresponds to Kivalo's Case (*c*).

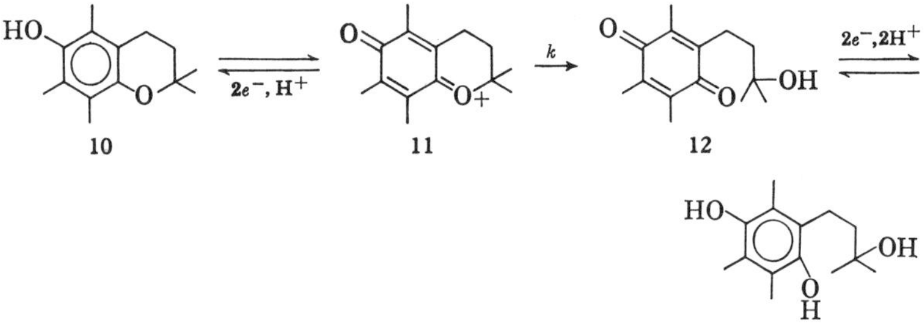

$$\mathbf{10} \underset{2e^-,\,H^+}{\rightleftharpoons} \mathbf{11} \xrightarrow{k} \mathbf{12} \underset{}{\overset{2e^-,2H^+}{\rightleftharpoons}} \text{(hydroquinone)}$$

Orthoquinones also form reversible systems. The cathodic $E_{1/2}$ of *o*-benzoquinone (**13**) is identical with the anodic $E_{1/2}$ of catechol (**14**) and in excellent agreement with the potentiometrically determined E^0 (47). Also, $dE_{1/2}/d\text{pH} = -0.06$ v., as required, although the log plot slope implies $\alpha n = 1.3$ (48), possibly because of a slight tendency to semiquinone formation. 1,2-Naphthoquinone (**15**) and its hydroquinone behave similarly, but pyrogallol (**16**) undergoes irreversible oxidation. According to oscillopolarography, the initial oxidation is reversible, but the primary product is very unstable and reacts further (47). Adrenalin (**17**) shows a reversible anodic wave (49), but the product reacts to form a product which is further oxidized to adrenochrome (**18**).

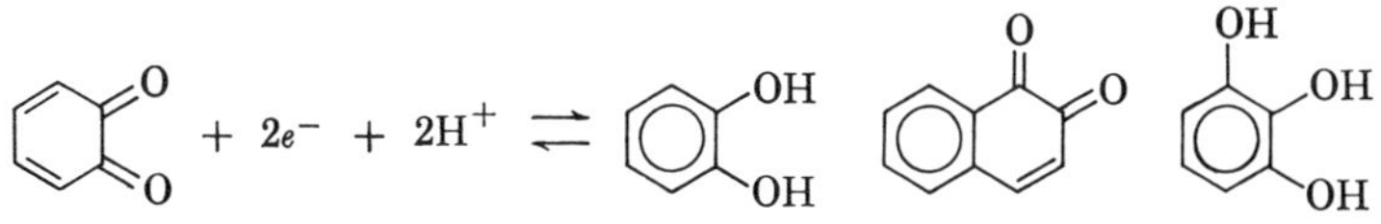

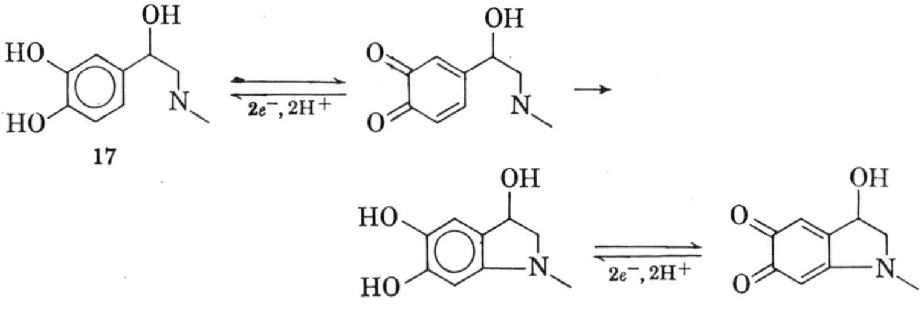

Quinoneimines behave similarly. For example, 2-methyl-4-amino-α-naphthol (**19**) undergoes a reversible, two-electron oxidation, and chemical oxidation leads to a quinone monoimine (**20**) which is reduced at the same $E_{1/2}$, if the cathodic wave is scanned before the imine is hydrolyzed (50). The $E_{1/2}$–pH plot shows a kink at the pK_a of the naphthol. The reducible portion of riboflavin (**21**) is similar to a quinoneimine, and it is reduced at an $E_{1/2}$ equal to its E^0, although there is an adsorption prewave in acid (51).

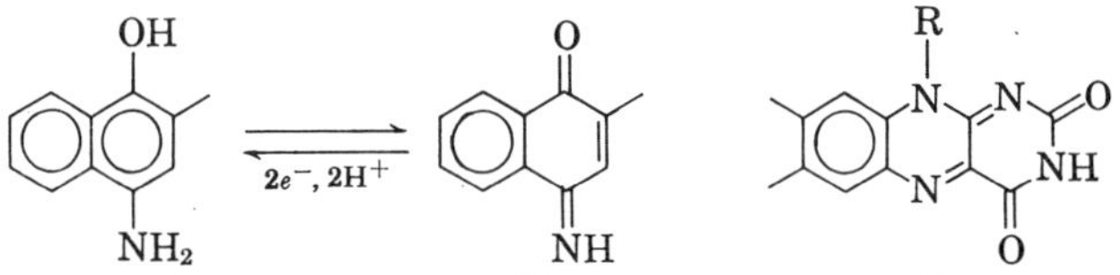

The semiquinone cations of substituted phenylenediamines are well-known as Würster's salts (**22**). The radical cation of *p*-phenylenediamine has been detected by ESR even in aqueous solution (52). The analogous enetetramine, tetrakis(dimethylamino)ethylene (**23**)

shows two reversible anodic waves in aprotic media, corresponding to stepwise oxidation to the radical cation, whose ESR spectrum can be recorded, and to the di-cation, whose salts can be crystallized (53).

22 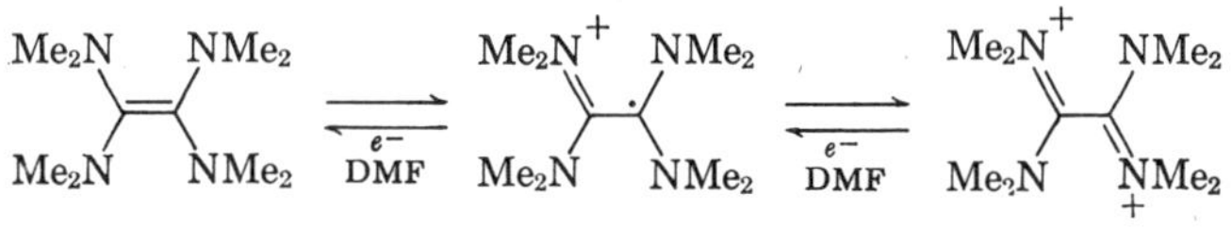 **23**

In alkali, rhodizonic acid (**24**) is reduced in a reversible four-electron step, but in acid there are apparently two waves of equal height (54). Rhodizonic acid gives no anodic wave, although the expected product, $(CO)_6$, is stable as a hexahydrate. The hexahydrate is reducible only at very negative potentials, which is further confirmation of the dodecahydroxycyclohexane structure (**25**). Croconic acid (**26**) and

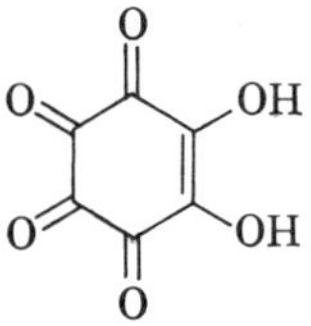

24 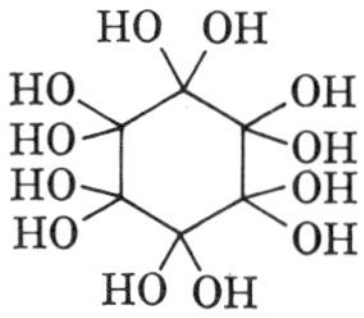 **25** 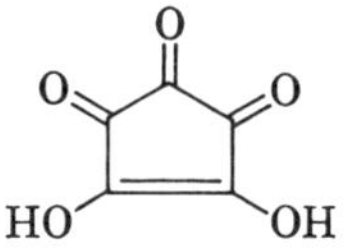 **26**

its anion are reduced in a four-electron wave, but in weakly acid solution a kinetic wave, limited by the rate of protonation of the anion, is observed (55).

2. Ene-diols and α-Diketones

Ene-diols and α-diketones are structurally similar to hydroquinones and quinones, but they generally give irreversible polarographic waves. The electron transfer is reversible, but the product undergoes irreversible chemical reaction. Oxidation of an ene-diol produces an α-diketone, which is then hydrated. The height of the cathodic wave is very small, since it is a kinetic wave limited by the rate of dehydration of this hydrate. Reduction of an α-diketone produces an ene-diol, which tautomerizes to a ketol, which is not re-oxidizable to the diketone at the same $E_{1/2}$.

The ene-diol, "reductone" (α,β-dihydroxyacrolein) (**27**) undergoes two-electron oxidation, with an $E_{1/2}$–pH dependence expected of a reversible reaction, but the product (**28**) is hydrated. Coumarindiol (**29**) behaves similarly (56). It has been suggested (57) on the basis of

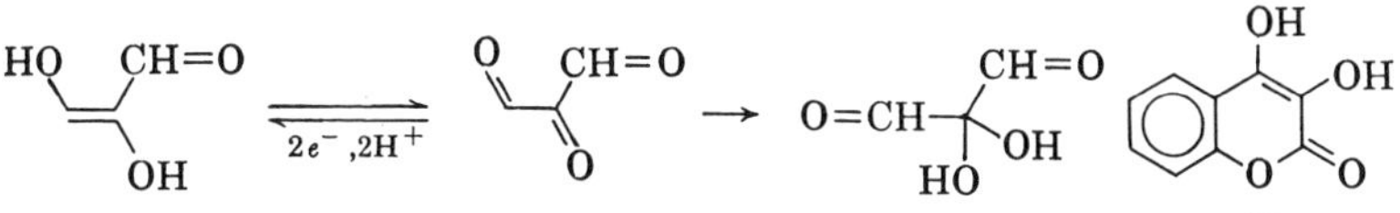

polarography that the stable form of 2,2′-pyridoin is an ene-diol (**30**) which is irreversibly oxidized to a hydrate of 2,2′-pyridil (**31**).

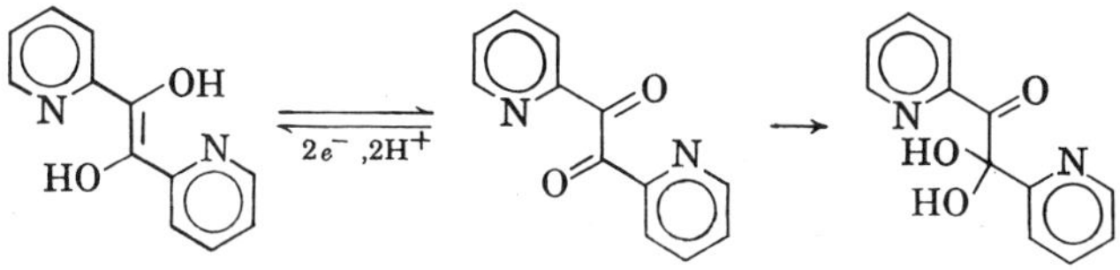

Ascorbic acid (**32**) is oxidized to a diketolactone (**33**) which is rapidly hydrated to dehydroascorbic acid (**34**). In base, **34** rearranges to

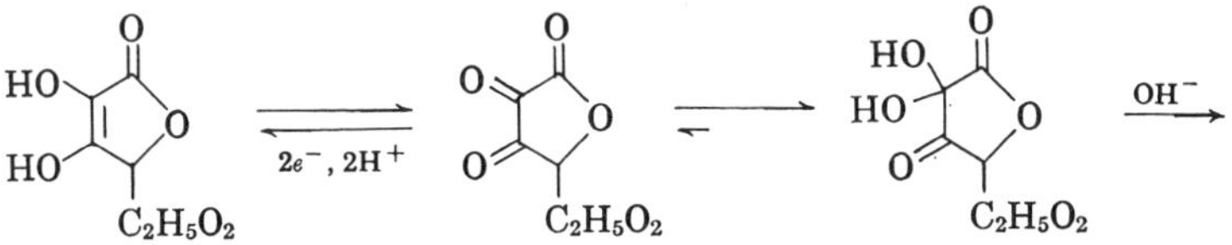

another ene-diol (**35**), which is further oxidized at higher potentials. The reduction wave of **34** is a kinetic wave (i_k independent of h, rises with temperature), although it is much larger than expected, and cpe regenerates ascorbic acid (**32**) (58).

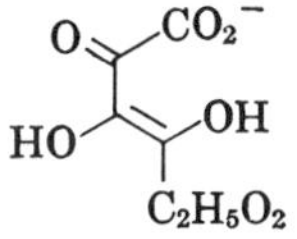

Polarographic reduction of biacetyl (**36**, R = Me) has occasionally shown two waves, but it has been suggested (59) that the splitting is due to inadequate buffering. The wave height corresponds to two-electron reduction to the ene-diol (**37**, R = Me) or its anion, which rearranges to acetoin (**38**, R = Me). Camphorquinone (**39**) behaves similarly at pH < 11, but in strong base, a reversible one-electron reduction to a semiquinone analog is observed (60).

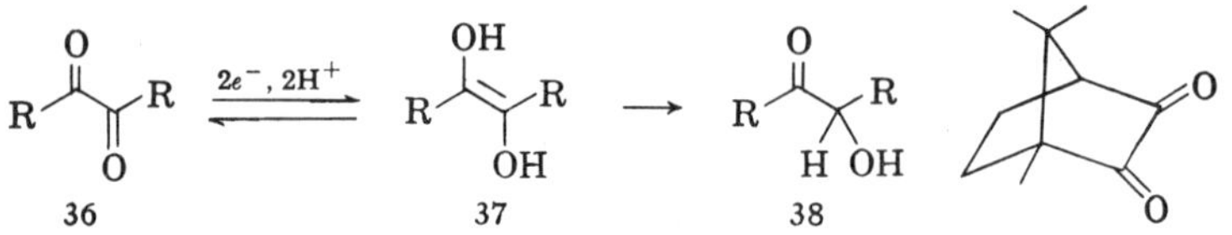

Benzil (**36**, R $= \phi$) gives a two-electron reduction wave, with a second wave at the $E_{1/2}$ of benzoin (61). In contrast, methyl phenyl diketone gives two waves in acid, and cpe on the first plateau produces a dimeric diketopinacol (62). Oscillopolarography (63) of benzil indicates that reduction to the ene-diol (**37**, R $= \phi$) is reversible, but rearrangement to benzoin (**38**, R $= \phi$) is irreversible. The rearrangement is base-catalyzed, and similar results are obtained for *p*-acetylacetophenone (**40**, R $=$ Me) and *p*-benzoylbenzophenone (**40**, R $= \phi$) (64). In DMF, benzil (**41**) shows two one-electron waves, with the first reversible, and the second irreversible, presumably because the dianion (**43**) is sufficiently basic to remove a proton irreversibly from the solvent (64). Cpe on the first plateau produces the green radical anion (**42**), which regenerates benzil on air oxidation. In the presence of LiCl, a single, reversible two-electron reduction is observed, leading to the stable lithium salt of the dianion (**43**) (64).

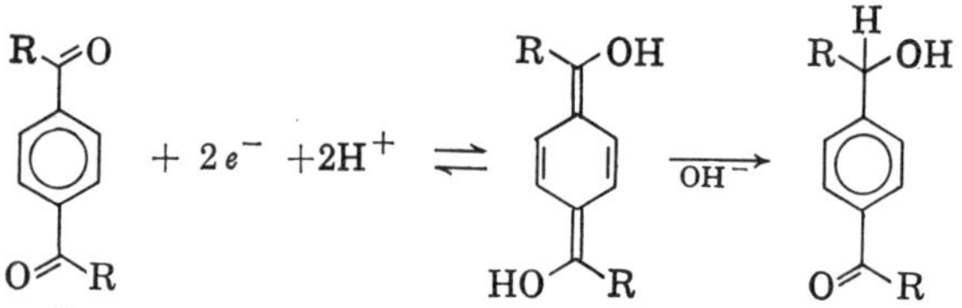

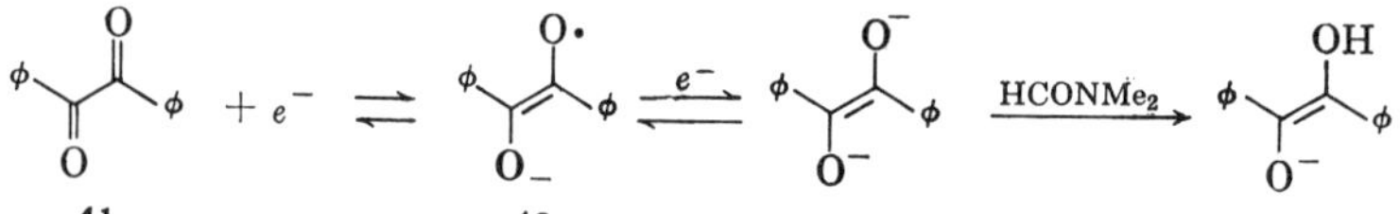

Ninhydrin (**44**), a hydrated triketone, shows two two-electron waves (65). The first wave presumably corresponds to

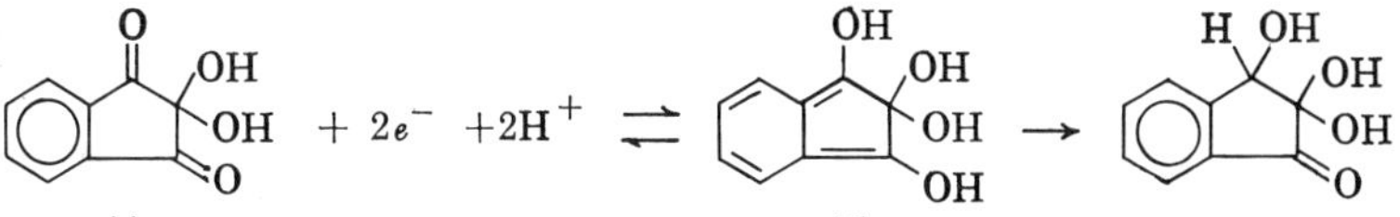

44 + $2e^- + 2H^+ \rightleftharpoons$ 45 $\rightarrow$

and it is claimed that the intermediate (**45**) is stable in alkali as its mono- or dianion.

In summary, quinones and hydroquinones form a reversible oxidation-reduction system, since the addition of two electrons, and the formation of two oxygen–hydrogen bonds, proceed sufficiently rapidly to maintain equilibrium at the electrode surface. The polarographic behavior of the seemingly similar α-diketones and ene-diols is irreversible because breaking carbon–hydrogen or carbon–oxygen bonds in the reverse reaction is slow. Reduction of an α-diketone produces an ene-diol which tautomerizes to an α-ketol, which cannot be reoxidized to the diketone sufficiently rapidly to maintain equilibrium. Oxidation of an ene-diol produces an α-diketone, which forms a hydrate; reduction back to the ene-diol requires prior dehydration to the free α-diketone, and the dehydration does not proceed sufficiently rapidly to permit equilibrium to be maintained.

C. AZO COMPOUNDS

The system azobenzene–hydrazobenzene has been found to be reversible. From pH 2 to 6, the anodic and cathodic waves have the same $E_{1/2}$, a slope of $E_{1/2}$ vs. pH of -0.059 v., and a log plot slope corresponding to $n = 2$ (66). At higher pH, the two $E_{1/2}$'s are unequal, but the irreversibility is ascribed to adsorption effects, and disappears at very low concentrations (67). Also, the reversibility requires adequate buffering. In poorly buffered solutions, a kinetic wave is observed for the reduction, whose rate is limited by the rate at which protons can be supplied (68). In strong acid, hydrazobenzene is further reducible to aniline

$$\phi N{=}N\phi + 2e^- + 2H^+ \rightleftharpoons \phi NHNH\phi \xrightarrow[\text{higher } a_{e^-}]{2e^-,\ 2H^+} 2\phi NH_2$$

cis-Azobenzene is reduced slightly more readily than *trans*, but both give the same product, hydrazobenzene, so that the reduction of the *cis* cannot be reversible, although the log plot slope suggests a reversible two-electron reduction (66).

Azobenzenes with hydroxy and amino substituents are reduced to the corresponding anilines on cpe, since the initially formed hydrazobenzene (e.g. **46**) decomposes to aniline and a quinone-imine, which is very readily reduced. The limiting current often indicates only two-electron reduction, but cpe and cpc indicate four-

$$H_2N\text{-}C_6H_4\text{-}N{=}N\text{-}\phi \underset{}{\overset{2e^-,\,2H^+}{\rightleftharpoons}} H_2N\text{-}C_6H_4\text{-}NH\text{-}NH\text{-}\phi\ (\mathbf{46}) \xrightarrow{k}$$

$$\phi NH_2 + HN{=}C_6H_4{=}NH \underset{\text{fast}}{\overset{2e^-,\,2H^+}{\rightleftharpoons}} H_2N\text{-}C_6H_4\text{-}NH_2$$

electron reduction. The decomposition of the intermediate hydrazobenzene (**46**) is usually slow compared to the drop time, but fast compared to the time required for complete electrolysis (69). With some azo dyes—amaranth (70) and Eriochrome Violet B (71)—the decomposition is so rapid that the diffusion current corresponds to immediate four-electron reduction to two anilines.

Azoxybenzene (**47**) undergoes four-electron reduction to hydrazobenzene. Azobenzene is presumably an intermediate, but the reduction of azoxybenzene is irreversible, and requires potentials so cathodic (ca. 0.3 v. more negative than $E_{1/2}$ of azobenzene) that the azobenzene is immediately reduced further (72).

$$\underset{\mathbf{47}}{\phi N{=}\overset{\overset{O}{\uparrow}}{N}\phi} \xrightarrow{2e^-,\,2H^+} \phi N{=}N\phi \underset{\text{fast}}{\overset{2e^-,\,2H^+}{\rightleftharpoons}} \phi NHNH\phi$$

Polarograms of aryldiazonium salts show two irreversible waves, with the height of the second almost three times that of the first. The limiting current corresponds to a four-electron reduction to the arylhydrazine (73), which has been detected chemically after cpe. By microscale cpc on the first plateau, $n = 1$, and on the second plateau, $n_{\text{total}} = 4$ (74). The $E_{1/2}$ of the first wave is independent of pH, whereas $dE_{1/2}/d\text{pH}$ of the second is -0.05 v. (75). The dependence of the first limiting current on h indicates an adsorption phenomenon (74), and it has been suggested that the mercury surface complexes and stabilizes an intermediate $\phi N_2^{\cdot}$ radical (76). Cpe on the first plateau has led to either diphenylmercury (73) or a nitrogen-containing tar (75). Apparently the nature of the first product depends upon whether stirring removes $\phi N_2^{\cdot}$ from the electrode into the solution, where it polymerizes. The mechanism may be

$$\phi N_2^+ + e^- \rightarrow \phi N_2^{\cdot}(Hg) \xrightarrow[\text{higher } a_{e^-}]{3e^-,\,3H^+} \phi NHNH_2$$

$$\text{polymer} \leftarrow \phi N_2^{\cdot}(Hg) \rightarrow N_2 + \phi\cdot \rightarrow \phi_2 Hg$$

At pH above 5, diazoacetophenone, $\phi COCH{=}\overset{+}{N}{=}N^-$, is reduced in an irreversible six-electron process to ammonia and α-aminoacetophenone, $\phi COCH_2NH_2$, which gives another wave at more negative potentials (77). In acid, the height of the first wave drops to about one-third its value in alkali, and the reactive species is phenacyldiazonium ion, $\phi COCH_2N_2^+$, which is reduced in a two-electron process to acetophenone and nitrogen. The mechanism of the two-electron process is similar to polarographic reduction of halides.

Hydroxyl-1,3-diphenyltriazene, $\phi N(OH){-}N{=}N\phi$, is reduced in a six-electron process to phenylhydrazine and aniline (78).

Finally, the stable free radical, diphenylpicrylhydrazyl (**48**, Ar = 2,4,6-$(O_2N)_3C_6H_2$) undergoes both reversible one-electron oxidation to a stable cation and reversible one-electron reduction to a stable anion (79).

$$\phi_2\overset{+}{N}{=}NAr \underset{\text{MeCN}}{\overset{e^-}{\rightleftharpoons}} \underset{\mathbf{48}}{\phi_2N{-}\dot{N}Ar} \underset{\text{MeCN}}{\overset{e^-}{\rightleftharpoons}} \phi_2N{-}\bar{N}Ar$$

D. NITROSO COMPOUNDS

The cathodic wave of nitrosobenzene (**49**, Ar = ϕ) has the same $E_{1/2}$ as the anodic wave of N-phenylhydroxylamine (**50**, Ar = ϕ), both waves being reversible two-electron processes (80). α-Nitrosonaphthalene (**49**, Ar = 1-$C_{10}H_7$) behaves similarly (81). At pH < 4, the slope of the $E_{1/2}$—pH plot for nitrosobenzene increases from that

$$\underset{\mathbf{49}}{ArNO} + 2e^- + 2H^+ \rightleftharpoons \underset{\mathbf{50}}{ArNHOH} \overset{H^+}{\rightleftharpoons} ArNH_2^+OH$$

required for a two-proton process to a slope indicative of a three-proton process, leading to the conjugate acid of phenylhydroxylamine (32). In DMF, two one-electron waves are observed, the first corresponding to a reversible formation of $\phi NO^{\cdot-}$. Addition of benzoic acid shifts $E_{1/2}$ of the second wave to less negative potentials, until it merges with the first, since the greater proton availability decreases the stability of the intermediate radical, according to the mechanism (82):

$$\phi NO + e^- \rightleftharpoons \phi NO^{\cdot-} \overset{e^-}{\rightleftharpoons} \phi NO^{=}$$

$$\phi NO^{=} + 2HA \rightleftharpoons \phi NHO^- + A^- + HA \rightleftharpoons \phi NHOH + 2A^-, \text{ reversible}$$

$$\phi NO^{=} + HCONMe_2 \rightarrow \phi NHO^- + CO + Me_2N^-, \text{ irreversible}$$

In acid, aromatic hydroxylamines are further reduced to anilines, by an irreversible two-electron process (32).

$$ArNHOH + 2e^- + 2H^+ \rightarrow ArNH_2 + H_2O$$

Aliphatic nitroso compounds, such as nitrosocyclohexane, which are dimeric, undergo irreversible six-electron (per dimer, according to i_d) reduction to the symmetrical dialkylhydrazine. The reduction does not proceed via prior dissociation to the monomer, which would be reduced to the hydroxylamine in a four-electron process (two electrons per monomer) (84). The azo-aliphatic is presumably an intermediate.

$$(RNO)_2 + 4e^- \rightarrow RNHN(OH)R \rightarrow RN{=}NR \xrightarrow{2e^-} RNHNHR$$

Nitrosophenols (quinone oximes) and nitrosoanilines are reduced reversibly to the hydroxylaminophenols or hydroxylaminoanilines. But these are dehydrated to a quinoneimine, which is rapidly reduced further. Most of the conclusions about nitrosophenols are derived from experiments on the more readily available nitrophenols. Nitroaromatics are generally reduced irreversibly at potentials more cathodic than those required to reduced nitrosoaromatics, which are intermediates in the reduction. The nitrosoaromatics are reduced immediately on being formed, but polarography of nitrophenols and nitroanilines provides a method of generating nitrosophenols and nitrosoanilines *in situ* for further reduction. *m*-Nitrophenol, *o*- and *p*-nitroanisole (85), *m*-nitroaniline, and *p*-nitro-*N*,*N*-dimethylaniline (86) are reduced only to the hydroxylamine, so that the intermediate nitroso compounds must undergo normal two-electron reduction. However, *p*-nitrophenol (**51**) (87), *o*-nitrophenol (85), the nitroresorcinols (88), *p*-nitroaniline (89), and α-nitroso-β-naphthol (90) can be reduced directly in a single six-electron step to the amine.

OH–C$_6$H$_4$–NO$_2$ (51) + $2e^-$ + $2H^+$ ⟶ OH–C$_6$H$_4$–NO $\underset{\text{fast}}{\overset{2e^-,\ 2H^+}{\rightleftharpoons}}$ OH–C$_6$H$_4$–NHOH (52) $\xrightarrow[\text{H}^+ \text{ or OH}^-]{k}$ O=C$_6$H$_4$=NH (53) $\underset{\text{fast}}{\overset{2e^-,\ 2H^+}{\rightleftharpoons}}$ OH–C$_6$H$_4$–NH$_2$

52 $\xrightarrow{2e^-,\ 2H^+ \text{ (higher } a_{e^-})}$ OH–C$_6$H$_4$–NH$_2$

The transformation of the hydroxylaminophenol (**52**) into the quinoneimine (**53**) is catalyzed by acids and bases. Only if the dehydration reaction is sufficiently fast does the reduction proceed to the amine in a single step. If the reaction is slow compared to the drop time, the hydroxylamine is stable at the potentials required to reduce the nitro or nitroso compound, and the reduction of the hydroxylamine occurs only at more negative potentials. Thus, anthraquinone monooxime gives one four-electron wave in acid, but the hydroxylamine is stable in base (91). In acid or base, *o*-nitrophenol is reduced to *o*-aminophenol. Near neutrality, two waves are observed, the second representing reduction of that *o*-hydroxyphenylhydroxylamine which has not been converted to *o*-quinoneimine and reduced immediately in the first wave. The second wave attains a maximum height of half the height of the first wave near pH 6, where the chemical reaction is slow compared to the drop time. In the other solutions, acid or base catalysis speeds up the dehydration (85). Chronopotentiometric methods have been used to estimate a value of 20 sec.$^{-1}$ for the rate constant k of the *ortho* isomer at pH 6.2 (92) and a value of 1 sec.$^{-1}$ for k of the *para* isomer at pH 4.8 (93). In acid, there is a further element of irreversibility in the reduction of *p*-nitrosophenol, which is present largely as quinone mono-oxime. The limiting current is independent of h, indicative of a kinetic current limited by the rate of formation of the conjugate acid, *p*-$HOC_6H_4NH^+{=}O$ (94), or perhaps limited by the rate of formation of the nitroso tautomer.

Another case of irreversible reduction of aromatic nitroso compounds is that of *o,o'*-dinitrosobiphenyl (**54**) (95). Reversible two-electron transfer presumably leads to *o*-(*o*-nitrosophenyl)phenylhydroxylamine) (**55**), which is transformed into an azoxy compound (**56**) reducible only at higher a_{e^-}.

$$\mathbf{54}\ \xrightleftharpoons[]{2e^-,\ 2H^+}\ \mathbf{55}\ \rightarrow\ \mathbf{56}\ \xrightarrow[\text{higher } a_{e^-}]{4e^-}$$

In acid solutions, secondary nitrosamines are reduced in an irreversible four-electron process to the unsymmetrical dialkylhydrazine (96). In neutral and basic solutions, the limiting current drops

$$R_2NN{=}O + 4e^- + 4H^+ \rightarrow R_2NNH_2$$

to half its value in acid, and the $E_{1/2}$ is independent of pH. The reduction is still an irreversible process, but leads to the secondary amine and nitrous oxide, which was isolated after cpe (97). It has also been

$$R_2NN{=}O + 2e^- + 1^1/_2H_2O \rightarrow R_2NH + {}^1/_2N_2O + 2OH^-$$

claimed that the two-electron reduction leads to a stable dialkylaminohydroxylamine, R_2NNHOH (98), although such reduction would not be expected to have an $E_{1/2}$ independent of pH.

Aromatic *N*-nitrosohydroxylamines are reduced in three two-electron steps to the aryl hydrazine. In base and strong acid, the first

$$ArN(OH)N{=}O + 4e^- \rightarrow ArN(OH)NH_2 \xrightarrow{2e^-} ArNHNH_2$$

two steps occur simultaneously, but $\phi N(OH)NH_2$ is stable and reduced only at higher a_{e^-}. The four-electron reduction in base is pH-independent (99). Above pH 10, four-electron reduction leads to the aromatic hydrocarbon plus nitrogen plus five hydroxide ions, whose

$$ArNONO^- + 4e^- + 3H_2O \rightarrow ArH + N_2 + 5OH^-$$

presence was confirmed by titration after cpe in unbuffered medium (100).

Hydroxamic oximes (e.g. **57**) give a reversible two-electron anodic wave, with $dE_{1/2}/d\text{pH} = 0.057$ v. at pH below 5.75 or above 10.75, and = −0.087 v. at intermediate pH. These kinks occur at the pK_a's of the oxidant and reductant. The hydroxamic oxime (**57**)

$$CH_3C(=N{-}O^-)(N{=}O) + H^+ \underset{pK_a\ 5.75}{\rightleftharpoons} CH_3C(=NOH)(N{=}O) \overset{2e^-,\,2H^+}{\rightleftharpoons}$$

$$\underset{\mathbf{57}}{CH_3C(=NOH)(NHOH)} \underset{pK_a\ 10.75}{\rightleftharpoons} CH_3C(=N{-}O^-)(NHOH) + H^+$$

also gives a two-electron reduction wave to acetamide oxime, $CH_3C(NH_2){=}NOH$ (101).

The stable free radical, di-*t*-butylnitroxyl (**58**, R = *t*-Bu) gives reversible one-electron oxidation and reduction waves in acetonitrile. The radical species is prepared by cpe of 2-nitro-2-methylpropane, Me_3CNO_2 (102).

$$R_2\overset{+}{N}{=}O + e^- \underset{MeCN}{\rightleftharpoons} \underset{\mathbf{58}}{R_2N{-}O\cdot} \underset{MeCN}{\overset{e^-}{\rightleftharpoons}} R_2N{-}O^-$$

In the absence of possibilities for further chemical reaction, monomeric nitroso compounds are reversibly reduced to the corresponding hydroxylamine. The addition of two electrons, and the protonation on nitrogen and oxygen, proceed quite rapidly. But, as with the previously considered systems—α-diketones, ene-diols, azo compounds—the incursion of subsequent, chemically irreversible reactions renders the electrochemical reaction irreversible.

V. Examples of Irreversible Systems

With the few exceptions discussed above, most organic compounds undergo irreversible polarographic reduction and oxidation. The reactions do not involve electron transfer alone, but are further complicated by formation or breakage of a new bond. The few reversible cases above are reversible because O–H and N–H bonds are formed and broken (ionically) sufficiently rapidly to maintain equilibrium at the electrode surface. In order to break some other kind of bond at a rate comparable to the rate of diffusion of the substrate to the electrode, the electrode potential must be much more reducing (oxidizing, for an anodic process) than the standard potential. This overpotential raises the activity of the electron to a level at which it has sufficient "push" to break a bond. Of course, if the irreversibility is due to irreversible formation of some bond subsequent to the electron transfer, and if the possibility of bond formation is eliminated by a suitable choice of solvent, the electron transfer step itself may be reversible, and accessible for further study.

A. HYDROCARBONS

1. Aromatic Hydrocarbons

Much insight into the products of polarographic reduction can be obtained from the complete polarograms of a series of such hydrocarbons. Benzene gives no polarographic wave in the accessible potential region, but naphthalene shows a single wave of height corresponding to a two-electron reduction. Since styrene and 1,2-dihydronaphthalene are reducible, the latter cannot be the product, and it can be concluded that 1,4-dihydronaphthalene is the product, as in Birch reduction. Anthracene likewise gives a single two-electron wave, indicating reduction to 9,10-dihydroanthracene. Phenanthrene gives two two-electron waves, with the second at the same $E_{1/2}$ as the single waves of biphenyl, fluorene, and 9,10-dihydrophenanthrene. Therefore, reduction at the 9,10 positions leads to the latter compound as primary product. Pyrene (**59**) shows three two-electron waves, with the second and third at the same $E_{1/2}$'s as the two waves of phenanthrene; therefore, initial reduction has occurred at the 4,5-positions:

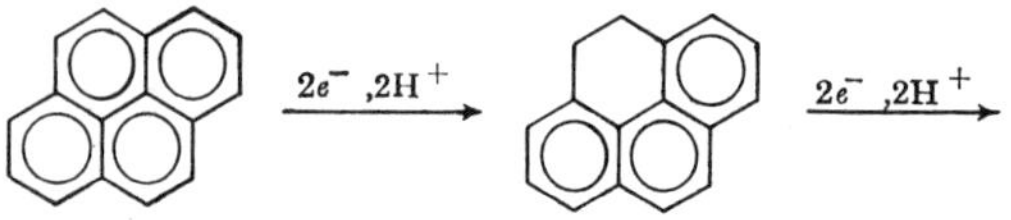

$2e^-, 2H^+$ $2e^-, 2H^+$

59

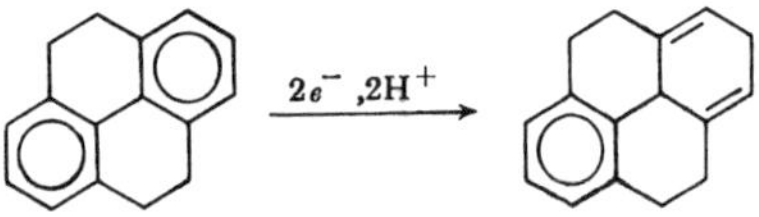

$2e^-, 2H^+$

Chrysene (**60**) is reduced in three poorly resolved two-electron waves. 1,2-Benzanthracene (**61**) is reduced in two two-electron waves, with the second at the same $E_{1/2}$ as the wave of naphthalene (103).

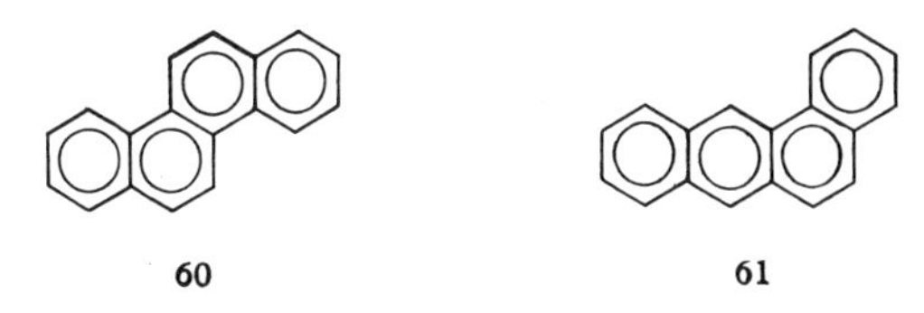

In 75% dioxane, the half-wave potentials of these reductions are independent of pH, although the acid region is not accessible because hydrogen discharge precedes reduction of the hydrocarbon. The slope of the log plot is consistent with $n_a = 1$, or $\alpha n = 1$.

Hoijtink (104) has explained these results on the basis of the following mechanism: (S = substrate = aromatic hydrocarbon)

$$S + e^{-} \rightleftharpoons S^{\dot{-}} \xrightarrow[H_2O]{k} SH\cdot \xrightarrow[fast]{e^-} SH^- \xrightarrow{H^+} SH_2$$

The initial electron transfer to S is rapid and reversible, but the radical anion, $S^{\dot{-}}$ is protonated irreversibly. According to Kivalo's Case (*c*), $E_{1/2}$ depends upon the rate constant k, and to explain the independence of $E_{1/2}$ and pH, it is necessary to assume that water is the most important proton donor, rather than hydronium ion. Then $E_{1/2}$ would depend upon k, but k would be independent of pH. The electron affinity of the radical, SH·, is greater than that of S, since for alternant hydrocarbons, the lowest vacant molecular orbital (LVMO) of S is antibonding, but the LVMO of SH· is nonbonding. Therefore, electron activities capable of converting S to $S^{\dot{-}}$ are capable of converting SH· to SH^- very rapidly. Thus, as long as the protonation of $S^{\dot{-}}$ is rapid compared to the drop time, two-electron reduction will be observed, even though the potential-determining step involves only one electron. The electron transfer is reversible, but the overall process is irreversible because ionization of SH· to $S^{\dot{-}}$ and H^+ is too slow to maintain equilibrium at the electrode.

For alternant hydrocarbons, there is a good correlation of $E_{1/2}$ with Hückel molecular orbital (MO) calculations of the energy of the LVMO (105). This correlation is not surprising, since the E^0 of the initial electron transfer should be closely related to the electron affinity of S, and $E_{1/2}$ depends upon E^0. Although $E_{1/2}$ also depends upon k, k should also correlate roughly with MO calculations. Hückel MO calculations have also been applied to determine the position of highest electron density in $S^{\dot{-}}$ and in SH^-, in order to determine the sites of protonation in the two successive steps. From the predicted products of two-electron reduction, it is possible to deduce the total course of reduction. Thus, pyrene (**59**) is expected to be reduced first to a 4,5-dihydro compound, then to a 4,5,9,10-tetrahydro compound, in agreement with experiment (106).

Most aromatic hydrocarbons have a higher electron affinity than the dihydro derivative formed on polarographic reduction. As a result, the electron activity must be increased in order to reduce the latter, and each polarographic wave involves but two electrons. There are a few exceptions, such as coronene (**62**), in which the dihydro aromatic has a greater electron affinity, according to MO calculations. For such compounds, the dihydro aromatic is further reduced as soon as it is formed, and the first polarographic wave involves four (or more) electrons (104).

As the protonating power of the solvent is decreased, the radical anion, $S^{\cdot-}$, becomes more stable (104). If the water content of the dioxane is decreased from 25% to 4%, the lifetime of $S^{\cdot-}$ becomes longer than the drop time, and it merely diffuses away into the solution, where it eventually abstracts a proton and either dimerizes or disproportionates. Then the polarographic wave is seen to represent reversible transfer of only one electron. At higher a_{e^-} it is possible to add another electron to form $S^{=}$. The simple MO treatment considers the electron affinities of S and $S^{\cdot-}$ to be equal, so that S and $S^{\cdot-}$ would be expected to add an electron at the same potential. But this treatment neglects electron repulsion, which makes it more difficult to add a second electron to an orbital which already contains one electron. This behavior has been verified experimentally; potentiometric titration of dibiphenyleneethylene (**63**) with sodium biphenyl shows a break in the curve, corresponding to a stability of the mono-

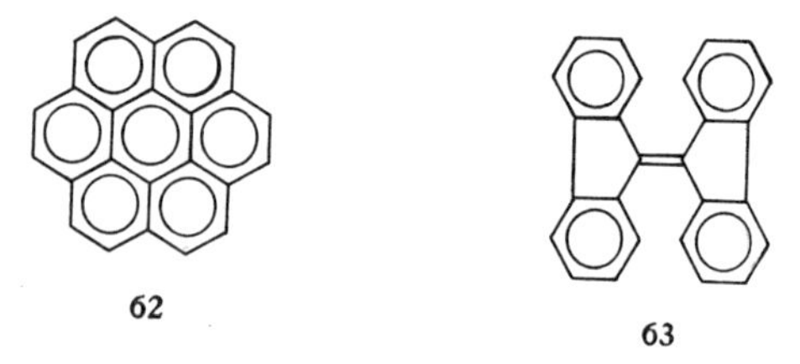

62 **63**

anion to disproportionation (106). Therefore $S^{=}$ is formed only at more negative potentials. This second step is irreversible, since $S^{=}$ is sufficiently basic to abstract a proton even from a solvent of low proton availability. Addition of hydriodic acid increases the rate of pretenation, and one two-electron wave is again observed (104).

In DMF, aromatic hydrocarbons show two one-electron waves (107). The addition of water or phenol (43) causes the second wave to shift to less negative potentials, until it merges with the first to give a single two-electron wave, by the above mechanism. Cpe in the absence of water forms a stable solution of the hydrocarbon radical anion. For example, naphthalene (**64**) gives the green $C_{10}H_8^{\cdot -}$, although only naphthalene could be isolated, since the anion migrates to the anode, and is re-oxidized (108). In the presence of CO_2, 1,4-dihydronaphthalene-1,4-dicarboxylate (**65**) was produced. Presumably the mechanism is the same as in protic solvents, except that CO_2 takes the place of protons. One-electron cpe of phenanthrene (**66**)

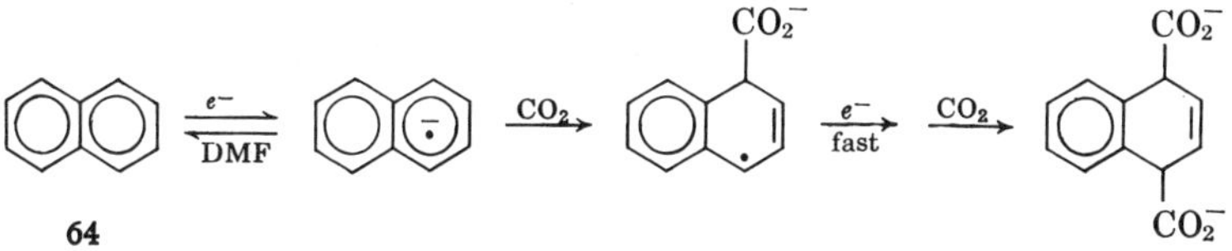

64 ⇌ (e⁻ / DMF) → (CO_2) → (e⁻, fast) → (CO_2) → 65

in DMF yields 9,9′-bi(9,10-dihydro)phenanthryl (**67**), the dimer of the one-electron reduction product. In the presence of CO_2, *trans*-9,10-dihydrophenanthrene-9,10-dicarboxylate (**68**) is formed (108).

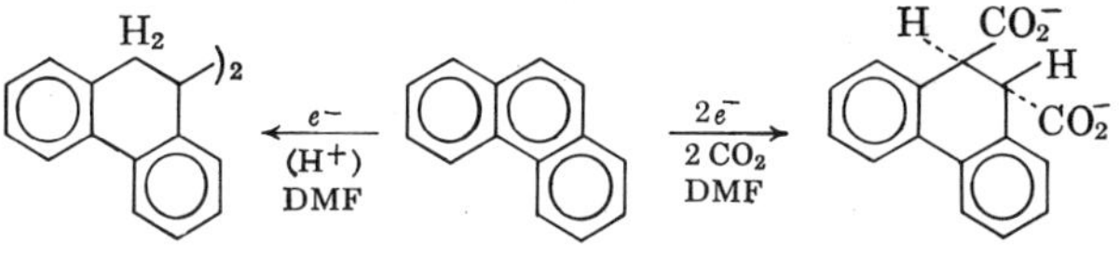

67 ← (e⁻, (H⁺), DMF) 66 → (2e⁻, 2 CO_2, DMF) 68

Radical anions of anthracene, biphenyl, chrysene (**60**), fluoranthene (**69**), and benzophenanthrene (**70**) have been generated by cpe in DMF (109).

69

70

Azulene (**71**) is a non-alternant hydrocarbon whose reduction mechanism differs from the above. Even in protic media, the first wave has a height corresponding to only one-electron reduction. Furthermore, $E_{1/2}$ is pH-independent, and the log plot slope indicates reversible one-electron transfer (110). Apparently the radical anion of azulene is not protonated during the lifetime of the drop. There are further waves at more negative potentials, but these represent further irreversible reduction. In DMF there are two one-electron waves (43). The second represents formation of the dianion (**72**), which abstracts a proton to give a nonreducible monoanion (**73**). The first wave is unaffected by added phenol, but the second doubles

e^- DMF; e^- higher a_{e^-}; $HCONMe_2$ or ϕOH

71 **72**

ϕOH; e^- fast

H_2 H_2 H_2

73

reducible at still higher a_{e^-}

in height and a third wave appears. The first wave of fluoranthene (**69**), also a non-alternant hydrocarbon, also involves only one electron, even in aqueous dioxane (111).

Cyclooctatetraene (COT) in ethanol behaves in a fashion similar to aromatic hydrocarbons, in that the limiting current is that of a two-electron reduction, and the $E_{1/2}$ is independent of pH. However, the log plot suggests $n_a = 2$, rather than $n_a = 1$, that is, that the electron transfer is a reversible two-electron process, rather than a reversible one-electron transfer, followed by irreversible steps (112). Vinyl-COT behaves similarly, and the reductive attack takes place at the COT portion of the molecule, rather than at the vinyl group (cf. styrene, below) (113). Cpe of COT leads to a mixture of 1,3,6- and 1,3,5-cyclooctatriene (114). The behavior of COT differs from that of aromatic hydrocarbons because the electron affinity of $COT^{\cdot-}$ is greater than that of COT. The first electron enters an antibonding MO of COT, but the second electron enters a non-bonding MO of a planar $COT^{\cdot-}$. Phrased differently, $COT^{\cdot-}$ is an unstable "semiquinone," subject to a Jahn–Teller distortion which requires that the odd electron be in an antibonding orbital, and which tends to disproportionate to COT and $COT^{=}$. In protic solvents, the dianion abstracts

$$COT + 2e^- \rightleftharpoons COT^{=} \xrightarrow{2HA} COT{-}H_2 + 2A^-$$

two protons from the solvent to produce the triene. In acetonitrile, COT undergoes reversible two-electron reduction to give the stable $COT^{=}$. In 96% dioxane, oscillopolarographic studies have shown that the two-electron reduction is reversible, but that the dianion reacts irreversibly with water (115). These authors have also discussed the factors governing the disproportionation of $COT^{\cdot-}$.

2. *Olefins and Acetylenes*

The reduction of olefins is generally similar to that of aromatic hydrocarbons. Unconjugated olefins are not reducible in the accessible potential range, but a double bond conjugated with an aromatic ring or another double bond gives a well-defined reduction wave. Examples include styrene, β-methylstyrene, stilbene, 1,1-diphenylethylene, triphenylethylene, tetraphenylethylene, and substituted fulvenes (103,111,116). Even allene, butadiene, vinylacetylene, and divinylacetylene are reducible, although i_d is less than that expected for even one-electron reduction (117). In 75% dioxane, a two-electron reduction is observed, with $E_{1/2}$ independent of pH, and a log plot indicative of reversible one-electron transfer. The mechanism may be indicated for 1,1-diphenylethylene:

$$\phi_2C{=}CH_2 + e^- \rightleftharpoons \phi_2\dot{C}{-}CH_2^- \xrightarrow[H_2O]{k}$$

$$\phi_2\dot{C}{-}CH_3 \underset{fast}{\overset{e^-}{\rightleftharpoons}} \phi_2\bar{C}CH_3 \xrightarrow[H_2O]{} \phi_2CHCH_3$$

and the reasoning (104), comparing the electron affinities of S and SH·, is still applicable. Thus, dibiphenyleneethylene (**63**) undergoes two-electron electrolytic hydrogenation of the double bond, and there is another polarographic wave at the same $E_{1/2}$ as that of fluorene (111).

Cpe of 1,2-dipyridylethylene, PyCH=CHPy, leads to dipyridylethane. According to oscillopolarography, the 4,4′-isomer undergoes reversible electron transfer in neutral and basic solution, but the radical anions of the other isomers (and the 4,4′ in acid) are protonated too rapidly for any reversibility to be observed (118).

As the number of ethylenic linkages in the 1,ω-diphenylpolyenes increases, the mechanism of reduction shifts toward addition of the second electron before protonation (119), since as the negative charge of the radical anion is spread over more atoms, the molecule is slower to protonate, while the electron repulsion opposing addition of a second electron decreases.

The non-alternant hydrocarbon, fulvene (**74**), shows two reversible one-electron reduction waves in 75% dioxane, since the monoanion (**75**) is apparently stable to protonation during the lifetime of the drop, and a higher electron activity is required to add a second proton (120) (cf. azulene, above).

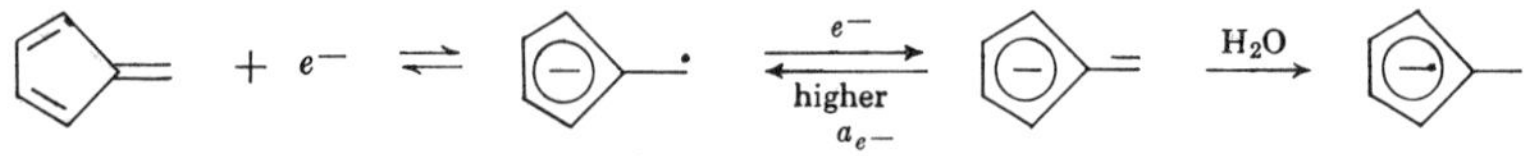

$$\mathbf{74} + e^- \rightleftharpoons \mathbf{75} \xrightarrow[\text{higher } a_{e^-}]{e^-} \ \xrightarrow{H_2O}$$

Tetraphenylbutatriene (**76**), a cumulene, undergoes two-electron reduction to 1,1,4,4-tetraphenylbutadiene (**77**), which is further reduced to both *trans*-1,1,4,4-tetraphenylbutene-2 (**78**) and 1,1,4,4-tetraphenylbutene-1 (**79**), the latter of which is further reduced to 1,1,4,4-tetraphenylbutane (121).

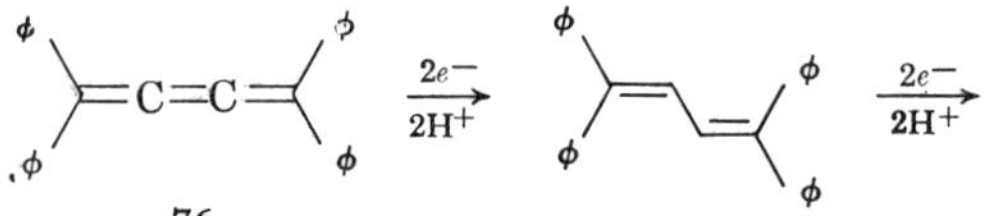

$$\phi_2C{=}C{=}C{=}C\phi_2\ (\mathbf{76}) \xrightarrow[2H^+]{2e^-} \mathbf{77} \xrightarrow[2H^+]{2e^-} \mathbf{78} + \mathbf{79} \xrightarrow[2H^+]{2e^-}$$

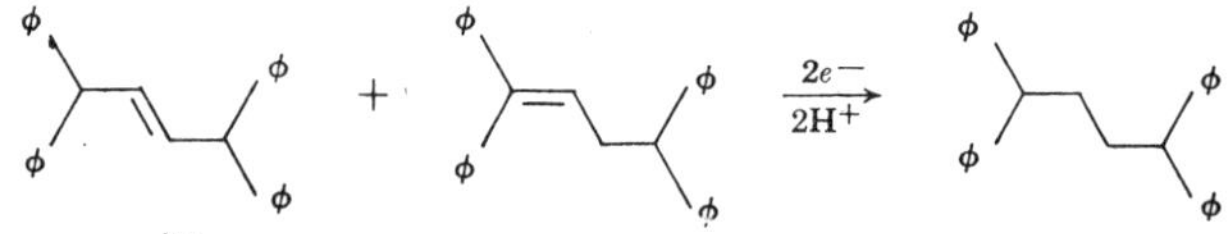

Acetylenes are reduced slightly less readily than the corresponding olefins. For example, $E_{1/2}$ of diphenylacetylene is -2.20 v., and $E_{1/2}$ of stilbene is -2.14 v. Therefore, a potential capable of reducing an acetylene to an olefin is also capable of reducing the olefin. Thus the wave heights of phenylacetylene and diphenylacetylene are twice those of styrene and stilbene (116).

In DMF, stilbene (**80**) shows two one-electron waves (107). The first forms the radical anion, which is stable during the life of the drop, but eventually dimerizes or disproportionates. The second wave corresponds to addition of a second electron to form the dianion (**81**), which is a sufficiently strong base to abstract a proton from the solvent. In the presence of CO_2, *meso*-diphenylsuccinic acid (**82**) is formed; CO_2 serves as a proton analog, and adds to the radical anion

$$\underset{\mathbf{80}}{\phi CH{=}CH\phi} \underset{DMF}{\overset{e^-}{\rightleftarrows}} \phi\dot{C}H{-}\bar{C}H\phi \overset{e^-}{\rightleftarrows} \underset{\mathbf{81}}{\phi\bar{C}H\bar{C}H\phi} \xrightarrow{HCONMe_2} \phi\bar{C}HCH_2\phi$$

82: $\phi(COOH)CH{-}CH(COOH)\phi$

to produce a readily reduced radical species. Addition of 20% water causes the second wave to merge with the first (122). In contrast to aromatic hydrocarbons in 96% dioxane, stilbene shows only a single, two-electron wave, indicating that the mono-anion of stilbene is protonated in this solvent more rapidly than are the radical anions of aromatic hydrocarbons (123). Cpe preparations of radical anions of several substituted ethylenes have been reported (109).

In DMF, diphenylacetylene shows two one-electron waves. Cpe in the presence of CO_2 produces diphenylfumaric acid, diphenylmaleic acid, and *meso*-diphenylsuccinic acid (108).

3. Hydrocarbon Cations

In strong acid ($BF_3 \cdot H_2O$ in F_3CCOOH) polycyclic aromatic hydrocarbons show two waves. The first has been attributed to one-electron reduction of the radical cation of the hydrocarbon, $S^{+\cdot}$, produced by air oxidation. The second wave is that of the protonated hydrocarbon, SH^+. The second $E_{1/2}$ is not independent of the nature of the hydrocarbon, as would have been expected from the fact that the electron enters a vacant non-bonding MO. However, self-consistent field (SCF) calculations give a good correlation with $E_{1/2}$ (124).

In methanesulfonic acid, trityl cation (**83**) undergoes two-electron reduction to triphenylmethane. The log plot slope suggests reversible one-electron transfer, consistent with the mechanism

$$\underset{\mathbf{83}}{\phi_3C^+} + e^- \rightleftharpoons \underset{\mathbf{84}}{\phi_3C\cdot} \xrightarrow[\text{fast}]{H^+} {}^1/_2\phi_3CH + {}^1/_2\phi_3C^+$$

In the presence of water, two waves are observed, with a total height corresponding to two-electron reduction. Water decreases the rate of acid-catalyzed disproportionation, so that the radical is stable during the life of the drop, and merely dimerizes. More negative potentials are required to reduce the radical (**84**) to the anion, which rapidly adds a proton (125). In liquid SO_2, trityl cation undergoes one-electron reduction to triphenylmethyl radical (**84**), which dimerizes (126).

Tropylium ion (**85**) shows two waves, with the $E_{1/2}$ of the first independent of pH. At low concentrations, the i_d of the first wave is proportional to $h^{1/2}$, but at higher concentrations, i_d becomes dependent on the first power of h, indicative of an adsorption phenomenon. The log plot slope indicates that the reduction is irreversible. Cpc indicates that one-electron reduction to bitropyl (**86**) occurs, and $E_{1/2}$ depends upon concentration, as expected for a reaction leading to

H, OH $\underset{}{\overset{H^+}{\rightleftharpoons}}$ (+) $\overset{e^-}{\rightleftharpoons}$ (•) $\rightarrow$ 1/2 (H, H)

85 **86**

dimeric product. The i_l of the second wave is independent of h, indicative of a kinetic current limited by the rate of the reaction (127):

$$H^+ + TrOH \xrightarrow{k} Tr^+ + H_2O$$

In acetonitrile, triphenylcyclopropenium cation (**87**) undergoes one-electron reduction to a dimer, and the reduction requires more reducing (by 0.2 v.) potentials than does the reduction of trityl cation. Oscillopolarography shows that the electron transfer is reversible, and that with sufficiently rapid potential reversal, the resulting cyclopropenyl radical (**88**) can be re-oxidized before it dimerizes (128).

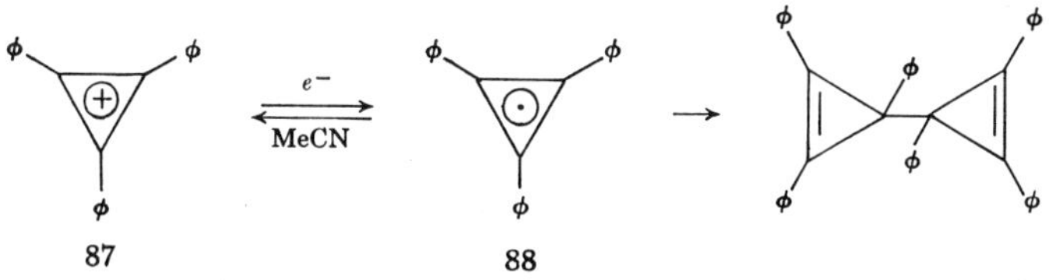

87 88

4. *Anodic Polarography*

The anodic polarography of hydrocarbons has received much less attention. In acetonitrile, anthracene exhibits a two-electron oxidation wave, whose $E_{1/2}$ is independent of the concentration of added perchloric acid, but is shifted to less positive values by addition of pyridine. Salts of 9,10-dihydroanthracene-9,10-dipyridinium ion (**89**) can be isolated after cpe in the presence of pyridine (129). In ethanol, anodic cpe of anthracene leads to bianthrone (**90**) (130).

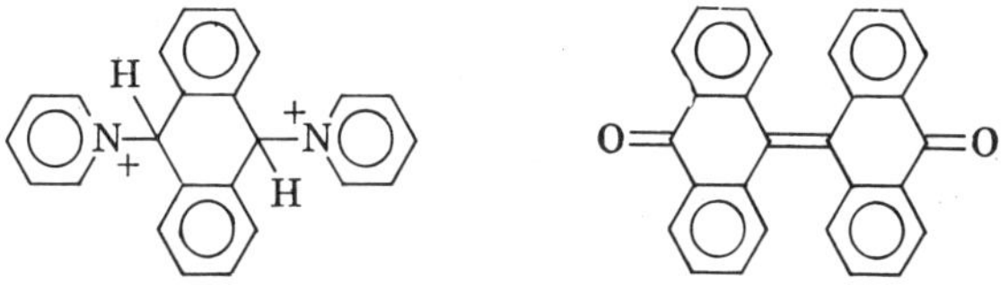

89 90

In acetonitrile, cyclohexadiene-1,3 is oxidized to benzene at a platinum anode; both cycloheptatriene and bitropyl (**86**) form tropylium ion (**85**), all by two-electron processes (131).

In acetic acid, cpe of *trans*-stilbene produces *meso*-1,2-diphenyl ethylene glycol diacetate. In wet acetic acid, *threo*-1,2-diphenylethylene glycol monoacetate is formed. The reaction proceeds through the cation (**91**), which is also an intermediate in solvolyses (132).

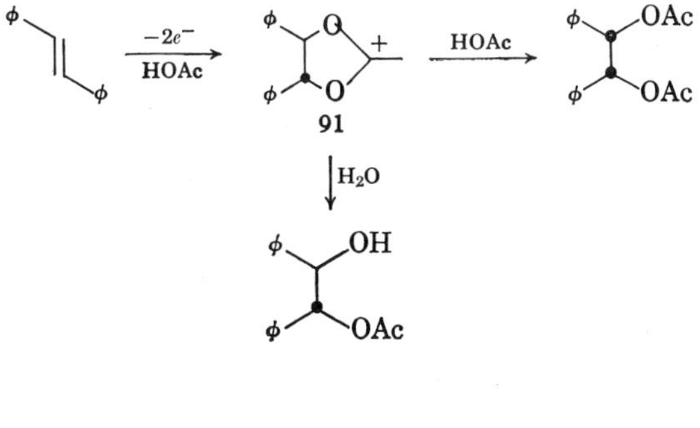

91

Such oxidations presumably proceed via production of hydrocarbon radical cations, rather than via acetoxy radicals, which are formed only at more oxidizing potentials (133). The mechanism for the oxidation of naphthalene may be

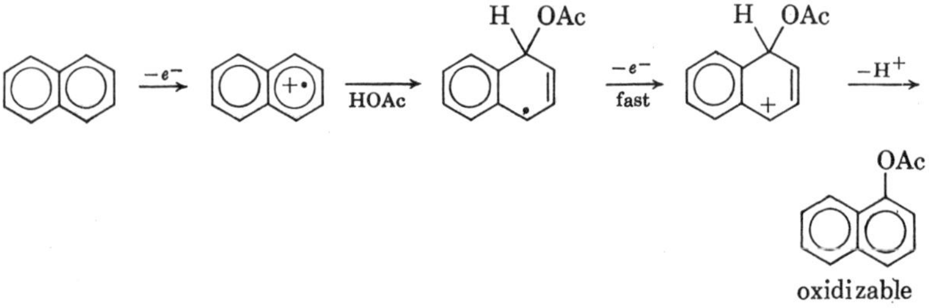

H OAc H OAc

$\xrightarrow{-e^-}$ $\xrightarrow[\text{HOAc}]{}$ $\xrightarrow[\text{fast}]{-e^-}$ $\xrightarrow{-H^+}$

OAc

oxidizable

The oxidation of stilbene would be similar, except addition of a nucleophile occurs in the last step, rather than proton loss. Stepwise removal of two electrons is likely, by analogy with reduction of aromatic hydrocarbons (104). Removal of the second electron, from a nonbonding MO, should be easier than removal of the first, from a bonding MO.

In methanol, electrolysis (not cpe) of stilbene produces 1,2-diphenyl-1,2-dimethoxyethane. The *trans* isomer produces a greater amount of the *dl* product, and the *cis* isomer produces a greater amount of the *meso* product, indicating a preference for *cis* methoxylation (134). The reaction presumably proceeds through the radical cation of stilbene, rather than by formation of methoxyl radicals. *Cis*-methoxylation is probably preferred because the substrate orients itself with its plane parallel to the electrode surface. Attack of solvent on the cationic species then occurs from the side opposite to the electrode.

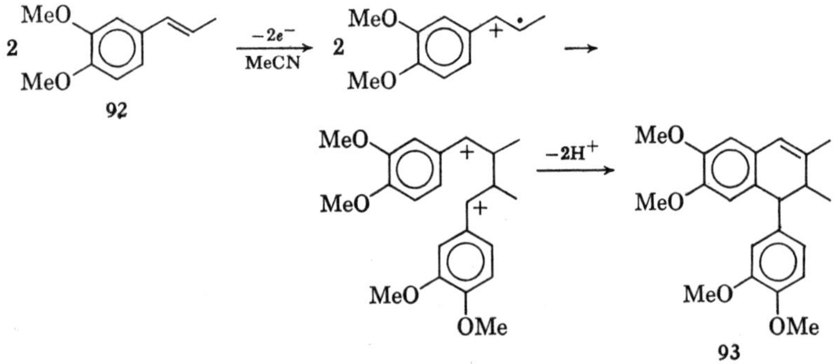

MeO MeO 2 $\xrightarrow[\text{MeCN}]{-2e^-}$ 2 MeO MeO →

92

MeO MeO $\xrightarrow{-2H^+}$ MeO MeO

MeO OMe MeO OMe

93

In the presence of pyridine, the substituted β-methylstyrene (**92**) undergoes two-electron oxidation and adds two pyridine molecules to give a dipyridinium salt. In acetonitrile alone, **92** undergoes an interesting one-electron oxidation to a dimeric product (**93**) (135).

B. CARBONYL COMPOUNDS

1. *Aromatic Ketones and Aldehydes*

In acid, aromatic ketones and aldehydes show two waves, each of a height corresponding to one-electron reduction. The log plot slopes indicate that both waves are irreversible (136). Cpe on the first plateau usually produces the pinacol, and cpe on the second plateau usually produces the carbinol. Thus, acetophenone can be reduced to its pinacol (137). At pH 1.3, benzaldehyde is reduced to hydrobenzoin. At pH 1.9, benzophenone is reduced to benzpinacolone, the rearrangement product of benzpinacol (61); at pH 3.6, both benzpinacol and benzhydrol can be isolated (138). Cpe of *p*-(dimethylamino) acetophenone yields the pinacol, although different isomers are formed in acid and base (139). Fluorenone and α-tetralone are also reduced to their pinacols (140).

$$R_2C{=}O + e^- + H^+ \rightarrow R_2\dot{C}OH \rightarrow {}^1/_2R_2C(OH)C(OH)R_2$$

For the first waves of benzophenone (136) and acetophenone (141), $dE_{1/2}/d\text{pH} = -0.06$ v., implying one proton is involved. Therefore this wave corresponds to the addition of one electron and one proton, to form an uncharged ketyl radical, which dimerizes irreversibly. Cpc verifies the one-electron nature of this wave for acetophenone (141) and xanthone (**94**) (142). Elving and Leone (141) have en-

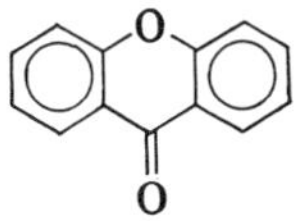

94

visioned the protonation as occurring in the electric field of the electrode, rather than as a prior equilibrium, which might be expected to give a kinetically controlled wave, limited by the rate of protonation. The electric field polarizes the C=O so as to facilitate proton addition to the oxygen. This is the behavior to be expected if the carbonyl group orients itself relative to the electrode with its carbon atom closer to the electrode than is the oxygen. Likewise, the ensuing protonation of the carbonyl oxygen facilitates electron addition. This simultaneous, synergetic addition of both a proton and an electron leads to a transition state in which an O—H bond is partially formed, and an electron is partially transferred. The shape of the wave does not fit either eq. (15) (rate-limiting electron transfer) or eq. (18) (reversible electron transfer, followed by rate-limiting dimerization), but indicates that neither step is strictly rate-limiting, and the rate of re-oxidation of the ketyl radical is comparable to its rate of dimerization (143). Oscillopolarography of benzaldehyde shows an anodic peak on reversing the polarity, suggesting that the intermediate ϕCHOH· radical can be re-oxidized before it dimerizes (144).

In acid, $E_{1/2}$ of the second wave of benzophenone depends only very slightly on pH (143). This wave represents addition of an electron to the ketyl to form ϕ_2COH^-. The slight dependence of $E_{1/2}$ on pH may be attributed to a partially reversible transfer of this second electron, followed by proton abstraction from both water and hydronium ion. This case differs from the reduction of hydrocarbons, for which only water is effective as proton donor. But ketone reductions can be carried out in more acidic solutions, where protons are present in sufficiently large concentrations to compete with water. Ketone reduction also differs from hydrocarbon reductions in that the neutral radical has a lower electron affinity than the substrate, in apparent contradiction to Hoijtink's MO argument (104). But this behavior is not surprising if the substrate is considered to be a partially protonated ketone. Then the electron affinities will be in the order $R_2COH^+ > R_2COH\cdot > R_2C{=}O > R_2C\cdot{-}O^-$, and $ArH_2^+ > ArH_2\cdot >$ $ArH > ArH^{\overline{\cdot}}$ (43).

As the pH is increased, the first wave shifts to more negative potentials more rapidly than does the second, until the two waves merge near pH 6.5 (benzophenone), and produce a single two-electron wave, whose $dE_{1/2}/d$pH is -0.03 v., approximately the average of the values of the two separate waves (138). In these solutions, the reduction process is a two-electron reduction directly to the carbinol, verified by cpe (141). But the hydrogen-ion concentration is too low to protonate the ketone simultaneously with electron transfer, and the initial product is the ketyl anion, which is sufficiently basic to abstract a proton from the solution. As mentioned above, the ketyl radical does have a greater electron affinity than does the starting ketone, so it is rapidly reduced to R_2COH^-, which irreversibly abstracts a proton from the solution.

For the normal esters of phthalaldehydic acid (**95**, R = H) (145) and *o*-benzoylbenzoic acid (**95**, R = ϕ) (146), the carbinol initially

R, O, COOR′ **95** + $2e^-$ + H^+ → R, O^-, COOR′ **96** →

R, O, O **97** + $R'O^-$

formed in the reduction (**96**) enters into ester interchange and forms the phthalide (**97**), whose reduction wave is observed at more negative potentials. This ester interchange adds a further element of irreversibility to the reduction. The behavior of the acids (**95**, R′ = H) is more complex, because of their ability to ionize and to form a cyclic pseudo-acid. The polarographic behavior of the cyclic esters is discussed in the section on cleavage of single bonds.

At still higher pH, i_d of the combined wave of ketones decreases and approaches the height of a one-electron wave, again verified by cpc (141). A new wave appears at more negative potentials, such that the sum of the heights of the two waves remains constant at the height of a two-electron reduction (136). The first wave still represents reduction of ketone to a ketyl anion, which is not protonated during the life of the drop, but merely diffuses away from the electrode, adds a proton, and dimerizes. Thus, cpe of acetophenone at pH 12.3 produces acetophenone pinacol (141). The second wave at more negative potentials represents addition of another electron to the ketyl anion, to form a dianion which rapidly abstracts a proton to form the carbinol. Metal ions shift the $E_{1/2}$ of this second wave to less negative potentials, as might be expected if the metal ion complexes with the ketyl anion to form a metal ketyl simultaneously with the addition of the second electron. Alternatively, the effect of metal ions may be considered an effect of ionic strength which facilitates reaction of an electron with an anion (141).

In DMF, cpe of benzophenone (**98**) produces a solution of the stable, blue ketyl anion, $\phi_2CO^{\overline{\cdot}}$. In the presence of CO_2, benzilic acid (**99**) is formed, by addition of one CO_2, then a second electron, and then a second CO_2, the first of which is lost on workup. In the presence of ethyl iodide, ethyldiphenylcarbinol (**100**) is formed. At more nega-

$$\underset{\mathbf{98}}{\phi_2CO} + e^- \underset{DMF}{\rightleftharpoons} \phi_2CO^{\overline{\cdot}} \xrightarrow{CO_2} \phi_2\dot{C}{-}OCO_2^- \xrightarrow{e^-} \xrightarrow{CO_2} \phi_2C(OCO_2^-){-}CO_2^- \xrightarrow[-CO_2]{H^+} \underset{\mathbf{99}}{\phi_2C(OH){-}COOH}$$

$$\phi_2CO^{\overline{\cdot}} \xrightarrow{EtI} \underset{\mathbf{100}}{\phi_2CHOEt}$$

tive potentials in DMF alone, a second wave is observed, due to the addition of a second electron to form the dianion, $\phi_2CO^=$ (147). Addition of benzoic acid or phenol produces a new wave at less negative potentials, attributed to reduction of a hydrogen-bonded form (43). Cpe of fluorenone and 4,5-benzfluorenone produces the ketyl anions, studied by ESR (42). In pyridine, benzophenone shows only a single one-electron wave, due to reduction to the ketyl anion (148).

In acetonitrile containing pyridine, anodic cpe of *p*-methoxybenzyl alcohol produces anisaldehyde (149). Anodic cpe of benzpinacol

$$p\text{-}MeOC_6H_4CH_2OH - 2e^- \xrightarrow[MeCN]{} p\text{-}MeOC_6H_4CH{=}O + 2H^+$$

(150) and fluorenone pinacol (151) in alkaline solution produces benzophenone and fluorenone, respectively. The oxidation is irreversible, and $E_{1/2}$ does not depend upon concentration, so that the mechanism does not involve prior, reversible dissociation to two ketyl radicals. The oxidation is facilitated by strong base, and the mechanism is considered to be a slow electron transfer from the monoanion of the pinacol.

$$Ar_2C(HO){-}CAr_2(O^-) - e^- \rightarrow Ar_2C{=}O + Ar_2COH\cdot \xrightarrow[fast]{-e^-,\ -H^+} 2Ar_2C{=}O$$

2. Unsaturated Ketones and Aldehydes

The behavior of α,β-unsaturated carbonyl compounds is similar to that of the aromatic systems discussed above, except that Michael-like additions can occur.

Acrolein (152), crotonaldehyde (153), and homologous polyene aldehydes, $CH_3(CH{=}CH)_xCHO$, (154) show two waves of equal height, with the second at $E_{1/2}$'s corresponding to reduction of a saturated aldehyde. The height of the first wave indicates only one-electron reduction (154,155), and a $dE_{1/2}/d$pH of -0.06 v. indicates that one proton is also involved. The log plot slopes vary with pH and from compound to compound, but a reversible electron transfer, followed by irreversible steps, is not excluded. The presence of the second wave at more negative potentials shows that the initial product of the first step is not a pinacol, which is not reducible in the accessible potential range. A suggested mechanism for the reduction of crotonaldehyde is

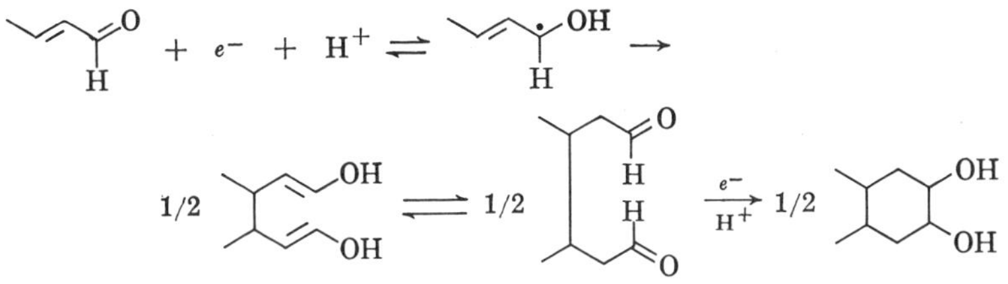

However, the nature of the second step has not been firmly established; it seems more likely that it involves addition of a second electron to the intermediate radical, in analogy to the reduction of aromatic aldehydes.

α,β-Unsaturated ketones undergo one-electron reduction in acid. At pH 1.3, cpe of mesityl oxide (**101**) produces the dimeric ketone (**102**) (61). In alkaline solution, two-electron reduction (hydrogenation of the double bond) is assumed, by analogy with the behavior of chalcones, discussed below.

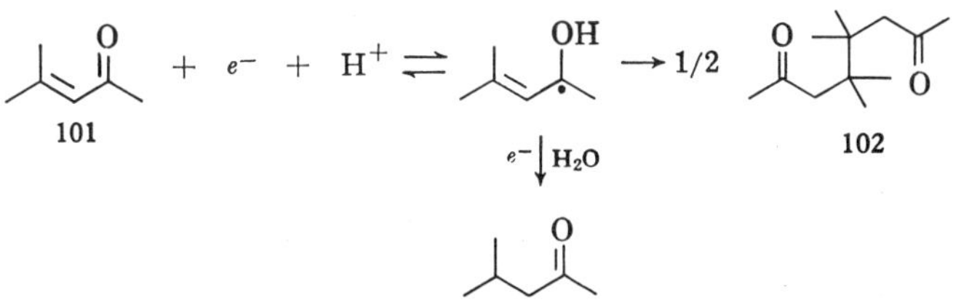

Some α,β-unsaturated steroid ketones undergo one-electron reduction and dimerize to a pinacol, presumably because of steric hindrance to dimerization through the β positions. Thus, cpe of prednisolone

(**103**) in acid produces a product without the UV absorption of an α,β-unsaturated ketone, which would be the expected product from 1,4-reduction of one of the two adjacent double bonds (156). Also, there is no further wave of an α,β-unsaturated ketone, although in alkali there is a further wave, attributed to addition of a second electron to the ketyl radical, rather than dimerization of the ketyl. Cpe of progesterone (**104**) and an androstadiene-1,4-one-3 (**105**) also leads to pinacol formation, since lead tetra-acetate oxidation regenerates

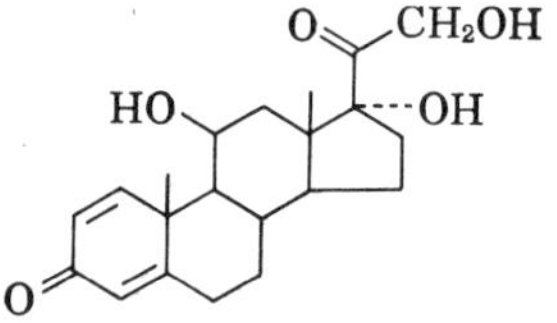

103 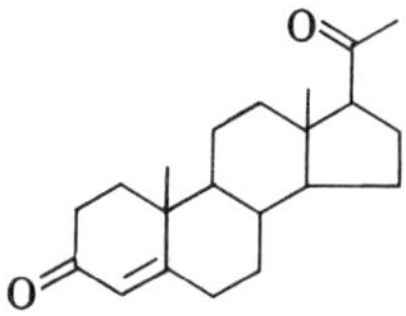 **104** 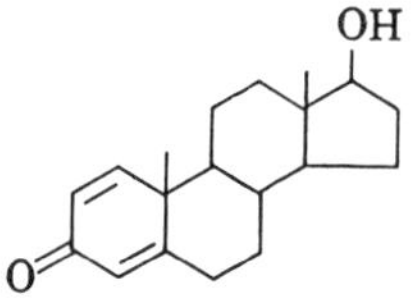 **105**

the original ketone. Cpe in acid produces a different pinacol from the one produced in alkali, since the pinacol may be either *dl* or *meso* (157). Zuman (158) has suggested that different pinacols arise because in acid the dimerizing radical is uncharged, whereas in alkali, the negative charges on the oxygens of the ketyl anions repel each other in dimerization. The mechanism of the polarography of santonin (**106**) is presumably similar (159).

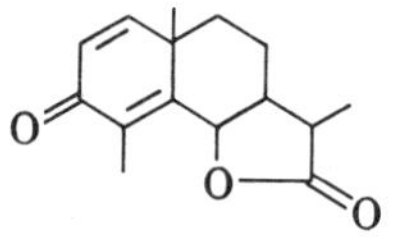

106 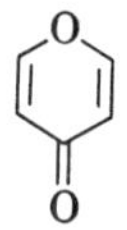 **107** 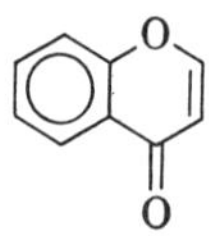 **108** 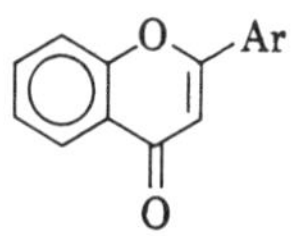 **109**

γ-Pyrone (**107**) undergoes two-electron reduction in alkaline solution (160), and chromone (**108**) undergoes irreversible two-electron reduction to an alcohol (161). Flavones (**109**) generally give two one-electron reduction waves, according to the i_d (162). The log plot slope indicates reversible transfer of the first electron. The absence of a third wave at more negative potentials indicates that the two-electron reduction does not hydrogenate the double bond, since there is no carbonyl remaining. The one-electron reduction is again attributed to pinacol formation, since steric hindrance prevents 4,4′-dimerization.

Cpe of methyl vinyl ketone (**110**) casts considerable doubt on the generality of the above mechanism. The electrolysis product was analyzed for mercury, and found to be di-(β-acetylethyl-)mercury (**111**) whose formation was rationalized on the basis of the mechanism (163):

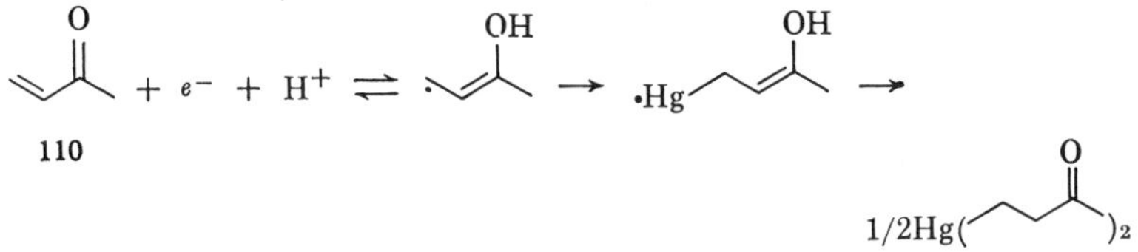

O OH OH
+ e^- + H^+ ⇌ → •Hg →
110
O
1/2Hg( )$_2$
111

What factors are operative in determining whether a radical dimerizes with or without inclusion of mercury is not clear.

Unsaturated ketones conjugated with an aromatic ring, such as chalcone (**112**, R = ϕ) and benzalacetone (**112**, R = Me), undergo one-electron reduction in acid, followed by dimerization of the radicals at the β position. These dimers have been isolated after cpe (61).

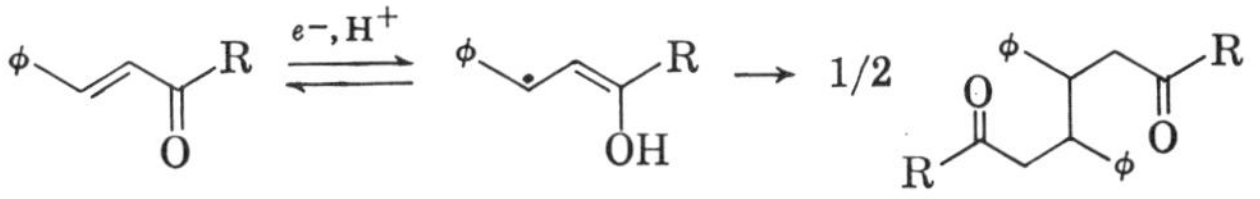

ϕ R $\xrightarrow{e^-, H^+}$ ϕ R → 1/2 ϕ R
O OH O O
112 R ϕ

In basic solution, two-electron reduction, hydrogenating the double bond, is observed, verified by cpe (61). In nearly neutral solution

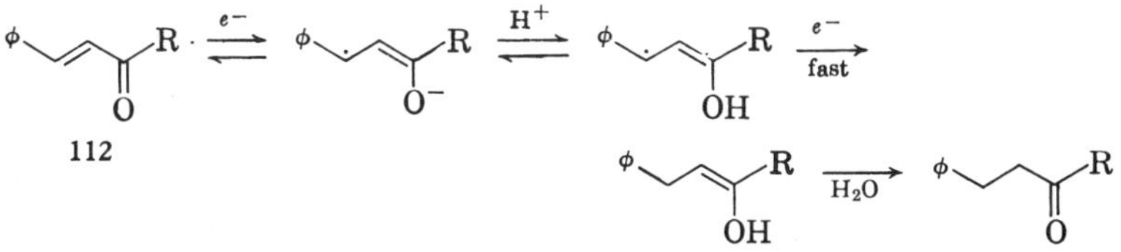

ϕ R $\overset{e^-}{\rightleftharpoons}$ ϕ R $\overset{H^+}{\rightleftharpoons}$ ϕ R $\xrightarrow[\text{fast}]{e^-}$
O O$^-$ OH
112
ϕ R $\xrightarrow{H_2O}$ ϕ R
OH O

both one-electron and two-electron processes occur. Both reduction products of chalcone (**112**, R = ϕ) still have a carbonyl group attached to a phenyl ring, and a further reduction wave for chalcone is observed near the $E_{1/2}$ of acetophenone. Cpe of dibenzalacetone (**113**) in acid produces a dimeric one-electron reduction product (**114**). In base, dibenzylacetone (**115**) is produced by a four-electron process (61).

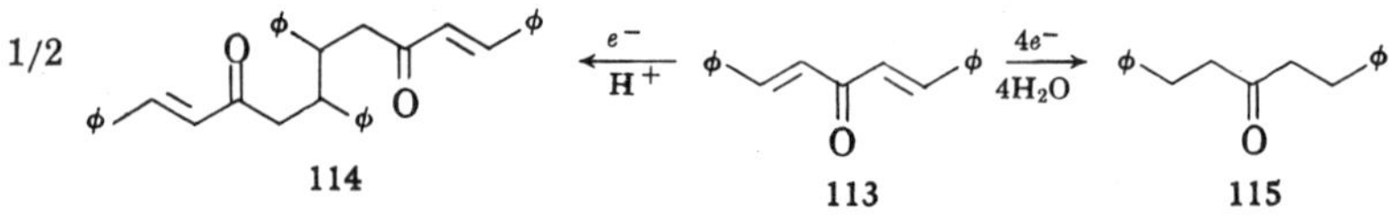

1/2 **114** $\xleftarrow[H^+]{e^-}$ **113** $\xrightarrow[4H_2O]{4e^-}$ **115**

Cpe of both *cis*- and *trans*-dibenzoylethylene produces dibenzoylethane (61).

$$\phi COCH{=}CHCO\phi + 2e^- + 2H^+ \rightarrow \phi COCH_2CH_2CO\phi$$

The polarograms of chalcones are very similar to those of aromatic ketones. Two waves are observed in acid, each one-electron, according to cpc. For the first, $dE_{1/2}/d\text{pH} = -0.06$ v., and for the second, $dE_{1/2}/d\text{pH} = -0.048$ v. The first wave approximates the shape of eq. (19). Near pH 7 the two waves merge, and the dependence on pH decreases. At pH 11 the merged wave divides into two waves. The mechanism of this reduction is entirely analogous to the reduction of benzophenone, except that the one-electron product is a 4,4′-dimer, rather than a pinacol arising from 2,2′-dimerization (164).

In DMF, cpe of chalcone produces only polymeric material. In

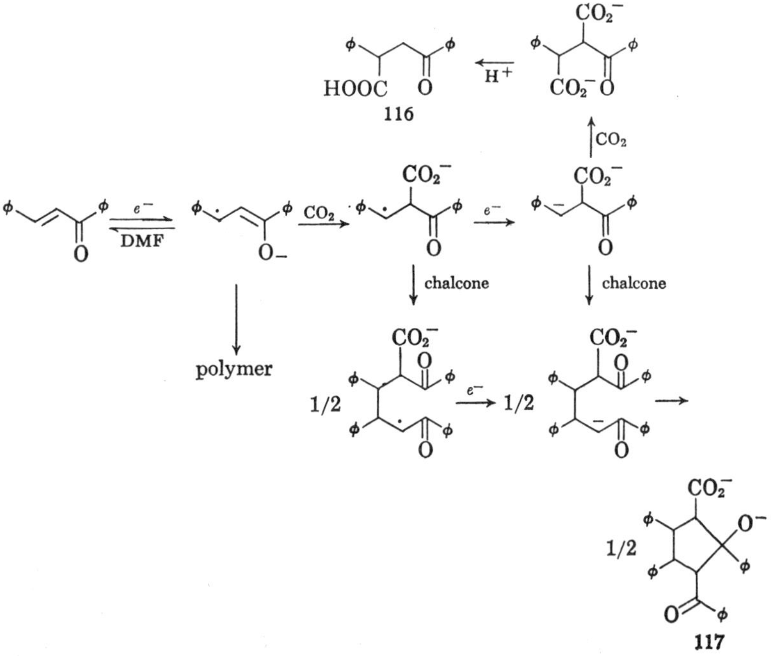

CO_2^- HOOC O **116** $\xleftarrow{H^+}$ CO_2^- O

CO_2 CO_2^-

$\xrightleftharpoons[DMF]{e^-}$ O– $\xrightarrow{CO_2}$ O $\xrightarrow{e^-}$ O

polymer chalcone chalcone

CO_2^- O 1/2 O $\xrightarrow{e^-}$ 1/2 CO_2^- O O $\longrightarrow$

CO_2^- O^- 1/2 O **117**

the presence of CO_2, both a monomeric (**116**) and a dimeric (**117**) acid can be obtained. Only a monomeric acid was obtained from cpe of benzalacetone in DMF in the presence of CO_2. Cpe of crotonaldehyde and cinnamaldehyde (**112**, R = H) in DMF gave only tars, even in the presence of CO_2 to trap unstable radical anion intermediates (147).

trans-Benzoylacrylic acid shows two two-electron waves, with the second at the same $E_{1/2}$ as acetophenone. The first wave involves reduction of the C—C double bond, to form β-benzoylpropionic acid (165).

$$\phi COCH{=}CHCOOH + 2e^- + 2H^+ \rightarrow \phi COCH_2CH_2COOH$$

Esters of *cis*- and *trans*-β-aroylcrotonic acid also undergo two-electron reduction, resulting in hydrogenation of the double bond. A second wave is observed for the reduction of the carbonyl group of the resulting keto-acid (166).

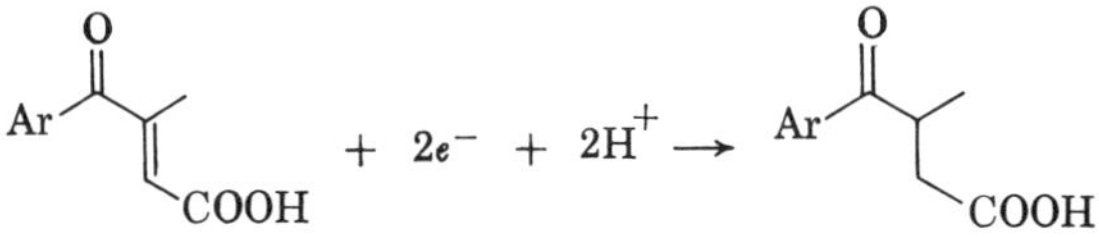

Ar–CO–C(CH₃)=CH–COOH + 2e⁻ + 2H⁺ → Ar–CO–CH(CH₃)–CH₂–COOH

The results of the polarography of coumarin (**118**) conflict among several workers (167–169). However, it seems reasonable that there is a one-electron reduction to a 4,4′-dimer, since cpe yields two high-melting lactones, perhaps the *meso* and *dl* dimers (167). Above pH 10, the current decreases, since the lactone ring is opened to a nonreducible anion. β-Methoxycoumarin (**119**) undergoes two-electron

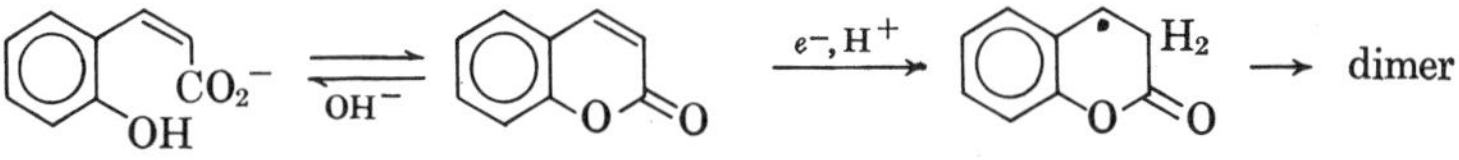

[Scheme: open-chain hydroxycinnamate (OH, CO_2^-) ⇌ (OH⁻) coumarin **118** —(e^-, H^+)→ radical (H_2) → dimer]

118

reduction, to hydrogenate the double bond; the β-hydroxycoumarin (**120**, R = H) enolizes in the other direction and is nonreducible (161).

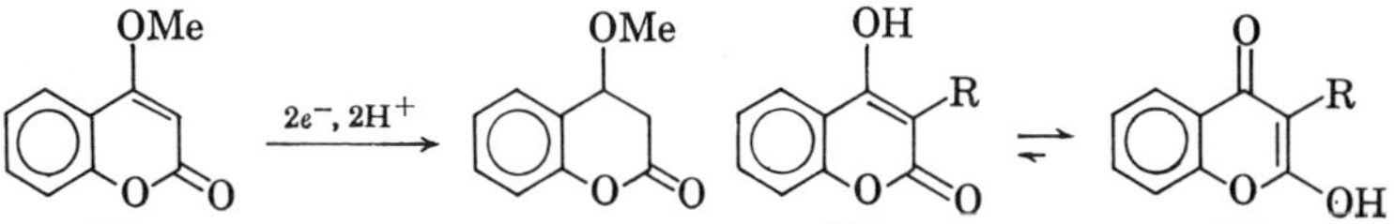

[Scheme: **119** (OMe) —($2e^-$, $2H^+$)→ 4-methoxy-3,4-dihydrocoumarin; **120** (OH, R) ⇌ **120** (O, R, OH)]

119 **120** **120**

Methyl acetylacrylate (**121**) undergoes two-electron reduction of the double bond, with the transfer of the first electron apparently reversible, according to the log plot slope. The product keto-ester is not further reducible (170).

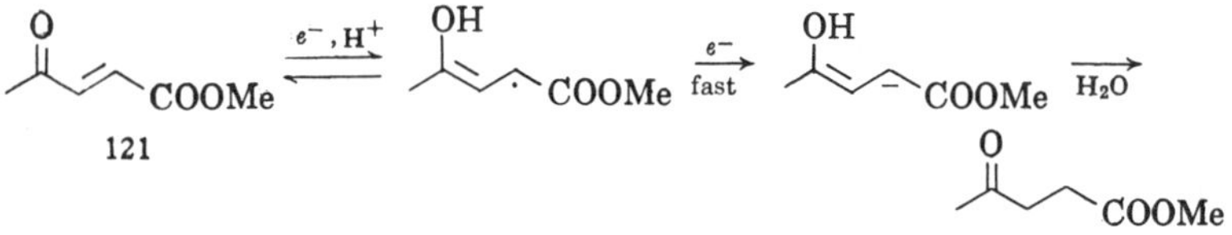

In acid, phenyl vinyl ketone (**122**) shows two one-electron waves, which merge at higher pH, followed by waves of propiophenone at more negative potential. The height of this last wave is a minimum at intermediate pH, since the initial product of the first two-electron process is the enol (**123**) of propiophenone, which is not reducible, but must be transformed into the keto form (**124**) in order for a third wave to be observed. This transformation is catalyzed by acids and bases (171).

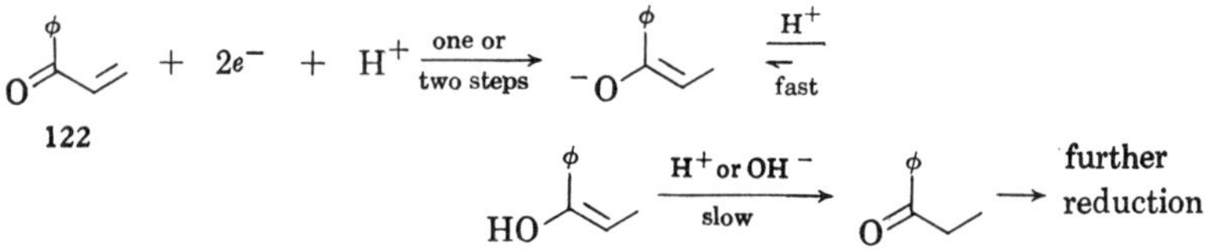

Cpe of tetracyclone (tetraphenylcyclopentadiene) (**125**) produces tetraphenylcyclopentenone (**126**) in base and nearly equal quantities of tetraphenylcyclopentenone (**126**) and tetraphenylcyclopentanone (**127**) in acid (172). Two one-electron waves are observed in acid, but another cpe on the second plateau produced only tetraphenylcyclopentenone (**126**) (173).

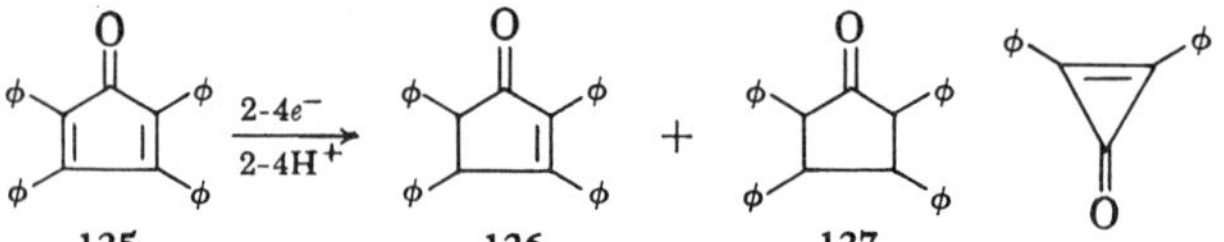

Diphenylcyclopropenone (**128**) in neutral solution shows two waves, each two-electron according to cpc. By analogy with tetracyclone

(**125**), it is claimed (174) that the first wave represents hydrogenation of the C—C double bond, and the second represents reduction of the resulting cyclopropanone. However, tetracyclone is hardly a reasonable reference compound.

Phenylbenzoylacetylene (**129**) shows two waves, with the first dependent on pH, and the second independent. According to the i_d's, they are one- and two-electron reductions, respectively, and the suggested mechanism is (175)

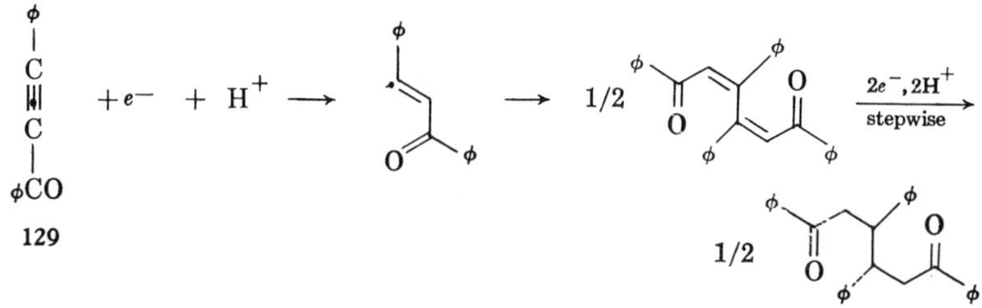

129

Perinaphthenone (**130**) shows two waves of equal height, each one-electron according to i_d. For the first, $E_{1/2}$ is nearly independent of pH, and the log plot slope suggests reversible electron transfer; the second wave involves a proton, and is irreversible (176).

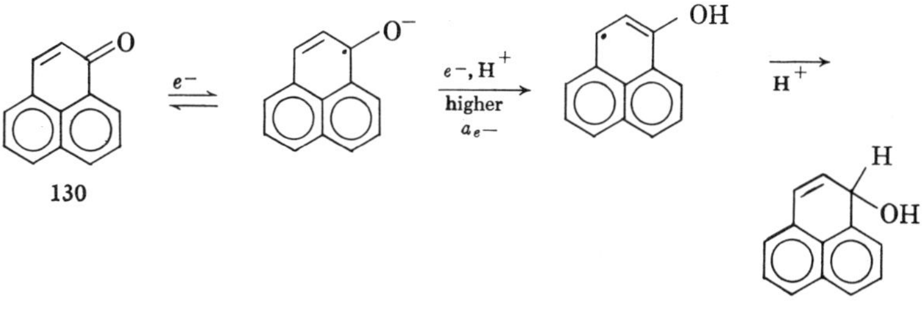

130

The polarographic behavior of tropone (**131**) is similar to that of benzaldehyde, and that of tropolone (**132**) resembles that of salicylaldehyde. Tropone (**131**) shows two two-electron waves, with the second corresponding to reduction of an unsaturated ketone (177).

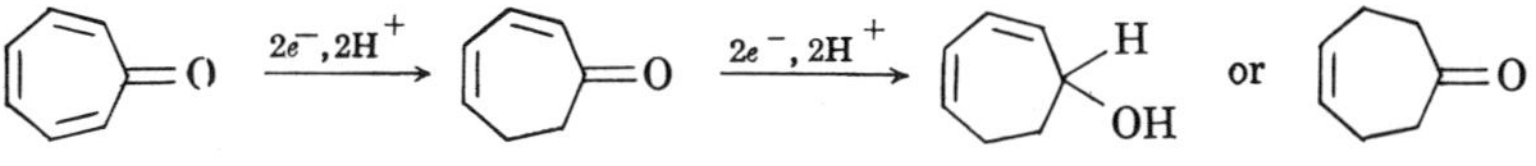

131

Tropolone (**132**) shows two waves at intermediate pH, and only one in acid or alkali. The total height remains fixed at that of a one-electron reduction. The two waves are of equal height at pH = pK_a, indicating that the two waves are due to separate reduction of tropolone and its conjugate base, with the recombination of anion with protons too slow to regenerate tropolone during the life of the drop. In acid, $dE_{1/2}/d\text{pH} = -0.055$ v., suggesting simultaneous protonation and electron transfer (178).

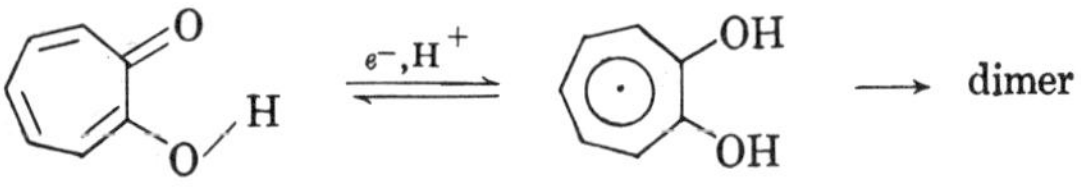

O H $\xrightleftharpoons{e^-, H^+}$ OH OH $\longrightarrow$ dimer

132

In basic solutions, β-methyltropolone (**133**) shows a further one-electron wave, pH-independent, attributed to (179).

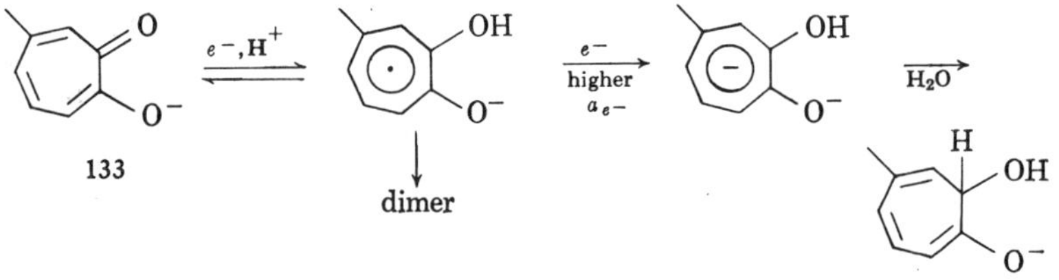

O O^- $\xrightleftharpoons{e^-, H^+}$ OH O^- $\xrightarrow[\text{higher } a_{e^-}]{e^-}$ OH O^- $\xrightarrow{H_2O}$ H OH O^-

133

dimer

Tropolone methyl ether (**134**) behaves similarly to tropolone, except the former is more readily reduced, since the hydrogen bonding in tropolone inhibits protonation simultaneous with electron transfer. Likewise, β-tropolone (**135**) is more readily reduced than α-tropolone (180).

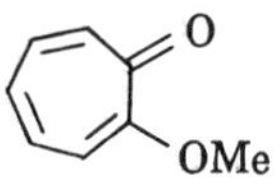

O OMe

134

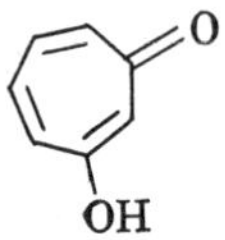

O OH

135

The enol form of a β-diketone behaves polarographically as an α,β-unsaturated ketone. Acetylacetone (**136**, X = Me) and ethyl acetoacetate (**136**, X = OEt) give two waves, originally attributed to separate reduction of keto and enol forms (181). However, the keto form would not be expected to be reducible, and both waves must be due to the enol. By analogy with unsaturated ketones, the first wave represents one-electron reduction to the diketopinacol, and the second represents reduction to the keto-alcohol. Dibenzoylmethane shows

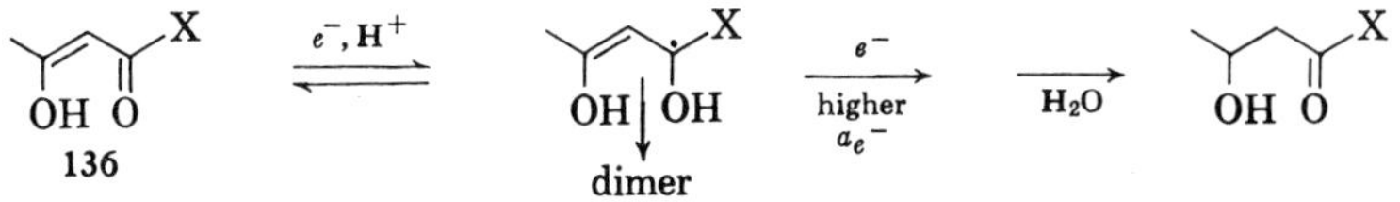

136

two one-electron waves, with a wave at more negative potential due to reduction of the keto alcohol (**137**) formed by two-electron reduc-

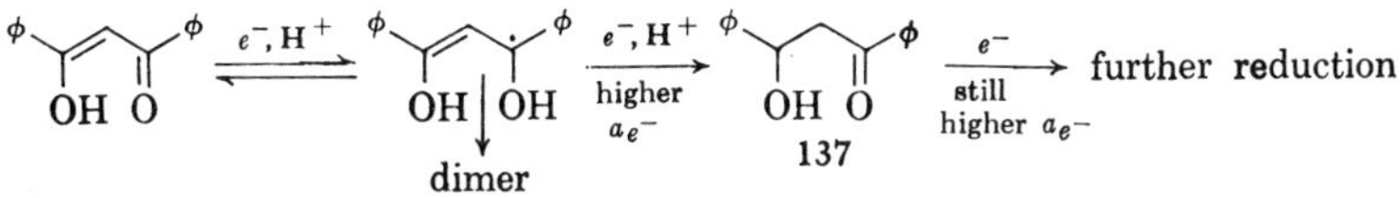

137

tion (61). Aureomycin (**138**) (182) and terramycin (**139**) (183) undergo four-electron reduction, presumably of the two enolized β-diketone groups of each. Thenoyltrifluoroacetone (**140**) shows a multitude of

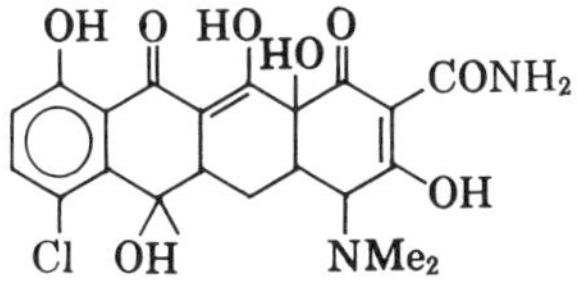

138

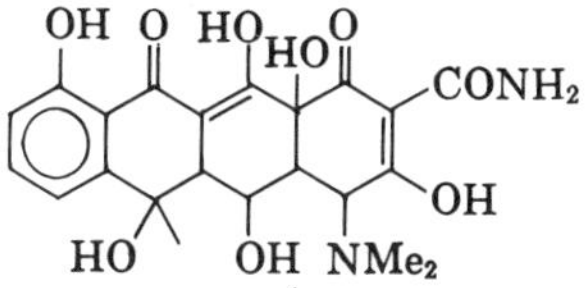

139

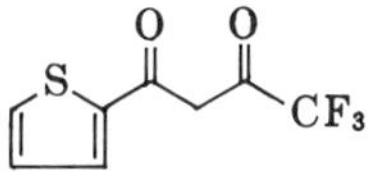

140

waves, but in acid the principal wave represents one-electron reduction of the carbonyl group adjacent to the thiophene ring, to form a pinacol, behavior similar to that of acetophenone. The carbonyl adjacent to the trifluoromethyl group is predominantly hydrated, and therefore unreactive (184).

3. Ketones and Aldehydes Showing Kinetic Waves

In aqueous solution, formaldehyde is present almost entirely as its nonreducible hydrate, $H_2C(OH)_2$, and the polarographic limiting current is governed by the rate at which dehydration to the free aldehyde can take place. In accord with the kinetic nature of this limiting current, i_k is independent of h (185) and increases by about 10% per degree, until near 80°C., the rate of dehydration in base becomes sufficiently great that the current is limited only by diffusion (186). The dehydration is general-base-catalyzed, and presumably also general-acid-catalyzed (185). As a result of the base catalysis, the limiting current increases with pH, and reaches a maximum near pH 13. Beyond pH 13, the current again decreases because formation of the anion, $HOCH_2O^-$, decreases the effective concentration of formaldehyde. At high temperatures, in alkali, the diffusion current corresponds to two-electron reduction to methanol. In alkali, $dE_{1/2}/dpH = -0.06$ v., similar to the behavior of other aldehydes (187), so that the overall mechanism may be written

$$H_2C(OH)_2 \overset{k_f}{\rightleftharpoons} H_2C{=}O \overset{e^-\ H^+}{\rightleftharpoons} H_2COH\cdot \overset{e^-}{\rightarrow} H_2COH^- \xrightarrow[H_2O]{} H_3COH$$

Oscillopolarography indicates that the intermediate radical is reoxidizable (188).

Acetaldehyde (**141**, R = H) and propionaldehyde (**141**, R = Me) (189) are much less extensively hydrated than is formaldehyde, and the dependence of i_l on temperature is much less marked. However, the limiting current does depend upon pH, and in unbuffered solutions, i_k is independent of h. The $E_{1/2}$ is shifted to more negative values by increasing the pH, and the electrode reaction is again two-electron reduction to the alcohol. Glycolaldehyde (**141**, R = OH) is much more extensively hydrated, and behaves quite similarly to formaldehyde (189). Glyoxal (**142**) also gives a kinetic current, with i_l independent of h, and a temperature coefficient of almost 10% per degree. The current is limited by the rate at which the dihydrate of glyoxal is dehydrated to dihydroxyacetaldehyde, the mono-hydrate. Cpc indicates $n = 3$, but the authors have suggested that this value represents an average value, rather than reduction exclusively to erythritol (**143**) (190).

$RCH_2CH{=}O$	$O{=}CHCH{=}O$	$HOCHCH_2OH$ / $HOCHCH_2OH$
141	**142**	**143**

Even in so-called anhydrous DMF, there is sufficient water present to hydrate aldehydes. The limiting current is quite small, but is proportional to $h^{1/2}$, so it is probably only the unhydrated aldehyde which is being reduced. Addition of excess phenol increases i_d sixfold, and $E_{1/2}$ shifts slightly to less negative values, as expected if the aldehyde is now present as a hydrogen-bonded form, and all the aldehyde present is capable of reduction (191). Another possibility is that general-acid-catalysis produces the aldehyde sufficiently rapidly that even the hydrate can contribute to i_d.

Aldoses likewise give a small wave for the reduction of the aldehyde group, but the current is a kinetic current limited by the rate at which the nonreducible cyclic hemiacetal opens to the aldehyde form (192).

α-Keto acids give two waves in solutions of intermediate pH, and only one wave in acid or base. In acid the single wave is due to reduction of the carbonyl group of the free acid, and in base the single wave is due to reduction of the carbonyl group of the anion. At intermediate pH, the wave doubling is due to separate reduction of acid and anion. The acid is reduced at less negative potentials than is the anion, because a —COOH group is more strongly electron-withdrawing than is a $—CO_2^-$ group. However, the height of the first wave is greater than expected on the basis of the concentration of the acid, since the acid is replenished at a finite rate from the anion. The rate of regeneration of the acid determines the kinetic component of the first wave, and pK', the pH at which both waves are of equal height, is greater than pK_a. Pyruvic (**144**, R = CH_3) and phenylglyoxylic (**144**, R = ϕ) acids show this behavior, but their esters undergo only the normal two-electron reduction of the keto group (193). There is a further complication for these acids, since the carbonyl group of pyruvic acid is extensively hydrated, and so are the carbonyl groups of phenylglyoxylic acid and its anion, and glyoxylic acid (**144**, R = H) and its anion. Thus the heights of these limiting currents are also governed by the rates of dehydration (194). 2-Ketogulonic acid (**145**) shows two waves at intermediate pH, with both limiting cur-

RCOCOOH
144

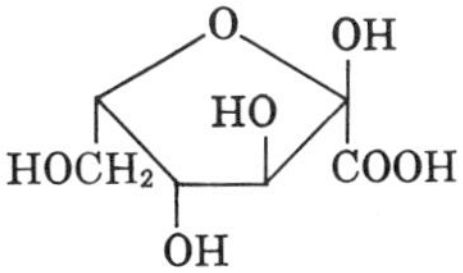

O OH
HO
$HOCH_2$ COOH
OH
145

rents independent of h, and the total current increasing with pH. Both the acid and the anion are present predominantly as a cyclic hemiketal, and the currents are governed by both the rate of ring-opening (general-base-catalyzed) and the rate of recombination of the anion with protons (195).

Finally, acetone shows an anodic wave, which is a kinetic current limited by the rate of enolization. The product is (hydroxymercuri)-acetone (196).

$$CH_3COCH_3 \overset{k_f}{\rightleftharpoons} CH_3C(OH){=}CH_2 \xrightarrow[Hg,\ H_2O]{-2e^-} CH_3COCH_2HgOH$$

4. Acids and Derivatives

Aliphatic and aromatic acids with no reducible groups give a polarographic wave, but this is usually due merely to a reduction of protons to hydrogen. Only a few cases of polarographic reduction of a carboxylic acid group are known, although Allen (7) has discussed many macroscale reductions. Cpe of β-naphthol-3-carboxylic acid in the presence of *p*-toluidine produces a small yield of the toluidine Schiff base of β-naphthol-3-aldehyde (**146**, Ar = *p*-MeC_6H_4) (197). Cpe of

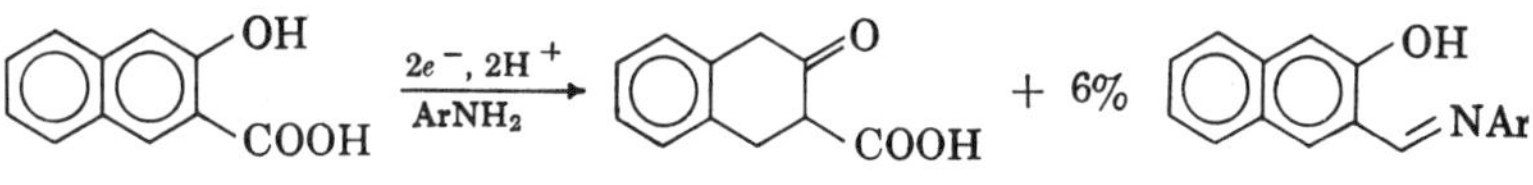

$$\text{(3-hydroxy-2-naphthoic acid)} \xrightarrow[ArNH_2]{2e^-,\ 2H^+} \text{(tetrahydro-3-oxo-2-naphthoic acid)} + 6\%\ \textbf{146}$$

146

isonicotinic acid and *N*-ethyl-isonicotinic acid produces good yields of the aldehyde; reduction stops at the aldehyde stage, since

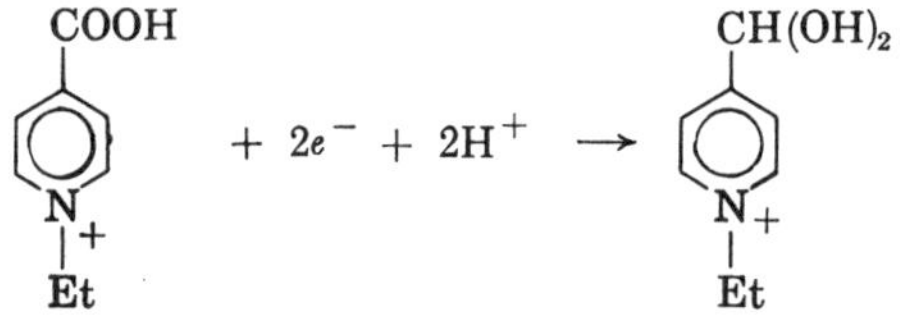

$$\text{(4-COOH-N-ethylpyridinium)} + 2e^- + 2H^+ \rightarrow \text{(4-}CH(OH)_2\text{-N-ethylpyridinium)}$$

these aldehydes are extensively hydrated and not further reducible (198). Isonicotinamide (**147**, R = H), thioisonicotinamide (**148**), isonicotinanilide (**147**, R = ϕ) (199), and thiobenzamide, $\phi CSNH_2$,

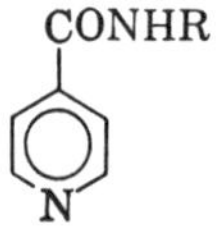

147 (4-CONHR-pyridine) **148** (4-$CSNH_2$-pyridine)

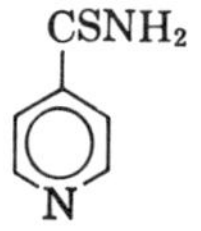

(200) are also reducible. Benzamidine undergoes four-electron reduction to benzylamine, isolated from cpe; $E_{1/2}$ is nearly independent of

$$\phi C(NH_2)_2^+ + 4e^- + 3H^+ \rightarrow \phi CH_2NH_2 + NH_3$$

pH, indicating that the potential-determining step is addition of a single electron, to form the radical, $\phi C(NH_2)_2^{\cdot}$ (201). Acid chlorides show a reduction wave in anhydrous acetone (202).

Unsaturated acids undergo reduction leading to hydrogenation of the double bond, with the carboxylic acid group acting as an activator similar to a ketone group. Thus, cinnamic acid shows a wave of height corresponding to two-electron reduction to hydrocinnamic acid (203).

$$\phi CH{=}CHCOOH + 2e^- + 2H^+ \rightarrow \phi CH_2CH_2COOH$$

In acidic solution, acrylic acid, methacrylic acid (204), methyl methacrylate (**149**) (205), and acrylonitrile (206) undergo reduction-induced polymerization at a mercury cathode.

COOMe (**149**) $\xrightarrow{e^-}$ COOMe $\xrightarrow{H^+}$ COOMe $\rightarrow$ polymerization

Substituted benzalmalonic esters and nitriles (**150**, R = H, ϕ, alkyl, X = CN,COOEt) undergo two-electron reduction in acid, but at pH above 6 the wave splits into two, with the first independent of pH (207):

150 $\underset{}{\overset{e^-}{\rightleftharpoons}}$ (→ dimer; e^-, higher a_{e^-}) $\xrightarrow{H^+}$ $\underset{\text{fast}}{\overset{e^-}{\rightarrow}}$ $\xrightarrow{H^+}$

In DMF, two one-electron waves are observed; addition of phenol increases the height of the first at the expense of the second (207).

In barium acetate or tetraethylammonium chloride supporting electrolytes, methyl phenylpropiolate shows two two-electron waves, according to cpc. $E_{1/2}$ of the first wave is independent of pH (208), and the suggested mechanism is

$$\phi C{\equiv}CC(=O)OMe + e^- \rightarrow \phi\dot{C}{=}C{=}C(O^-)OMe \xrightarrow[\text{fast}]{e^-,\,2H^+} \phi CH{=}CHCOOMe \xrightarrow[\text{higher } a_{e^-}]{2e^-,\,2H^+} \phi CH_2CH_2COOMe$$

Use of the higher tetraalkylammonium salts as supporting electrolyte shifts the first wave to more negative potentials. These higher homologs are more strongly adsorbed on the negatively charged electrode surface. The surface layer hinders electron transfer to the triple bond, but apparently not to the double bond, whose orientation requirements are different (209).

Ethyl esters of maleic (**151**) and fumaric (**152**) acids undergo two-electron reduction to diethyl succinate (210). Two waves of unequal

COOEt / COOEt (**151**) $\xrightarrow{2e^-,\ 2H^+}$ COOEt / COOEt $\xleftarrow{2e^-,\ 2H^+}$ EtOCO / COOEt (**152**)

height occur at high pH, but only one is observed in acid. The $E_{1/2}$ of the first wave of diethyl fumarate depends upon pH. The log plot slope varies with pH, but indicates that at pH 7.6, one electron may be involved in a reversible step, to form a radical anion which adds a proton. Below pH 5.8, $E_{1/2}$ of the first wave of diethyl maleate depends upon pH, but at higher pH, the $E_{1/2}$ of the first wave is independent of pH, suggesting formation of a stable radical anion, which can dimerize before adding a proton. The log plot slope again varies with pH, but a reversible one-electron reduction may be operative at high pH. An unusual feature is that in acid the thermodynamically less stable maleate species (**151**) is less readily reduced than is the more stable fumarate species (**152**), although both form the same intermediate radical. Elving and Teitelbaum (210) suggest that the transition state for maleate reduction requires a planar configuration, and is quite unfavorable. Such behavior is not unusual; in 80% ethanol, *trans*-stilbene is reduced at −2.13 v., but *cis*-stilbene is reduced at −2.18 v. (211). At higher pH, where the reduction of diethyl maleate becomes pH-independent, whereas that of diethyl fumarate is still pH-dependent, fumarate becomes less readily reduced, since the two are reduced by different mechanisms (210). However, it is not clear why a pH-independent reduction of diethyl fumarate does not occur at high pH.

The polarographic behavior of maleic and fumaric acids is still more complicated, as a result of ionization of one or two protons, and the possibility of kinetic contributions to the waves, governed by the rate of recombination of anions with protons (210). It is reasonable to ascribe the doubling of the waves to successive one-electron reductions, with a metastable intermediate radical, by analogy with the behavior of the esters, although the possibility of separate reduction of acid and anion has not been eliminated. It is surprising that the mono-ethyl esters of fumaric and maleic acids show only a single wave over the accessible pH range (212); both possibilities for doubling of the wave are still present. Cpe might distinguish these two possibilities. At all pH values, maleic acid is more readily reduced than fumaric.

Maleic hydrazide behaves in a fashion similar to that of maleic acid, although a total of three waves are observed over the accessible pH range (213). Cpe produces succinic hydrazide; reductive hydrogenation of the double bond occurs (214). No anodic wave is observed, indicating that the substrate is not present as the tautomeric 2,3-diazahydroquinone (**153**) (213).

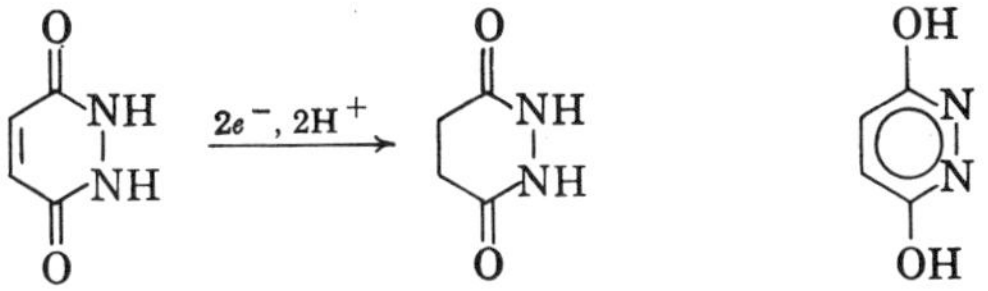

O NH NH O $\xrightarrow{2e^-,\ 2H^+}$ O NH NH O

OH N N OH

153

Maleic mono-*N*-methylamide undergoes two-electron reduction in acid, but above pH 6 two one-electron waves are observed, with the first attributed to reduction to a dimer (215).

One remarkable feature of these reductions is that cpe of dimethylmaleic anhydride (**154**) and dimethylfumaric acid (**155**) produce *dl*- (**156**) and *meso*- (**157**) α,α'-dimethylsuccinic acid, respectively, demonstrating stereospecifically *trans* hydrogenation (212).

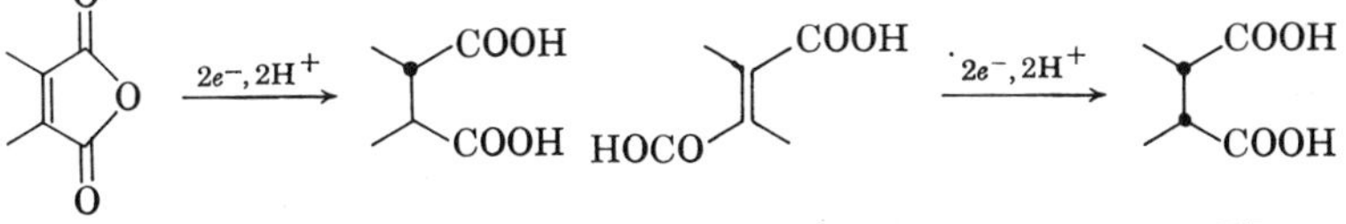

O O O $\xrightarrow{2e^-,2H^+}$ COOH COOH HOCO COOH $\xrightarrow{2e^-,2H^+}$ COOH COOH

154 **156** **155** **157**

Electrolysis at constant current (not cpe) of 2,3-diphenyl-*trans*,-*trans*-muconic acid (**158**, R = ϕ) produces mostly *trans*-3,4-diphenyl-$\Delta^{3,4}$-dehydroadipic acid (**159**, R = ϕ), with a small amount of the *cis* isomer (216), corresponding to 1,4-hydrogenation. Cpe of both *cis,cis-* and *trans,trans*-muconic acid (**158**, R = H) leads to *trans*-$\Delta^{3,4}$-dehydroadipic acid (**159**, R = H) (217).

$$\mathbf{158} \xrightarrow{2e^-,\ 2H^+} \mathbf{159}$$

Diethyl acetylenedicarboxylate gives two closely spaced waves in acid, with the second at the same $E_{1/2}$ as the wave of diethyl fumarate. Cpc indicates the first wave to be a two-electron reduction; $E_{1/2}$

$$\text{EtOCOC}\equiv\text{CCOOEt} \xrightarrow{2e^-,\ 2H^+} \text{EtOOC-CH=CH-COOEt} \xrightarrow[\text{higher } a_e-]{2e^-,\ 2H^+} \text{EtOOC-CH}_2\text{CH}_2\text{-COOEt}$$

depends upon pH. Cpc of the monoethyl ester (**160**, R = Et) indicates three-electron reduction, and diethyl *dl*-α,α'-dimethylsuccinate (**161**, R = Et) was obtained from cpe (212). The suggested mechanism is

$$\underset{\mathbf{160}}{\text{HOOC-C}\equiv\text{C-COOR}} \xrightarrow{e^-,\ H^+} \underset{\mathbf{162}}{\text{HOOC-CH=}\dot{\text{C}}\text{-COOR}} \xrightarrow{-CO_2} \text{CH}_2\text{=}\dot{\text{C}}\text{-COOR} \rightarrow$$

$$\frac{1}{2}\ \text{ROCO-C(=CH}_2\text{)-C(=CH}_2\text{)-COOR} \xrightarrow[\text{fast}]{e^-,\ H^+} \xrightarrow[\text{fast}]{e^-,\ H^+} \frac{1}{2}\ \underset{\mathbf{161}}{\text{ROCO-CH(CH}_3\text{)-CH(CH}_3\text{)-COOR}}$$

Acetylenedicarboxylic acid (**160**, R = H) shows the same behavior—three-electron reduction to *dl*-α,α'-dimethylsuccinic acid (**161**, R = H), obtained from cpe—and the same mechanism is suggested. In further support of the mechanism, it was noted that decarboxylation cannot have occurred prior to the addition of the first electron, because propiolic acid, HC≡CCOOH, is not reducible in the accessible potential range. Also, decarboxylation must occur prior to dimerization of the one-electron reduction product (**162**), because butadiene-

1,2,3,4-tetracarboxylic acid is reducible without decarboxylation (212).

Esters of phthalic acid show two irreversible waves, four- and two-electron, respectively, according to i_d. The $E_{1/2}$ of the second wave is the same as that of phthalide (**97**, R = H), isolated from cpe on the first plateau (218). Cpe on the second plateau produced only a non-crystalline resin.

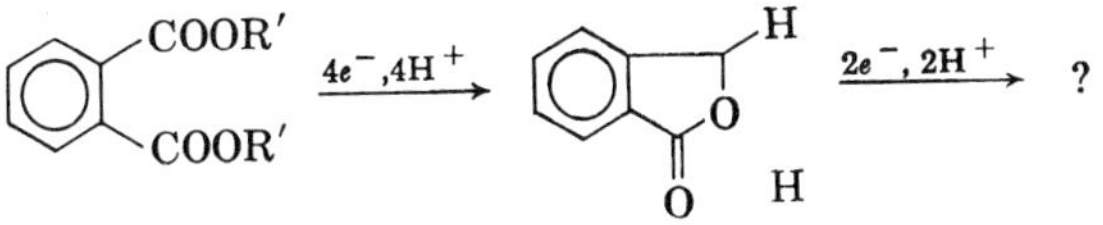

$$\text{C}_6\text{H}_4(\text{COOR}')_2 \xrightarrow{4e^-, 4H^+} \textbf{97} \xrightarrow{2e^-, 2H^+} ?$$

97

Phthalic acid gives a total of three waves over the pH range, with possible kinetic components due to reduction of the various acid and anion species (219). In DMF, substituted phthalic anhydrides and pyromellitic anhydride (**163**) undergo one-electron reduction to the radical anions (220).

163 **164**

Phthalimide undergoes two-electron reduction in acid. In base the wave splits into two. The splitting is not due to separate reduction of phthalimide and its anion, since *N*-alkylphthalimides behave similarly. The first wave must represent formation of a one-electron reduction product, which dimerizes (221). The wave of naphthalimide (**164**) splits into two even at pH values above 2. Further support for a two-electron reduction, rather than a four-electron reduction as found with phthalic anhydride, is afforded by the cpe of an *N*-alkyltetrachlorophthalimide (**165**, R = $CH_2CH_2NMe_2$) at a lead cathode, to produce a two-electron reduction product (**166**) (222).

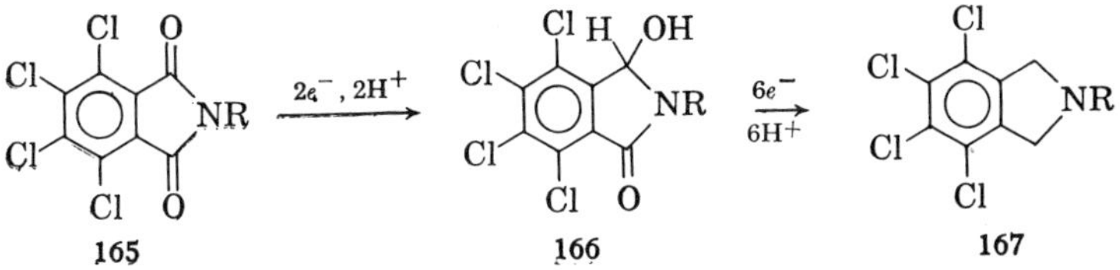

$$\textbf{165} \xrightarrow{2e^-, 2H^+} \textbf{166} \xrightarrow[6H^+]{6e^-} \textbf{167}$$

165 **166** **167**

At more negative potentials, further six-electron reduction produces the isoindoline (**167**) (222).

Cpe of the bis-furanone (**168**, *cis* or *trans*) in sulfuric acid produces 4,5-diketosuberic acid (**169**). In acetonitrile–water mixtures, $E_{1/2}$ depends upon pH (223).

$$\mathbf{168} \xrightarrow[H_2SO_4]{2e^-,\ 2H^+} \xrightarrow{H^+} HOCO(CH_2)_2CO\text{-}CO(CH_2)_2COOH\ \ \mathbf{169}$$

In absolute ethanol, the sodium salt of diethyl malonate undergoes one-electron oxidation and dimerization to tetraethyl ethanetetracarboxylate (224).

$$Na^+\ {}^-CH(COOEt)_2 - e^- \underset{EtOH}{\rightleftharpoons} \cdot CH(COOEt)_2 \longrightarrow 1/2\ (EtOCO)_2CH\text{-}CH(COOEt)_2$$

5. *Cyano Compounds*

Tetracyanoethylene (TCNE) (**170**) undergoes reversible one-electron reduction in acetonitrile (225). 1,1,2,2-Tetracyanocyclopropane (**171**) undergoes irreversible reduction to the anion of TCNE, as determined by its anodic wave and ESR spectrum. The fate of the CH_2 group was not established. At more negative potentials a second electron can be added to TCNE, although this dianion is sufficiently basic to abstract a proton irreversibly from the solvent (226). Tetracyanoquinodimethane (TCNQ) (**172**) undergoes stepwise two-electron

$$\underset{\mathbf{170}}{(NC)_2C{=}C(CN)_2} + e^- \underset{MeCN}{\rightleftharpoons} (NC)_2CC(CN)_2^{\overline{\cdot}} \longleftarrow \mathbf{171}$$

172

reduction, through a stable radical anion (226,227). In 90% acetic acid, the reduction is still reversible, although the product is $TCNQH_2$, rather than the dianion, since the radical anion is no longer stable to disproportionation (226).

Polarography of TCNE and other π-acids (acceptors) in DMF in the presence of excess hexamethylbenzene (HMB) has yielded interesting information about the charge-transfer complexes formed (228). Even in the presence of HMB, the log plot slopes for TCNE, TCNQ, chloranil (**173**), pyromellitonitrile (**174**), tetrachlorophthalic anhydride (**175**), and 2,3-dichloro-5,6-dicyanoquinone (**176**) indicate that

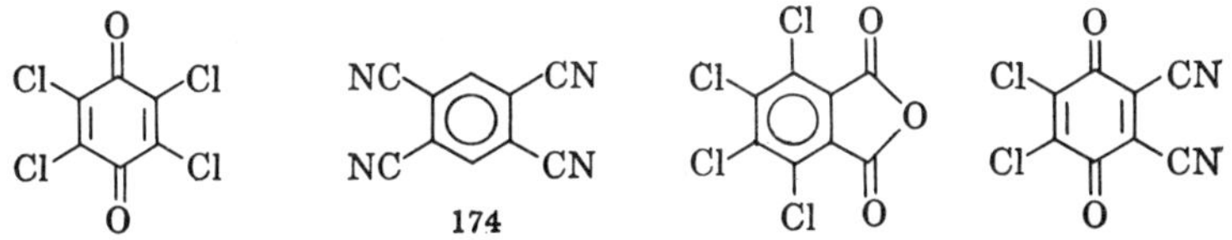

173 **174** **175** **176**

the reductions are reversible, and the limiting currents are of normal magnitude and proportional to $h^{1/2}$, so that dissociation of the donor-acceptor complexes is sufficiently rapid not to lead to complications.

$$\mathrm{D{-}A} \underset{\text{rapid}}{\rightleftharpoons} \mathrm{D + A} \underset{\text{DMF}}{\overset{e^-}{\rightleftharpoons}} \mathrm{A}^{\cdot-}$$

The effect of the donor HMB is only to shift the $E_{1/2}$ to more negative values, according to eq. (10), where E^0 is to be interpreted as a formal potential, as was done to obtain eq. (12). The slope of $-E_{1/2}$ vs. log [HMB] equals 0.059 times the number of molecules of donor per molecule of acceptor in the complex, and the relationship of $E_{1/2}$ to E^0 and [HMB] can be used to determine the equilibrium constant for complex formation.

In DMF, crotononitrile (**177**, R = Me) undergoes two-electron reduction to butyronitrile (**178**, R = Me), but at higher concentrations, one-electron reduction followed by dimerization is observed. Cinnamonitrile (**177**, R = ϕ) shows two one-electron waves (229).

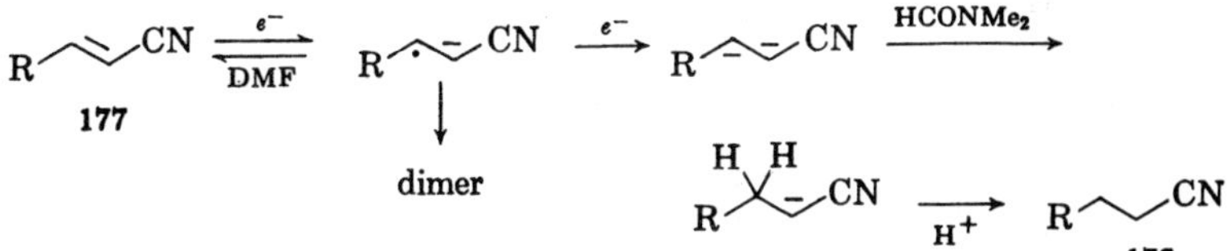

$$\text{R–CH=CH–CN } (\mathbf{177}) \underset{\text{DMF}}{\overset{e^-}{\rightleftharpoons}} \text{R}\dot{\text{C}}\text{H–}\bar{\text{C}}\text{H–CN} \xrightarrow{e^-} \text{R}\bar{\text{C}}\text{H–}\bar{\text{C}}\text{H–CN} \xrightarrow{\text{HCONMe}_2}$$

$$\text{R}\dot{\text{C}}\text{H–}\bar{\text{C}}\text{H–CN} \longrightarrow \text{dimer}$$

$$\text{RCH}_2\text{–}\bar{\text{C}}\text{H–CN} \xrightarrow{\text{H}^+} \text{R–CH}_2\text{–CH}_2\text{–CN } (\mathbf{178})$$

In DMF, benzonitrile (**179**) undergoes reversible one-electron reduction to a radical anion (**180**). Phthalonitrile (**181**) undergoes reversible, stepwise two-electron reduction, but the dianion (**182**) irreversibly abstracts a proton from the solvent and loses a cyanide anion to form benzonitrile (**179**), which adds an electron to form the radical anion (**180**) of benzonitrile, whose ESR spectrum is observed (230). *p*-Fluoro

181 ⇌ (e^- / DMF) ⇌ (e^- / higher a_{e^-}) **182** → ($HCONMe_2$) CN^- + **179** ⇌ (e^- / DMF) **180**

(**183**, X = F) and *p*-amino- (**183**, X = NH_2) benzonitrile add an electron to form a radical anion which dimerizes with the expulsion of fluoride or amide anion, and then adds another electron for each molecule of dimer (230).

2 **183** ⇌ ($2e^-$ / DMF) → $2X^-$ + dimer ⇌ (e^-) [dimer]$^{\cdot -}$

Tetramethylammonium pentacyanopropenide (**184**) undergoes reversible stepwise addition of two electrons, but the resulting trianion

184 ⇌ (e^- / DMF) ⇌ (e^- / higher a_{e^-})

185 → ($HCONMe_2$) CO + Me_2N^- + **186** + CN^-

(**185**) irreversibly abstracts a proton from DMF and expels cyanide anion to form a monoanion (**186**) (230).

Below pH 5, picolinonitrile (**187**) and isonicotinonitrile (**188**) show a four-electron reduction wave in acid (231). Cpe of the latter in 1*M* HCl produces 4-aminomethylpyridine (**189**) (232).

$$\textbf{188}\ (\text{4-cyanopyridine}) \xrightarrow{4e^-,\ 4H^+} \textbf{189}\ (\text{4-}CH_2NH_2\text{-pyridine}); \quad \textbf{187} = \text{2-cyanopyridine}$$

In acetonitrile, perchlorate anion undergoes one-electron oxidation to the perchloryl radical (**190**). This radical abstracts a hydrogen atom from the solvent, to form a radical which dimerizes to succinonitrile (**191**) (233).

$$ClO_4^- \xrightarrow{-e^-} \underset{\textbf{190}}{ClO_4^{\cdot}} \xrightarrow{CH_3CN} HClO_4 + CH_2CN\cdot \rightarrow \underset{\textbf{191}}{{}^1/_2NCCH_2CH_2CN}$$

6. Ketone Derivatives

a. Imines

Saturated ketones are reducible only at very negative potentials, but their derivatives are much more readily reducible. Thus, acetone shows a wave at -2.1 v., but this potential range is accessible only in solutions with tetraalkylammonium salts as supporting electrolyte. In ammoniacal solution, acetone imine is formed, and gives a wave at -1.6 v. (234). Cyclohexanone (**192**) in the presence of methylamine undergoes two-electron reduction, according to cpc, but the product isolated after cpe is not cyclohexanol, but *N*-methylcyclohexylamine (**193**) (235).

$$\underset{\textbf{192}}{C_6H_{10}{=}O} + MeNH_2 \rightleftharpoons C_6H_{10}{=}NMe \xrightarrow{2e^-,\ 2H^+} \underset{\textbf{193}}{C_6H_{10}(H)NHMe}$$

The mechanism of the reduction of imines is entirely analogous to the reduction of the corresponding ketone. Thus benzophenone anil, $\phi_2C{=}N\phi$, shows two one-electron waves in acid, which merge at higher pH. The $E_{1/2}$ of the first wave depends upon pH, but $E_{1/2}$ of the second does not. Cpe produces benzhydrylaniline, $\phi_2CHNH\phi$ (235).

The cyclic Schiff base salt, hydrastinine (**194**), undergoes two-electron reduction in acid (236).

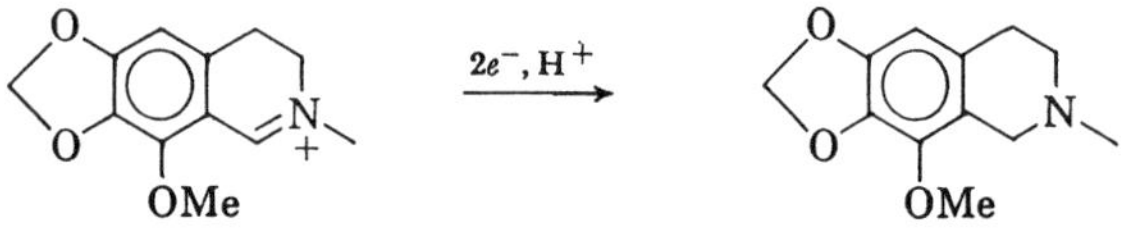

$$\mathbf{194}\ (\text{OMe}) \xrightarrow{2e^-,\ H^+} \text{(reduced amine, OMe)}$$

b. Oximes

Souchay and Ser (237) have reviewed the polarography of oximes. According to the wave height, the reduction consumes four electrons, to form the amine. The imine is a likely intermediate, but the hydroxylamine cannot be an intermediate, since it would be reduced only at much more negative potentials. In acid, an oxime is reduced more readily than its parent ketone, but in strong base the oxime is converted to the less readily reduced anion. The wave height decreases above pH 8, and a new wave appears at more negative potentials. Between pH 6 and 10, the height of the first wave is independent of h, indicative of a current limited by the rate of protonation of the oxime, to form its conjugate acid (237,238). The second wave is due to a pH-independent reduction of the oxime. The two waves are not due to the reduction of oxime and anion, separately, since the pK_a of oximes is near 11, and at these low pH values, the oxime is present only as free oxime. At pH below 6 the rate of protonation of the oxime is sufficiently high that the current is only diffusion-controlled. The mechanism may be written as

$$R_2C{=}NOH_2^+ + 2e^- + H^+ \rightarrow R_2C{=}NH \xrightarrow[\text{fast}]{2e^-,\ 2H^+} R_2CHNH_2$$

$$\Big\Updownarrow k_f[H^+]$$

$$R_2C{=}NO^- \overset{H^+}{\rightleftharpoons} R_2C{=}NOH \xrightarrow[\substack{\text{higher}\\ a_{e^-}}]{2e^-} \xrightarrow{2H^+} R_2C{=}NH \xrightarrow[\text{fast}]{2e^-,\ 2H^+} R_2CHNH_2$$

In 5–20% sulfuric acid, the oxime is converted entirely to its conjugate acid, and $E_{1/2}$ becomes independent of acidity (238).

Cpe of isobutyraldoxime produces isobutylamine; phenylacetaldoxime produces β-phenethylamine, and benzophenone oxime produces benzhydrylamine (235). Cpe of isonicotinaldoxime produces 4-aminomethylpyridine (**189**) (239). In support of the intermediacy of the imine, 2,4-dihydroxybenzophenone oxime shows two two-electron waves in acid, with the second at the same $E_{1/2}$ as that of the imine, isolated from cpe on the first plateau. The cyclic oxime benzoate (**195**, Ar = p-$MeOC_6H_4$) behaves similarly (240), and **196** was isolated from cpe on the first plateau.

$$\mathbf{195} \xrightarrow{2e^-,\ 2H^+} \mathbf{196} \xrightarrow[\text{higher } a_{e^-}]{2e^-,\ 2H^+} \text{(amino acid, } NH_2\text{, COOH)}$$

In contrast to α,β-unsaturated ketones, whose reduction leads to hydrogenation of the C—C double bond, α,β-unsaturated imines (the intermediates in the reduction of α,β-unsaturated oximes) are reduced with hydrogenation of the C—N double bond. Thus cinnamaldoxime (**197**, R = H) undergoes only four-electron reduction, according to cpc, to form primarily cinnamylamine (**198**, R = H), isolated from cpe. Cpe of benzalacetoxime (**197**, R = Me) produces 1-methylcinnamylamine (**198**, R = Me). During cpe in base, the intermediate benzalacetone imine (**199**, R = Me) is hydrolysed to benzalacetone (**112**, R = Me), which is hydrogenated to benzylacetone (235).

$$\mathbf{197}\ (\phi CH{=}CHC(R){=}NOH) \xrightarrow{2e^-,\ 2H^+} \mathbf{199}\ (\phi CH{=}CHC(R){=}NH) \xrightarrow{2e^-,\ 2H^+} \mathbf{198}\ (\phi CH{=}CHCH(R)NH_2)$$

$$\mathbf{199} \xrightarrow{OH^-} \mathbf{112}\ (\phi CH{=}CHC(R){=}O) \xrightarrow[\text{fast}]{2e^-,\ 2H^+} \phi CH_2CH_2C(R){=}O$$

Also, the oxime of testosterone propionate (**200**) shows two two-electron waves, although the imine intermediate formed by cpe on the first plateau hydrolyzes to the ketone during the cpe (235).

Mono-oximes of diketones are less readily reduced than the parent ketones. One four-electron wave is observed in acid or base, but two waves are observed at pH values near the pK_a of the oxime, corresponding to separate reduction of the oxime and its monoanion to an α-amino ketone (237).

The dioxime, dimethylglyoxime, is still less readily reduced than the mono-oxime, but the 1,3-dioxime (**201**) of cyclohexanetrione-1,2,3

OCOEt

HON

200

O

HON NOH

201

undergoes very facile four-electron reduction (237). In acid, dimethylglyoxime is largely present as its conjugate acid (**202**, R = Me), and undergoes eight-electron reduction to 2,3-diaminobutane. From the slope of the $E_{1/2}$–pH curve, one proton is involved in the first, irreversible step, and the mechanism suggested is (241)

$$\text{R–C}(=\overset{+}{\text{N}}\text{HOH})\text{–C(R)}=\text{NOH}\ (\mathbf{202}) + e^- + \text{H}^+ \longrightarrow \text{R–C}(\overset{+\bullet}{\text{N}}\text{HOH})=\text{C(R)}\ddot{\text{N}}\text{HOH} \xrightarrow[\text{fast}]{7e^-,\ 8\text{H}^+} \text{RCH}(\text{NH}_3^+)\text{–CH(R)NH}_3^+$$

Gelb and Meites (241) have also discussed the mechanism of α-furil dioxime (**202**, R = C_4H_3O) reduction in detail.

Benzamide oxime undergoes two-electron reduction to benzamidine (235).

$$\phi\text{C(NH}_2)\text{=NOH} + 2e^- + 2\text{H}^+ \rightarrow \phi\text{C(NH}_2)\text{=NH}$$

Mesoxalic acid, $OC(COOH)_2$, occurs predominantly as its nonreducible hydrate, $(HO)_2C(COOH)_2$, in solution, but its oxime and hydrazone show reduction waves of height corresponding to two-electron reduction (242). However, cpe indicates four-electron reduction to aminomalonic acid, $H_2NCH(COOH)_2$ (235), and the oxime diethyl ester also undergoes four-electron reduction (237).

c. Hydrazones

The polarographic behavior of hydrazone derivatives of ketones is closer to that of the parent ketone than to that of oximes. According to the limiting current, the reduction of Girard derivatives of ketones

(**203**) is a two-electron process. The log plot slope indicates that the potential-determining step is transfer of one electron, and one proton is also implicated, according to the pH-dependence of $E_{1/2}$ (243).

$$\underset{\textbf{203}}{R_2C{=}N{-}NHCOCH_2NMe_3^+} + e^- + H^+ \rightarrow R_2\dot{C}{-}NHNHCOCH_2NMe_3^+ \xrightarrow[\text{fast}]{e^-,\ H^+} \underset{\textbf{204}}{R_2CHNHNHCOCH_2NMe_3^+}$$

This two-electron reduction to the 1,2-disubstituted hydrazine (**204**) contrasts with the four-electron reduction of oximes, in which the first step is cleavage of the N—O bond.

An exception is the four-electron reduction of benzaldehyde thiosemicarbazone (**205**, R = $CSNH_2$) to benzylamine plus thiourea in acid, although in base only a two-electron reduction is observed (244).

$$\underset{\textbf{205}}{\phi CH{=}NNHR} + 4e^- + 4H^+ \rightarrow \phi CH_2NH_2 + RNH_2$$

Also, benzaldehyde and benzophenone phenylhydrazones (e.g., **205**, R = ϕ) and semicarbazones (e.g., **205**, R = $CONH_2$) undergo four-electron reduction to the amines, isolated from cpe. In base, cinnamaldehyde semicarbazone undergoes two-electron reduction to hydrogenate the C—C double bond. In acid, benzaldehyde azine

$$\phi CH{=}CHCH{=}NNHCONH_2 + 2e^- + 2H^+ \rightarrow \phi CH_2CH_2CH{=}NNHCONH_2$$

(**206**) is reduced to benzylamine (**207**), by a six-electron process. In base, the reduction can be stopped after only two-electron reduction to benzaldehyde benzylhydrazone (**208**), isolated from cpe; this latter can be further reduced at pH 5 to two moles of benzylamine (235).

$$\underset{\textbf{206}}{\phi CH{=}N{-}N{=}CH\phi} + 2e^- + 2H^+ \rightarrow \underset{\textbf{208}}{\phi CH{=}NNHCH_2\phi} \xrightarrow{4e^-,\ 4H^+} \underset{\textbf{207}}{2\phi CH_2NH_2}$$

The multiplicity and time-dependence of the waves of various hydrazone derivatives have been ascribed to equilibrium among hydrazone (**209**), azo (**210**), and ene-hydrazine (**211**) forms (245). The

$$\underset{\textbf{209}}{R_2CH{-}CH{=}NNHX} \rightleftharpoons \underset{\textbf{210}}{R_2CH{-}CH_2N{=}NX} \rightleftharpoons \underset{\textbf{211}}{R_2C{=}CH{-}NHNHX}$$

differences between the $E_{1/2}$'s of the various waves were used to determine equilibrium constants for interconversion of the various forms. However, no evidence has been found for the existence of the ene-hydrazine form in aprotic media (246). O'Connor (246) therefore ascribed the multiplicity of waves to adsorption and kinetic waves. It is certainly not justifiable to use differences in $E_{1/2}$ values of irreversible waves to determine relative energies of the various forms, and thereby determine equilibrium constants for equilibria among the various species, as the Russian workers have done, but the origin of the many waves is still not clear.

Glucosazones undergo four-electron reduction. The glucosazone formed with 1-methylphenylhydrazine (**212**) is reduced appreciably more readily than the ordinary osazone (**213**), which is stabilized by hydrogen bonding (247).

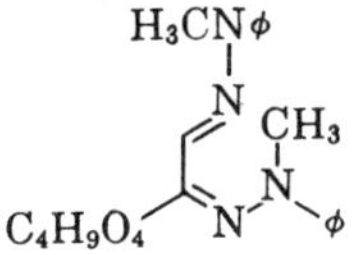

$H_3CN\phi$, N, CH_3, N, $C_4H_9O_4$, N, ϕ

212

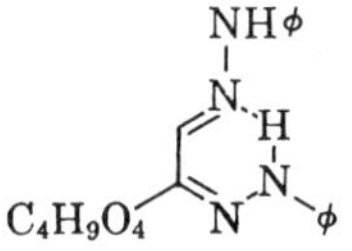

$NH\phi$, N, H, N, $C_4H_9O_4$, N, ϕ

213

C. NITRO COMPOUNDS

The polarographic behavior of nitro compounds has been reviewed by Seagers and Elving (248).

1. *Nitroaromatics*

Nitrobenzene shows a four-electron reduction wave, leading to formation of phenylhydroxylamine. Below pH 4 a two-electron wave is observed at more cathodic potentials, corresponding to further reduction of phenylhydroxylamine to aniline (32). Cpc verifies the four-electron nature of the reduction of nitrobenzene, *m*-nitroaniline, *m*-nitrophenol, *p*-nitroanisole, and 4-nitrotropolone (86). Nitrosobenzene is presumably an intermediate in the reduction of nitrobenzene but since it is reduced at less negative potentials than the $E_{1/2}$ of nitrobenzene, only a single four-electron wave is observed (249).

In methanesulfonic acid, the phenylhydroxylamine undergoes rearrangement to *p*-hydroxyaniline (125).

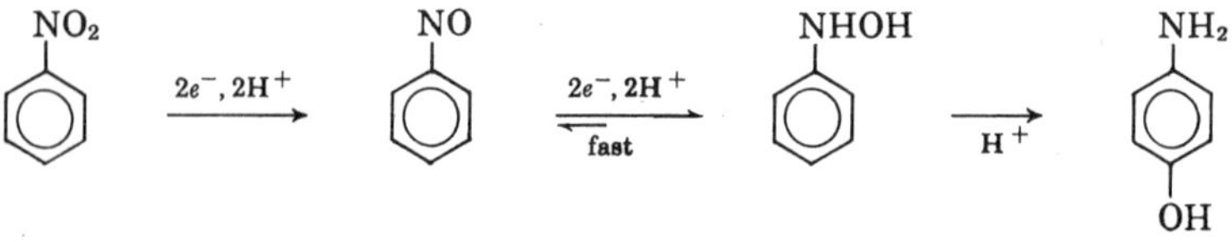

$$C_6H_5NO_2 \xrightarrow{2e^-,\,2H^+} C_6H_5NO \underset{\text{fast}}{\overset{2e^-,\,2H^+}{\rightleftharpoons}} C_6H_5NHOH \xrightarrow[H^+]{} p\text{-}HOC_6H_4NH_2$$

The mechanism of nitrobenzene reduction is not entirely clear. The log plot slope indicates that the reduction is irreversible, with less than one electron transferred at the transition state (32,86). Conclusive proof for the irreversibility is the fact that neither phenylhydroxylamine nor nitrosobenzene shows an anodic wave at the $E_{1/2}$ of nitrobenzene. The $E_{1/2}$–pH plot indicates that at least one proton is included in the transition state (32). Perhaps the potential-determining step is formation of the radical, $\phi NOOH\cdot$, which then rapidly adds another electron and loses OH^- to form nitrosobenzene, which is rapidly reduced further.

The unstable radical anion, $\phi NO_2^{\dot{-}}$, is implicated as another intermediate in nitrobenzene reduction. A catalytic wave is observed in the presence of periodate or of cupric tartrate, since the radical anion is re-oxidized to nitrobenzene by these species (250). Above pH 7, the ESR spectrum of the radical anion can be observed, even in aqueous solution (52). In the presence of camphor, the four-electron wave splits into a one-electron wave and a three-electron wave. This behavior is attributed to an impedance to further reduction of the radical anion, caused by an adsorbed layer of camphor on the electrode surface (251).

In acetonitrile, reversible one-electron reduction forms the radical anion, $\phi NO_2^{\dot{-}}$, whose ESR spectrum was observed (252). A second wave is observed at more negative potentials. The mechanism is presumably

$$\phi NO_2 \underset{\text{MeCN}}{\overset{e^-}{\rightleftharpoons}} \phi NO_2^{\dot{-}} \xrightarrow[\text{higher } a_{e^-}]{e^-} \phi NO_2^{=}$$

$$\xrightarrow{H_3CCN} \phi NOOH^- \rightarrow OH^- + \phi NO \xrightarrow{2e^-,\, H^+} \phi NHO^-$$

Similar results are observed for α- and β-nitronaphthalene and nitromesitylene in DMF (253). Addition of benzoic acid shifts the second, three-electron, wave to less negative potentials, until it merges with the first, since the rate of protonation of the dianion increases (82). Addition of water or ethanol has the same effect (254).

The reduction of polynitroaromatics occurs in stages, via the intermediate nitro-aromatic hydroxylamines. The presence of one nitro group facilitates the reduction of the other, but once one nitro group has been reduced, the substituent effect of the resultant hydroxylamino group impedes the reduction of the remaining nitro group.

m-Dinitrobenzene is reduced in two four-electron steps to *m*-di(hydroxylamino)benzene. Both *o*- and *p*-dinitrobenzene give a four-

$$O_2N\text{-}C_6H_4\text{-}NO_2 \xrightarrow{4e^-,\ 4H^+} O_2N\text{-}C_6H_4\text{-}NHOH \xrightarrow[\text{higher } a_{e^-}]{4e^-,\ 4H^+} HONH\text{-}C_6H_4\text{-}NHOH$$

electron wave followed by a wave at more negative potentials (32). Depending on pH, the second step may or may not be a complete eight-electron reduction to the phenylenediamine, since the intermediate di(hydroxylamino)benzene (e.g. **214**) can form the readily reducible quinoneimine oxime (**215**).

$$O_2N\text{-}C_6H_4\text{-}NO_2 \xrightarrow{4e^-,\ 4H^+} HONH\text{-}C_6H_4\text{-}NO_2 \xrightarrow[\text{higher } a_{e^-}]{4e^-,\ 4H^+} \underset{\mathbf{214}}{HONH\text{-}C_6H_4\text{-}NHOH} \xrightarrow[OH^-]{H^+\text{or}} \underset{\mathbf{215}}{HON{=}C_6H_4{=}NH} \xrightarrow[\text{fast}]{2e^-,\ 2H^+}$$

$$HONH\text{-}C_6H_4\text{-}NH_2 \xrightarrow[OH^-]{H^+\text{or}} HN{=}C_6H_4{=}NH \xrightarrow[\text{fast}]{2e^-,\ 2H^+} H_2N\text{-}C_6H_4\text{-}NH_2$$

At pH 12, the height of the first wave of *p*-dinitrobenzene corresponds to only a two-electron reduction, since the quinonoid dianion (**216**) is apparently stable, although reducible at more negative potentials (255). Cpc, however, indicates four-electron reduction, since the dianion disproportionates or decomposes to *p*-nitronitrosobenzene during the time required for cpc (256). Dinitrodurene behaves normally, because of steric hindrance to formation of the planar dianion (**217**) (257). 1,3,5-Trinitrobenzene shows three four-electro- waves (32). Above pH 11 the limiting current decreases as the substrate is converted to the nonreducible adduct with hydroxide ion (**218**) (258).

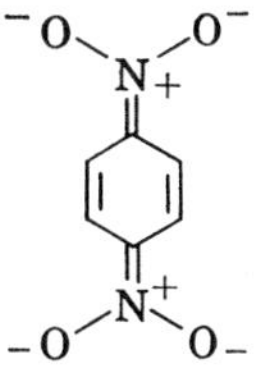

216 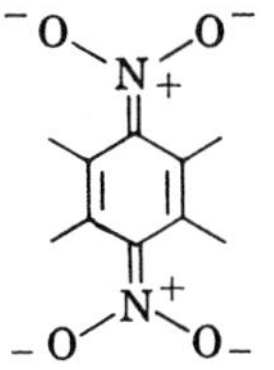 217 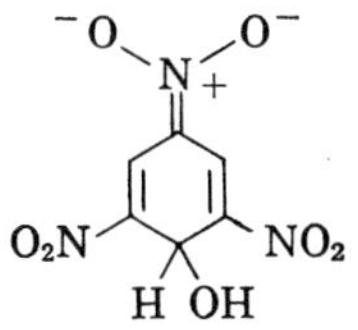 218

Nitroferrocene, $C_5H_5FeC_5H_4NO_2$, undergoes six-electron reduction (259). (Cf. nitrophenols below.)

The nitro group of the nitrophenylthiocyanates undergoes normal reduction, except for the *ortho* isomer at pH 3, which undergoes the cyclization (260):

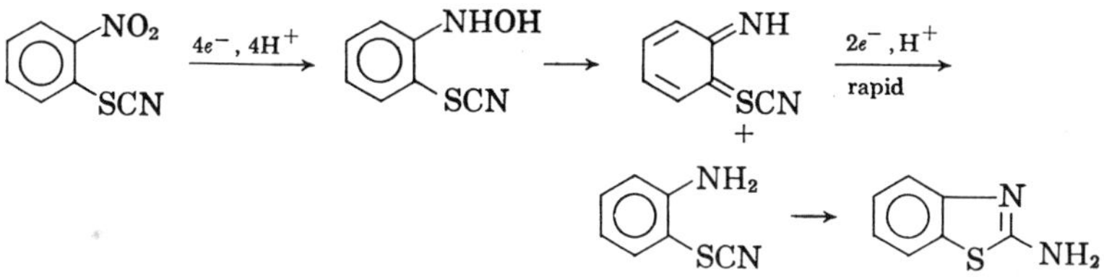

$4e^-, 4H^+$ $2e^-, H^+$ rapid

2. *Nitrophenols and Nitroanilines*

The unusual features of the polarography of nitrophenols and nitroanilines have been discussed in the section on nitroso compounds, which are intermediates in the reduction of nitro compounds. The *ortho* and *para* isomers (e.g. **51**) are capable of undergoing six-electron reduction all in one wave, since the intermediate hydroxylamine (e.g. **52**) can form the readily reducible quinone imine (e.g. **53**) or diimine. Only four-electron reductions are observed for the nitroacetanilides (**219**) (261).

At low concentrations in acid, picric acid (**220**) is reduced to 2,4,6-triaminophenol (262), since cpc indicates eighteen-electron reduction.

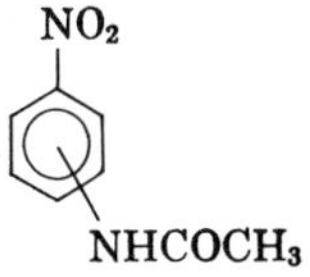

219 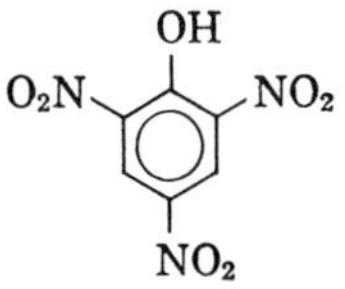 220

4-Nitrocatechol (**221**) is an exception to the rule that nitro compounds are less readily reduced than the corresponding nitroso compound. Above pH 9 it gives a two-electron wave, followed by a four-electron wave. Apparently the nitroso compound (**222**) is reduced only at more negative potentials (263).

$$\text{HO},\text{HO}\text{-}C_6H_3\text{-}NO_2 \xrightarrow{2e^-,\ 2H^+} \text{HO},\ O{=}C_6H_3{=}NOH$$

221 **222**

3. Aliphatic Nitro Compounds

In acid, the magnitude of i_d indicates that nitroalkanes (**223**) undergo four-electron reduction to the alkylhydroxylamine. The reduction is irreversible, and $E_{1/2}$ depends upon pH (264,265). In alkaline solution, i_d decreases because of formation of the non-reducible *aci*-nitro anion (**224**) (266). Cpc verifies the four-electron nature of the

$$RCH{=}NO_2^- \overset{H^+}{\rightleftharpoons} RCH_2NO_2 \xrightarrow{4e^-,\ 4H^+} RCH_2NHOH$$

224 **223**

reduction. The alkylhydroxylamine is not further reducible in the accessible potential range, although phenylhydroxylamine is reducible to aniline (267). ω-Nitro-acetophenone (**223**, R = φCO) and its oxime (**223**, R = φCNOH) undergo four-electron reduction to the hydroxylamine (268).

Nitro-olefins are more readily reduced than nitro-alkanes (267). ω-Nitrostyrene (**225**, R = φ) shows a four-electron reduction wave, followed by a two-electron wave. The first represents reduction to phenylacetoxime (**226**, R = φ), isolated from cpe on the first plateau; the second represents reduction to β-phenethylhydroxylamine (**227**, R = φ), isolated from cpe on the second plateau (269). 1-Nitropentene-1 (**225**, R = *n*-Pr) behaves similarly (270). But it is certainly surprising that under these circumstances, the oxime (**226**) undergoes only two-electron reduction, rather than the usual four-electron reduction to the amine.

$$RCH{=}CHNO_2 \xrightarrow{2e^-,\ 2H^+} RCH{=}CHN{=}O \underset{\text{fast}}{\overset{2e^-,\ 2H^+}{\rightleftharpoons}} RCH{=}CHNHOH$$

$$\rightleftharpoons RCH_2CH{=}NOH \underset{\text{higher } a_{e^-}}{\xrightarrow{2e^-,\ 2H^+}} RCH_2CH_2NHOH$$

225 **226** **227**

4. Gem-Dinitro Compounds

2,2-Dinitropropane (**228**) undergoes pH-independent reduction. The reduction is a two-electron process, according to the magnitude of i_d and cpc. One equivalent of nitrite ion was detected after cpe. 2-Nitroso-2-nitropropane (**229**) was eliminated as a possible product, since it is reduced at less negative potentials (271). In acid, nitrite does add to 2-nitropropane to form the blue 2-nitroso-2-nitropropane (**229**), but in the presence of azide, nitrite is destroyed and no blue color is observed. The 2-nitroso-2-nitropropane (**229**) is further reducible to nitrous acid plus acetone oxime (**230**), isolated from cpe. A second wave is observed at more negative potentials, corresponding to further reduction of acetone oxime to isopropylamine and of nitrous acid to hydroxylamine (272).

$$\underset{\mathbf{228}}{Me_2C(NO_2)_2} \xrightarrow{2e^-} NO_2^- + Me_2C{=}NO_2^- \xrightarrow[pH\ 2]{} \underset{\mathbf{229}}{Me_2C(NO)NO_2} \xrightarrow{2e^-} \underset{\mathbf{230}}{Me_2C{=}NO^-} + NO_2^-$$

In acid, 1,1-dinitroethane shows a six-electron reduction wave followed by a two-electron wave, corresponding to reduction to acetamide oxime, isolated from cpe on the second plateau. In alkali,

$$MeCH(NO_2)_2 \xrightarrow{6e^-,\,6H^+} \underset{\mathbf{57}}{MeC(NHOH){=}NOH} \xrightarrow[\text{higher } a_{e^-}]{2e^-,\,2H^+} MeC(NH_2){=}NOH$$

dinitromethane undergoes eight-electron reduction to formamide oxime (273).

$$H_2C(NO_2)_2 \xrightarrow[8H_2O]{8e^-} HC(NH_2){=}NOH$$

At pH 6, nitroform (trinitromethane) undergoes twelve-electron (according to cpc) reduction to dihydroxyguanidine, whose UV spectrum was observed after cpe. At pH 12, methyl nitroform is first

$$HC(NO_2)_3 \xrightarrow{12e^-,\,12H^+} HONHC({=}NH)NHOH$$

reduced to the *aci*-anion of 1,1-dinitroethane, which is protonated and reduced further at more negative potentials, as described above (274).

$$\mathrm{MeC(NO_2)_3} \xrightarrow{2e^-} \mathrm{MeC(NO_2)_2^-} + \mathrm{NO_2^-}$$

5. *Nitrate Esters*

n-Butyl nitrate undergoes two-electron reduction, according to the magnitude of i_d (265). The half-wave potential is independent of pH (275). Polynitrate esters of ethylene glycol, glycerin, and pentaerythritol also undergo reduction, with an i_d corresponding to two electrons per nitrate ester group (276).

$$\mathrm{RONO_2} + 2e^- \rightarrow \mathrm{RO^-} + \mathrm{NO_2^-}$$

D. CLEAVAGE OF SINGLE BONDS

1. *Carbon–Halogen Bonds*

The polarographic cleavage of carbon–halogen bonds has been reviewed by Elving (14) and by Elving and Pullman (16).

a. Alkyl halides

Von Stackelberg and Stracke (117) have made an extensive study of the polarography of various organic halides. They found that the ease of reduction is I > Br > Cl, and that polyhalogenation facilitates the reduction. The order of the halogens parallels their relative leaving group abilities in S_N2 displacements, but in contrast, the effect of α-halogen substituents is to retard S_N2 displacements. Vinyl and aryl halides are more difficult to reduce, but allylic halides undergo facile reduction. The half-wave potentials of several bromides are independent of pH (117,277). Cpe indicates that in almost all cases, polarographic reduction replaces the halogen atom by a hydrogen, except that *vic*-dihalides often lose two halide ions to give an olefin, without hydrogen uptake. Von Stackelberg and Stracke (117) proposed the mechanism

$$\mathrm{RX} + e^- \rightarrow \mathrm{X^-} + \mathrm{R\cdot} \xrightarrow[\text{rapid}]{e^-} \mathrm{R^-} \xrightarrow[\mathrm{H_2O}]{} \mathrm{RH}$$

According to the log plot slope, the reduction is irreversible, with less than one electron transferred at the transition state (278).

Elving (14) has distinguished three possible mechanisms, S_N2, S_N1, and radical, by analogy with solvolysis:

$$S_N2\text{: } RX + 2e^- \rightarrow X^- + R^-$$

$$S_N1\text{: } RX \rightleftharpoons X^- + R^+ \xrightarrow{2e^-} R^-$$

$$\text{radical: } RX + e^- \rightarrow X^- + R \rightarrow \text{further reaction}$$

The S_N2 mechanism is envisioned as an approach of the alkyl halide to the negatively charged electrode with the carbon end of the carbon–halogen dipole closer to the electrode. Then a pair of electrons effects a backside displacement of halide ion. In this context, the S_N1 mechanism does not differ significantly from the S_N2. In solution, S_N1 ionization of an alkyl halide does not occur sufficiently rapidly to account for the electrochemical reaction, so that the solvolytically generated carbonium cannot be an intermediate. Rather, the ionization is assisted by the electric field at the interface, which orients the carbon–halogen dipole with the halogen more distant from the electrode. The electric field further repels the halide ion and effects the solvolysis. Thus the only difference between these two mechanisms is that one conceives the displacement to be effected by a pair of electrons, whereas the other conceives the displacement to be effected by the electric field. But it is clear that as the carbonium ion is generated, it will pull electrons from the electrode, so that the two mechanisms become essentially equivalent.

The radical mechanism, modified so as to incorporate some of the features of the S_N mechanism, seems most satisfactory (14). The isolation of dimeric products from some halide reductions shows that the process does not always involve simultaneous transfer of a pair of electrons, but rather that they are transferred in separate steps. In many cases, the addition of the second electron is the predominant pathway, but there are some cases in which dimerization is the preferred pathway. For example, bromomaleic acid is reduced to a mixture of fumaric acid, maleic acid, and butadiene-1,2,3,4-tetracarboxylic acid (279).

Subsequent discussion of polarographic cleavage of single bonds is in terms of this modified radical mechanism. One electron is transferred in the potential-determining step, and the nature of the product depends upon whether the resultant radical adds a second electron or dimerizes. Thus, cpe of benzyl chloride produces bibenzyl, from coupling of the radicals (280). But at very negative potentials, cpe of benzyl bromide in DMF in the presence of CO_2 produces a small yield of phenylacetic acid, from carbonation of the anion (281).

$$\phi CH_2X + e^- \rightarrow X^- + \phi CH_2^{\cdot} \xrightarrow{e^-} \phi CH_2^- \xrightarrow[DMF]{CO_2} \phi CH_2CO_2^-$$
$$\phi CH_2^{\cdot} \downarrow \ {}^1/_2\phi CH_2CH_2\phi$$

The species, $\phi CH_2Cl^{\cdot -}$, has also been suggested as an intermediate in the polarography of benzyl chloride in aqueous lithium chloride solutions. At potentials more negative than $E_{1/2}$, i_d decreases from that for a two-electron reduction to that for a one-electron reduction. At these negative potentials, the electric field of the electrode repels the anion, $\phi CH_2Cl^{\cdot -}$, into the solution, where it loses chloride to form benzyl radical, which dimerizes. (Similar dips in the current-potential curves are observed in the reduction of anions.) Also, adsorption effects, attributed to surface-active bibenzyl, are observed. Tetramethylammonium ion adsorbs on the electrode surface and diminishes the effect of the electric field, so that the radical anion, $\phi CH_2Cl^{\cdot -}$, remains in the vicinity of the electrode surface, where it eliminates chloride and undergoes further reduction to toluene. Thus, in the presence of tetramethylammonium salts, the normal two-electron reduction is observed at all potentials (280).

Exceptions to this mechanism occur in the polarography of benzyl and allyl iodides, which give waves at the $E_{1/2}$ of benzylmercuric iodide and allylmercuric iodide, respectively. It was suggested that the formation of the organometallic is catalyzed by benzyl radical, formed in the reduction (282).

$$\phi CH_2I + e^- \rightarrow I^- + \phi CH_2$$

$$\phi CH_2 + Hg \rightarrow \phi CH_2Hg\cdot \qquad \phi CH_2Hg\cdot + \phi CH_2I \rightarrow \phi CH_2HgI + \phi CH_2^{\cdot}$$

Considerable controversy has arisen over the interpretation of steric effects on halide reductions. Since this discussion has direct bearing on the mechanism of halide reduction, it will not be deferred to the section on structure and reactivity.

It is clear that α-bromopropionic acid (**231**) is more readily reduced

$$\underset{\mathbf{233}}{Me_2CBrCOOH} > \underset{\mathbf{231}}{MeCHBrCOOH} > \underset{\mathbf{232}}{CH_2BrCOOH}$$

than is bromoacetic acid (**232**) (283), and that α-bromo-isobutyric acid (**233**) is still more readily reduced (284). Although the steric and inductive effects of an alpha methyl group would be expected to impede a nucleophilic process, apparently the stabilizing effect of methyl groups on the radical-like transition state is more important. It is less obvious that increasing the length of the alkyl chain in the α-bromoalkanoic acids should lead to a steady increase in the ease of reduction, even to a C_8 acid, although this observation is understandable in view of the greater adsorbability of the longer chain compounds (283), which would provide an increase in the entropy of activation for the process. What is surprising is the observation that anions of α-ethyl-α-bromoacetic acids (α-bromobutyric acids) are less readily reduced than expected from a monotonic increase of reducibility with chain length (284). Elving, Markowitz, and Rosenthal (284) have suggested that such anions are stabilized by hydrogen bonding of the ω-hydrogen of the ethyl group with the oxygen of the carboxylate, according to Newman's "Rule of Six," to form a six-membered ring (**234**). Stabilization of the starting material then makes the sub-

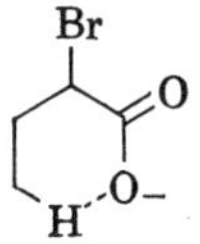

Br

O

O–

H

234

strate more difficult to reduce. The anomalous behavior is observed only for the anion, and not for α-bromobutyric acid or its ester, in which such hydrogen bonding would be less likely to stabilize the reactant.

More "normal" effects are observed in the polarography of alkyl bromides, whose behavior parallels S_N2 reactivities more closely (285). The ease of reduction is in the order ethyl > *n*-propyl > *i*-propyl, as expected for an S_N2-like process. Also, increasing the chain length in the normal alkyl bromides decreases the ease of reduction, but the effect persists only up to *n*-amyl bromide, beyond which all bromides are reduced at nearly the same $E_{1/2}$.

These authors object to invoking the hydrogen-bonded form, **234**, to explain steric effects in reduction of α-bromo acid anions, since the strength of a hydrogen bond involving an sp^3 carbon–hydrogen bond should be infinitesimal, and certainly less than the 5 kcal. inferred from the half-wave potentials (284). Lambert and Kobayashi (285) prefer to ascribe the steric effects to interference by the alkyl group with backside displacement by a very bulky electrode. However, they offered no explanation as to why a steric effect is observed in reduction of the anion, but not of the acid (286). Perhaps the effect of *alpha* ethyl groups is one of steric hindrance to backside displacement, but in the acid the hydrocarbon chain can swing away from the reaction center and toward the carboxylic acid group. In the anion, the hydrocarbon portion of the molecule is repelled by the lipophobic carboxylate group, and is forced to interfere with the reaction. Thus, the difficulty of reducing α-bromobutyrate may be ascribed to the *instability* of the conformation (**234**), rather than any stability.

Lambert and Kobayashi (287) have made a further extensive study of the steric effects on polarographic half-wave potentials of alkyl bromides. They found that the ease of reduction closely parallels S_N2 reactivities. Thus, ethyl bromide and cyclopentyl bromide are relatively easily reduced, but cyclohexyl bromide, and especially cyclobutyl, cyclopropyl, and neopentyl bromides are reduced with difficulty. These results are consistent with backside displacement of bromide by electrons from a very bulky electrode surface. The half-wave potentials do not correlate well with rates of abstraction reactions, indicating that the polarographic reduction does not take place by attack of an electron on the bromine, to displace the alkyl radical. Although most of the results may also be interpreted as reflecting the stability of the resulting radical, the greater reactivity of ethyl relative to isopropyl, and the very low reactivity of neopentyl, cannot be ascribed to radical stabilities. The only exception to this correlation with S_N2 reactivities is *t*-butyl bromide, which is quite readily reduced. The authors attribute this anomaly to the incursion of the pseudo-S_N1 reduction mechanism, whose ease is related to the stability of the *t*-butyl cation.

Apparently an S_N1-type reduction can also occur in strained systems in which backside displacement is difficult or impossible, although it is difficult to distinguish this type of process from an attack at the bromine to give a strained radical (288,289). For example, *exo*-norbornyl bromide is reduced more readily than *endo*-norbornyl bromide, paralleling their S_N1 solvolysis rates. Similarly, *cis*-4-*t*-butylcyclohexyl bromide is reduced about as readily as cyclohexyl bromide, since an axial bromide can be displaced by rearward attack. But the equatorial bromine of the *trans* isomer cannot be so displaced, and must undergo either pseudo-S_N1 reduction or attack on the bromine, and this reduction is more difficult. Bridgehead bromides—adamantyl, bicyclo-[2.2.2]-octyl, and bicyclo-[2.2.1]-heptyl—in which backside displacement is impossible, are nevertheless reducible, but with increasing amounts of difficulty. It seems unlikely that these reactions occur by S_N1-like processes, since these carbonium ions are difficult to form, especially the 1-norbornyl cation. Also, if these molecules orient themselves with the bromine most distant from the electrode, in order for the electric field to promote solvolysis, it seems that the low dielectric constant of the intervening hydrocarbon portion will divert the lines of force through the surrounding solvent, and moderate the effect of the electric field. It seems more reasonable to consider reduction of bridgehead bromides to occur by attack on the bromine, rather than by the usual S_N1-S_N2 back-side displacement. The difficulty of reducing these bromides is then to be ascribed to the decreased stability of the strained radicals which are produced.

One possible method for distinguishing between S_N1-like reduction and reductive attack at the bromine is on the basis of the transfer coefficient, α. If the reduction becomes more S_N1-like on going from primary alkyl to *t*-butyl and norbornyl, and then to equatorial cyclohexyl and bridgehead, α should decrease, since less electron has been transferred at the transition state. On the other hand, if reductive attack occurs on bromine, the transition state will be closer to product for reactions producing the more strained radicals (34), and α would be expected to increase as the radical becomes less stable.

The facile reduction of 1-bromotriptycene (288) may be attributed to the inductive effect of three phenyl groups on an essentially nucleophilic reaction.

Although $E_{1/2}$ for halide reductions is generally pH-independent, $E_{1/2}$ for reduction of halides capable of ionization, such as α-bromo acids (290), iodoacetic acid, iodobenzoic acids, and iodoanilines (291), depends upon pH. The pH-independent bond cleavage is complicated by pH-dependent protonation at another site. In acid solution, the acid form is reduced in a pH-independent process, and in basic solution, the basic form is reduced in a pH-independent process. But at intermediate pH values not too much greater than pK_a, the mechanism is such that the transition state includes one proton more

$$RCHXCO_2^- + e^- + H^+ \rightarrow X^- + R\dot{C}HCOOH \xrightarrow{e^-} RCH_2CO_2^-$$

than does the reactant. [See discussion following eq. (20).]

b. Activated halides

An adjacent carbonyl or nitro group facilitates reduction of alkyl halides, because of both the inductive effect and phenacyl-like participation in an S_N2-like process. This reduction does not proceed via the ketyl radical or nitro radical anion as intermediate, since the ketyl radical is formed only at much more negative potentials. For example, phenacyl chloride is reduced at −0.55 v., and an α-chloro substituent would not be expected to shift the $E_{1/2}$ for reduction of the ketone very much below $E_{1/2}$ for acetophenone, which is −1.26 v. at pH 4. However, the phenacyl participation may be viewed as a contribution of a ketyl resonance form to the S_N2-like transition state

$$\underset{\text{Cl}}{\overset{e^-(\text{Hg})\quad \text{O}}{\phi\text{—}\overset{\|}{\text{C}}\text{—CH}_2}} \leftrightarrow \underset{\text{Cl}^-}{\overset{(\text{Hg})\quad \text{O}}{\phi\text{—}\overset{\|}{\text{C}}\text{—}\dot{\text{C}}\text{H}_2}} \leftrightarrow \underset{\text{Cl}}{\overset{(\text{Hg})\quad \text{O}^-}{\phi\dot{\text{C}}\cdot\text{—CH}_2}}$$

Thus, phenacyl bromide (**235**, X = Br) shows two two-electron waves, with the second at the same $E_{1/2}$ as the wave of acetophenone. $E_{1/2}$ of the first wave is independent of pH (277). Clearly the first wave involves displacement of bromide by a single electron, to form the phenacyl radical (**236**) which rapidly adds another electron to give the enolate anion of acetophenone. Likewise, chloro-, bromo-, and iodoacetone each give a single, irreversible, pH-independent two-

$$\underset{\mathbf{235}}{\phi COCH_2X} + e^- \rightarrow X^- + \underset{\mathbf{236}}{\phi COCH_2\cdot} \xrightarrow{e^-} \phi COCH_2^-$$

electron wave, to form acetone and the halide ion (292). α-Bromocamphor (293) and 2-chlorocyclohexanone (294) behave similarly.

trans-4-*t*-Butyl-2-chlorocyclohexanone (**237**), with the chlorine axial, is more readily reduced than is the *cis* isomer (**238**), as expected

O Cl **237**

O Cl **238**

for back-side displacement on an α-keto halide. 2-Chlorocyclohexanone is reduced in a single wave at the same $E_{1/2}$ as the *trans-t*-butyl isomer (**237**), since the two conformers of 2-chlorocyclohexanone are in rapid equilibrium (295). Similar results are observed with α-chloro ketosteroids, where the stereochemistry is fixed (296).

Below pH 6, $E_{1/2}$ of the first wave of phenacyl fluoride (**235**, X = F) depends upon pH, but above pH 6, $E_{1/2}$ is independent of pH. This pH-dependence is attributed to a catalytic effect of a proton, which assists in pulling off the fluoride (297). Polarographic reduction of phenacyl chloride (**235**, X = Cl) shows acid catalysis only at pH below 3, since chloride is a weaker base than fluoride, and acid catalysis is less effective. The fact that the various halides show acid catalysis to an extent dependent upon their basicity is demonstration that the acid catalysis is not to be attributed to protonation of the carbonyl oxygen (298).

2-Bromo-2-nitropropane (**239**, X = Br) and 2-chloro-2-nitropropane (**239**, X = Cl) undergo very facile two-electron reduction to halide ion and 2-nitropropane, whose reduction wave is observed at more negative potentials. The $E_{1/2}$ of the first wave is independent of pH, suggesting the *aci* anion (**240**) of nitropropane as primary product.

$$\underset{\mathbf{239}}{Me_2CXNO_2} + 2e^- \rightarrow X^- + \underset{\mathbf{240}}{Me_2C{=}NO_2^-} \xrightarrow{H^+} Me_2CHNO_2$$

2-Bromo-2-nitrosopropane (**241**, X = Br) and 2-chloro-2-nitrosopropane (**241**, X = Cl) undergo two-electron reduction to halide ion and acetone oxime (**242**) by a pH-independent process (299). α-Bromo-α-

$$\underset{\mathbf{241}}{Me_2CXNO} + 2e^- \rightarrow X^- + Me_2C{=}NO^- \xrightarrow{H^+} \underset{\mathbf{242}}{Me_2C{=}NOH}$$

nitroacetophenone is reduced to α-nitroacetophenone (268). (Cf. the polarographic cleavage of nitrite ion discussed above.)

$$\phi COCHBrNO_2 + 2e^- \rightarrow Br^- + \phi COCH{=}NO_2^- \xrightarrow{H^+} \phi COCH_2NO_2$$

c. *Gem* polyhalides

The presence of one halogen on a carbon atom facilitates reductive cleavage of another. This activating effect is opposite to the effect of halogens on S_N2 reactivities, and may be due to the stabilizing effect of halogen on the electron-rich, radical-like transition state.

Thus a polyhalide gives a series of two-electron waves, corresponding to stepwise replacement of halogen by hydrogen. For example, carbon tetrachloride gives two irreversible two-electron waves, with the second at the same $E_{1/2}$ as the wave of chloroform (300). Like-

$$CCl_4 + 2e^- + H^+ \rightarrow Cl^- + HCCl_3 \xrightarrow[\text{higher } a_{e^-}]{2e^-, H^+} Cl^- + H_2CCl_2$$

wise, tribromoacetic acid shows three waves, with the second two at the same half-wave potentials as the two waves of dibromoacetic acid, and the third at the $E_{1/2}$ of the single wave of bromoacetic acid (301).

$$Br_3CCOOH \xrightarrow{2e^-, H^+} Br_2CHCOOH \xrightarrow{2e^-, H^+} BrCH_2COOH \xrightarrow{2e^-, H^+} CH_3COOH$$

The chloroacetic acids behave similarly (302). The first wave of trichloroacetic acid is indeed a two-electron reduction, according to cpc (303). 2,2,2-Trichloroethanol undergoes two-electron reduction to dichloroethanol. Above pH 11 the half-wave potential depends upon pH, since the alcohol is ionizable (304).

$$Cl_3CCH_2O^- \overset{H^+}{\rightleftharpoons} Cl_3CCH_2OH \xrightarrow{2e^-, H^+} Cl_2CHCH_2OH$$

Chloral hydrate (**243**) undergoes reduction according to the scheme (305)

$$\underset{\mathbf{243}}{Cl_3CCH(OH)_2} \xrightarrow{2e^-, H^+} \underset{\mathbf{244}}{Cl_2CHCH(OH)_2} \overset{k}{\rightarrow} Cl_2CHCH{=}O$$

$$\xrightarrow[\text{fast}]{2e^-, H^+} ClCH_2CH{=}O \xrightarrow[\text{fast}]{2e^-, H^+} CH_3CH{=}O$$

This mechanism is proposed on the basis of the kinetic nature of the reduction currents of dichloroacetaldehyde (**244**) and chloroacetaldehyde, and the fact that these two compounds are more readily reduced than is chloral hydrate. Under polarographic conditions, the dehydration of dichloroacetaldehyde hydrate is slow, so that only two-electron reduction is observed, but cpc elucidates the role of dehydration.

In DMF or acetonitrile, carbon tetrachloride undergoes two-electron reduction to Cl^- plus Cl_3C^-. This latter anion loses another chloride to form the carbene, CCl_2, which can be trapped with tetramethylethylene to form 1,1-dichloro-2,2,3,3-tetramethylcyclopropane. In the absence of a trapping agent, CCl_2 adds two electrons and two protons to form methylene chloride. Carbon tetrabromide, and

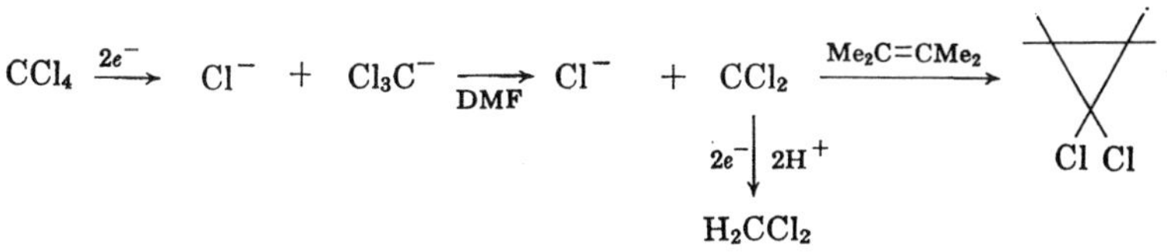

$$CCl_4 \xrightarrow{2e^-} Cl^- + Cl_3C^- \xrightarrow[DMF]{} Cl^- + CCl_2 \xrightarrow{Me_2C{=}CMe_2} \text{(tetramethyl-dichlorocyclopropane, Cl Cl)}$$

$$CCl_2 \xrightarrow{2e^-,\ 2H^+} H_2CCl_2$$

perhaps benzotrichloride, behave in similar fashion (306).

The trifluoromethyl group is not ordinarily reducible, except when present on a benzene ring with electron-withdrawing substituents. Thus, **245** and **246** are each reduced to the substituted toluene in a six-electron process, with liberation of three fluoride ions (307).

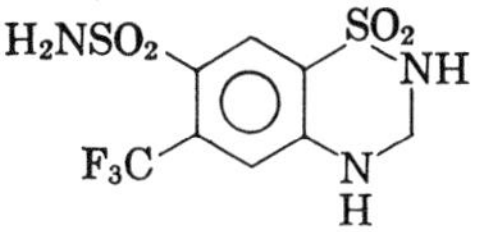

245: H_2NSO_2, F_3C, SO_2, NH, N, H

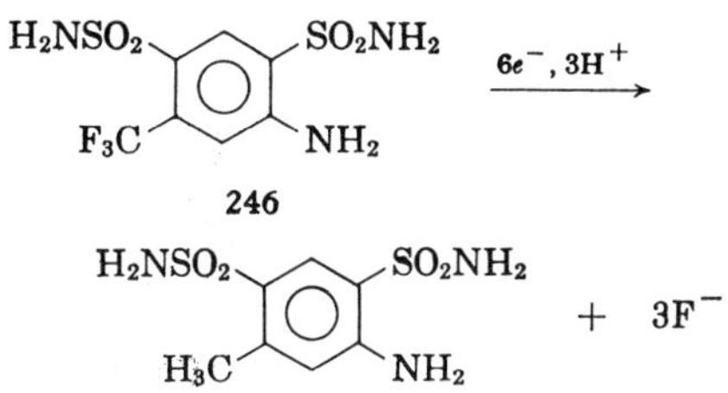

246: H_2NSO_2, SO_2NH_2, F_3C, NH_2 $\xrightarrow{6e^-,\ 3H^+}$ H_2NSO_2, SO_2NH_2, H_3C, NH_2 $+ \ 3F^-$

This is an exception to the rule that *gem* polyhalides are reduced in successive steps. Benzotrifluoride gives a reduction wave in DMF, but the nature of the product was not established, except that several electrons are consumed (308).

d. *Vic* dihalides

The behavior of *vic* dihalides differs from that of monohalides in that the carbanion resulting from two-electron reduction can eliminate a second halide ion to form an olefin. Thus, cpe of 1,2-dibromoethane produces both ethylene and ethane (117).

Cpe of benzene hexachloride indicates six-electron reduction to six chloride ions and benzene, identified after cpe (309). No protons are

$$C_6H_6Cl_6 + 6e^- \longrightarrow C_6H_6 + 6Cl^-$$

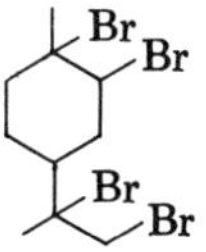

247: Br, Br, Br, Br

consumed in the process (310). An axial chlorine is more easily reduced (311), and the β isomer, with all chlorines equatorial, is most difficult to reduce (309). Heptachlorocyclohexane, with a pair of geminal chlorines, is even more readily reduced (311). 1-Bromo-2-chlorocyclopropane shows a reduction wave, but the nature of the product was not determined (312). Limonene tetrabromide (**247**) undergoes four-electron reduction to limonene (293). DDT (1,1-di-(*p*-chlorophenyl)-1,2,2,2-tetrachloroethane) (**248**, Ar = *p*-ClC_6H_4) undergoes pH-independent two-electron reduction to 1,1-di-(*p*-chlorophenyl)-2,2-dichloroethylene (**249**, Ar = *p*-ClC_6H_4) (313). Electrolysis at a lead cathode (not cpe) of 1,1,1-trifluoro-2-bromo-2-chloroethane, $F_3CCHBrCl$, produces a 70% yield of 1,1-difluoro-2-chloroethylene $F_2C{=}CHCl$ (314). Cpe of **250** produces *o*-cresoxyacetic acid (315).

$$\underset{\mathbf{248}}{Ar_2CCl{-}CCl_3} + 2e^- \longrightarrow \underset{\mathbf{249}}{Ar_2C{=}CCl_2} + 2Cl^-$$

$$\mathbf{250} \xrightarrow{4e^-} o\text{-}CH_3C_6H_4OCH_2CO_2^- + 3Cl^-$$

Both *meso*- and *dl*-diethyl α,α'-dibromosuccinate undergo two-electron reduction to diethyl fumarate, indicating that the reduction is not stereospecific. The half-wave potentials are independent of pH;

$$BrCH(COOEt){-}CH(COOEt)Br \xrightarrow{2e^-} Br^- + {}^{\ominus}CH(COOEt){-}CH(COOEt)Br \longrightarrow Br^- + EtOCO{-}CH{=}CH{-}COOEt$$

$E_{1/2}$ of the *meso* ester is +0.12 v., and $E_{1/2}$ of the *dl* ester is +0.05 v. Notice that the *meso* ester is the more easily reduced. In acid and base, both *meso*- and *dl*-dibromosuccinic acid are reduced to fumaric acid, but at pH 4 up to 70% maleic acid is formed from the *dl* acid (316). The radical resulting from one-electron reduction can undergo rotation about the central carbon–carbon bond. The radical from the *meso* acid (**251**) is formed in a stable conformation, and does not rearrange before adding a second electron and eliminating bromide, so that it forms the more stable fumaric acid. In acid and base, rotation occurs in the radical formed from the *dl* acid (**252**), to form the more stable conformation (**251**), which forms fumaric acid. But near pH 4, the radical from the *dl* acid is largely present as its monoanion (**253**), whose conformation is fixed by hydrogen bonding. When this radical

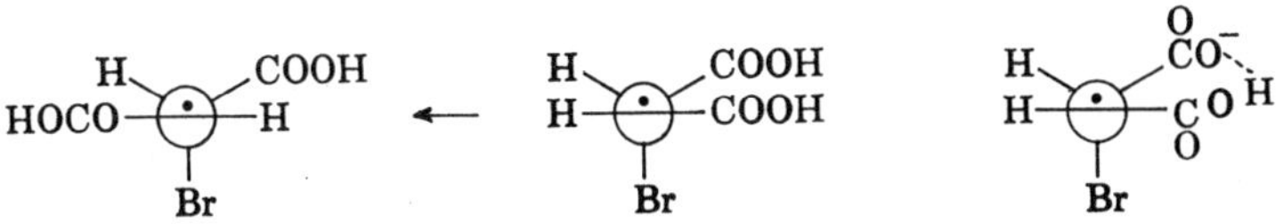

H, COOH, HOCO, H, Br — **251** ← H, COOH, H, COOH, Br — **252**; H, CO_2^-, H, CO_2H, Br — **253**

adds another electron and eliminates bromide, it forms maleic acid (316). 2,3-Dibromo-1,4-dihydroxybutane, $HOCH_2CHBrCHBrCH_2$-OH, undergoes pH-independent two-electron reduction. The *meso* isomer is reduced considerably more readily than is the *dl* (317).

The greater ease of reducibility of the *meso* dihalides suggests that *trans* elimination is preferred. This behavior is similar to the preferred *trans* dehalogenation of dihalides by metals and by iodide ion, and suggests a similarity of mechanism, except that the reduction at the electrode surface probably proceeds by stepwise transfer of the

e^- e^- Hg X X I^- X X

two electrons. It would be interesting to compare the half-wave potentials of the 2,3-dibromobicycloheptanes, (**254,255**) for which *cis* dehalogenation might be preferred, as a result of the impossibility of achieving a *trans* coplanar configuration.

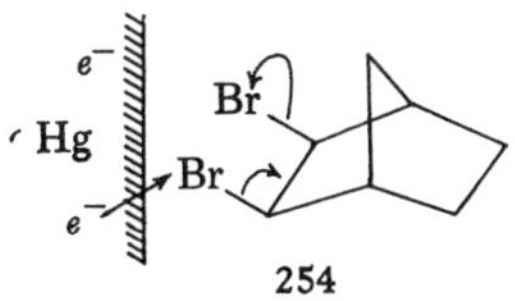

e^- Hg e^- Br Br **254** easier than e^- Hg e^- Br Br **255** ?

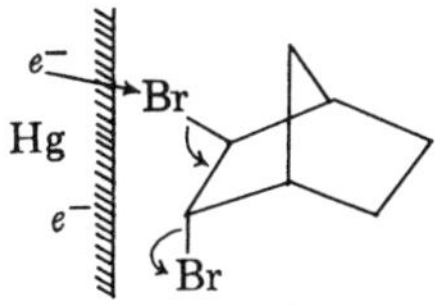

e. Aryl and vinyl halides

Aromatic and vinyl halides are more difficult to reduce than alkyl halides, but the process is still a two-electron reduction, to replace the halogen by hydrogen. Polyhaloaromatics undergo stepwise reduction (318,319).

Bromo- and chloro-pyridines undergo two-electron reduction, at potentials less negative than those required to reduce the corresponding halobenzene. In acid, $E_{1/2}$ is independent of pH, but at pH values above pK_a, $E_{1/2}$ depends upon pH, indicating that the reducible species is the pyridinium ion. Above pH 7, the limiting current decreases and represents a kinetic wave, limited by the rate of protonation of the halopyridine. The wave of the neutral halopyridine is observed

$$\text{X-C}_5\text{H}_4\text{N} \underset{}{\overset{k_f[\text{H}^+]}{\rightleftharpoons}} \text{X-C}_5\text{H}_4\text{NH}^+ \xrightarrow{2e^-, \text{H}^+} \text{C}_5\text{H}_5\text{NH}^+ + \text{X}^-$$

only at very negative potentials (244).

In DMF, the three iodonitrobenzenes show two waves. Cpe on the first plateau produces iodide ion, and cpe on the second plateau produces the radical anion of nitrobenzene. *o*-Bromonitrobenzene gives

$$IC_6H_4NO_2 \xrightarrow[\text{DMF}]{2e^-, H^+} I^- + C_6H_5NO_2 \underset{\substack{\text{higher} \\ ae-}}{\overset{e^-}{\rightleftharpoons}} C_6H_5NO_2^{\overline{\cdot}}$$

only a single wave, and cpe produces the radical anion of nitrobenzene, since potentials capable of cleaving bromide are also capable of adding an electron to the resulting nitrobenzene. Cpe of the *para* isomer produces radical anions of both nitrobenzene and *p*-bromonitrobenzene.

$$o\text{-BrC}_6\text{H}_4\text{NO}_2 \xrightarrow[\text{DMF}]{2e^-, \text{H}^+} \text{C}_6\text{H}_5\text{NO}_2 \underset{\text{DMF}}{\overset{e^-}{\rightleftharpoons}} \text{C}_6\text{H}_5\text{NO}_2^{\overline{\cdot}}$$

No halide is lost from *m*-nitrobromobenzene or the chloro- and fluoronitrobenzenes, since their radical anions are stable to loss of halide (320).

Above pH 5, cpe of isonicotinonitrile splits off cyanide ion and produces pyridine. This is an unusual example of polarographic cleavage

$$4\text{-NC-C}_5\text{H}_4\text{N} \xrightarrow{2e^-, \text{H}^+} \text{C}_5\text{H}_5\text{N} + \text{CN}^-$$

of a carbon–carbon bond (199,231), similar to the two-electron cleavage of cyanogen, N≡C—C≡N, to two CN^- ions (321).

Cpe of both diethyl bromofumarate (**256**) and diethyl bromomaleate (**257**) produces diethyl fumarate (**258**). Apparently there is free rotation about the double bond of some intermediate species, since the

EtOCO–CH=C(Br)–COOEt (**256**) $\xrightarrow{2e^-,\ H^+}$ EtOCO–CH=CH–COOEt (**258**) $\xleftarrow{2e^-,\ H^+}$ EtOCO–CH=C(Br)–COOEt (**257**)

maleate species isomerizes to the more stable fumarate on reduction. Bromofumaric acid is reduced to fumaric acid, but bromomaleic acid is reduced to a mixture of maleic acid, fumaric acid, and butadienetetracarboxylic acid, a dimeric species arising from coupling of two radicals. The yield of maleic acid is maximum at pH 4.2, where the conformation is held *cis* by hydrogen bonding in the monoanion, as in the reduction of *dl*-dibromosuccinic acid (279).

1,2-Dibromoolefins undergo two-electron reduction to form two bromide ions and the acetylene, which is ordinarily not further reducible at the potentials employed. *Trans* dibromides are reduced more easily than the *cis* isomers, again suggesting a preference for *trans* elimination (322). But *cis*-2,3-dichloroacrylonitrile (**259**) shows two two-electron waves, corresponding to stepwise replacement of chloride by hydrogen, to form acrylonitrile (**260**), whose reduction wave is observed at more negative potentials. The *trans* isomer (**261**) shows a single four-electron wave, corresponding to two-electron reduction to two chloride ions and propiolonitrile (cyanoacetylene) (**262**), which is immediately electrolytically hydrogenated to acrylonitrile. Apparently the anion (**263**) formed from two-electron reduction of the *cis* isomer merely adds a proton, but *trans* elimination of chloride from the anion (**264**) formed from the *trans* isomer is faster than proton uptake. 2,2,3,3-Tetrachloropropiononitrile undergoes facile two-electron reduction to *trans*-2,3-dichloroacrylonitrile (**261**), whose

259 $\xrightarrow{2e^-}$ Cl^- + **263** $\xrightarrow{H^+}$ ClCH=CH–CN $\xrightarrow{2e^-,\ H^+}$ CH_2=CH–CN (**260**) + Cl^-

261 $\xrightarrow{2e^-}$ Cl^- + **264** $\xrightarrow{-Cl^-}$ HC≡C–CN (**262**) $\xrightarrow[\text{fast}]{2e^-,\ 2H^+}$ CH_2=CH–CN (**260**)

reduction waves are observable at more negative potentials, as described above (323).

In DMF, 1,1,4,4-tetraphenyl-2,3-dihalobutadienes (**265**, X = Cl, Br) undergo two-electron reductive dehalogenation to the cumulene, 1,1,4,4-tetraphenylbutatriene (**76**), whose polarographic behavior has already been discussed (121).

$$\text{265} \xrightarrow[\text{DMF}]{2e^-} \phi_2C{=}C{=}C{=}C\phi_2\ (\text{76}) + 2X^-$$

m-Dibromobenzene shows two two-electron reduction waves, with the second at the $E_{1/2}$ of bromobenzene; stepwise replacement of halogen by hydrogen occurs. *m*-Bromochlorobenzene behaves similarly. *o*-Dibromobenzene (**266**, X = Br) shows only a single wave, of height corresponding to four-electron reduction. Two electron reduction produces the anion (**267**, X = Br) which eliminates a second bromide ion more rapidly than it adds a proton. The resulting benzyne (**268**) is rapidly hydrogenated electrolytically to benzene. *o*-Chlorobromobenzene (**266**, X = Cl) shows a small second wave, since protonation of the anion (**267**, X = Cl) produced by two-electron reductive cleavage of bromide is competitive with elimination of chloride to form benzyne. The small second wave represents reduction of chlorobenzene (**269**) which is formed by protonation of the *o*-chlorophenyl anion (**267**, X = Cl). Cpe of *o*-dibromobenzene (**266**, X =

$$\text{266} \xrightarrow{2e^-} Br^- + \text{267} \xrightarrow[\text{DMF}]{-X^-} \text{268} \xrightarrow[\text{fast}]{e^-} \cdot \xrightarrow[\text{fast}]{H^+} \cdot \xrightarrow[\text{fast}]{e^-,\,H^+} C_6H_6$$

$$\text{267} \xrightarrow{H^+} \text{269};\quad \text{268} \xrightarrow{\text{furan}} \text{270} \xrightarrow{H^+} \text{271}$$

Br) in DMF in the presence of furan produces naphthalene-1,4-epoxide (**270**), which is converted by acid to 1-naphthol (**271**), identified by vpc. The solution must be stirred, to remove benzyne from the electrode surface before it is reduced further (324).

f. Positive halogen compounds

Diphenyliodinium salts show three waves, with the third at the $E_{1/2}$ of iodobenzene (325). The first two are apparently reversible one-electron reductions, according to i_d and the log plot slopes (325). Comparison with the polarograms of phenylmercuric iodide and diphenyl mercury shows that neither of these is an intermediate. If the possibility of reversibility of the second wave is discounted, a possible mechanism is (326)

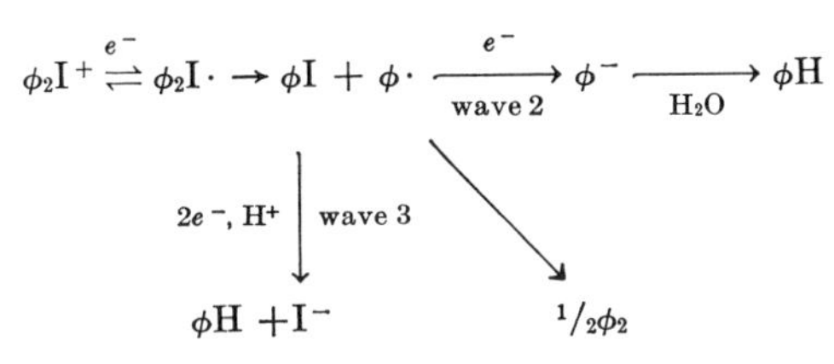

Iodosobenzene, ϕIO, shows two two-electron waves, corresponding to stepwise reduction, first to iodobenzene and then to benzene. Iodoxybenzene, ϕIO_2, undergoes four-electron reduction to iodobenzene, whose reduction wave is observed at more negative potentials (327). Iodosobenzene is presumably an intermediate, but it is reduced about as readily as is iodoxybenzene, so only one wave is observed. Iodosobenzene shows an anodic wave, but this anodic $E_{1/2}$ differs from the cathodic $E_{1/2}$ of iodoxybenzene, showing that this couple is not polarographically reversible, as is also clear from the fact that $E_{1/2}$ depends upon concentration, but not on pH (328).

2. Carbon–Oxygen Bonds

Polarographic cleavage of carbon–oxygen bonds is similar to that of carbon–halogen bonds, except that halogens are generally better leaving groups. Very little work has been done to establish the role of protons in facilitating the cleavage.

Pseudo-esters (cyclic esters) of *o*-benzoylbenzoic acid (**272**, R = ϕ) (146) and *o*-phthalaldehydic acid (**272**, R = H) are reduced in a two-electron process to the phthalide (**97**), whose reduction wave is observed at about the same potentials. $E_{1/2}$ is independent of pH, and

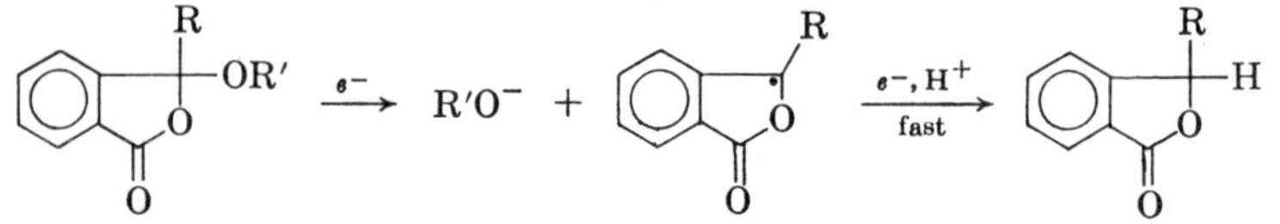

$$\textbf{272} \xrightarrow{e^-} R'O^- + \text{radical} \xrightarrow[\text{fast}]{e^-, H^+} \textbf{97}$$

the ease of reduction increases with the acidity of R′OH (R′ = Me < Et < ϕ < Ac). The reduction of these cyclic esters occurs at more negative potentials than does the reduction of the normal ester, whose carbonyl group is readily reduced. The cyclic amides behave similarly. The reduction involves splitting the C—OR′ bond because the intermediate radical is benzylic; splitting the C—OCO bond would require displacement of a bond in the plane of the benzene ring, and there could be no resonance stabilization of the transition state. Cpe of **272** (R = p-$MeOC_6H_4$, R′ = Et) on the first plateau produces the corresponding phthalide (**97**) (329).

Although at pH 8, benzoin (**273**) is reduced to hydrobenzoin, at pH 1.3, 1,2,3,4-tetraphenylbutanediol-2,3 (**274**) was isolated from cpe. Apparently the first step is polarographic cleavage of hydroxide, to produce desoxybenzoin (**275**), which is then reduced to the pinacol (**274**) (61). Various morphinone and codeinone derivatives (e.g.

$$\underset{\textbf{273}}{\phi COCHOH\phi} \xrightarrow{2e^-, H^+} OH^- + \underset{\textbf{275}}{\phi COCH_2\phi} \xrightarrow[\text{fast}]{e^-, H^+} \underset{\textbf{274}}{{}^1/_2\phi CH_2C(OH)\phi C(OH)\phi CH_2\phi}$$

276) which are cyclic α-aryloxy ketones show a two-electron reduction wave, ascribed to splitting apart the aryloxide (330). Germacrol (**277**) shows a two-electron reduction wave, and its polarogram differs

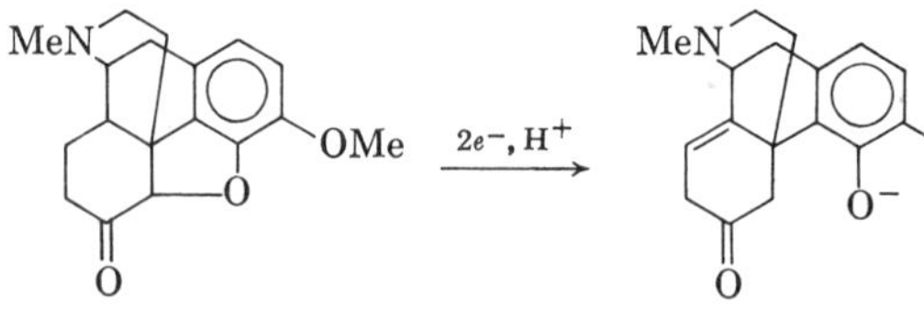

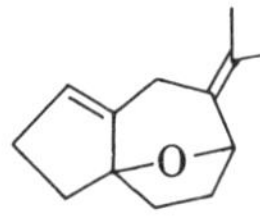

$$\textbf{276} \xrightarrow{2e^-, H^+} \text{(phenoxide product)} \qquad \textbf{277}$$

from that of germacrone, an α,β-unsaturated ketone arising from acid-catalyzed opening of the epoxide ring; apparently this wave represents cleavage of an allylic ether (331). 1-Phenylhydroxyacetone $\phi CHOHCOCH_3$, shows a reduction wave whose $E_{1/2}$ is independent of pH; since reduction of the carbonyl group is pH-dependent, this must represent cleavage of the hydroxide (332). Various steroidal α-ketols undergo two-electron reduction near -2.0 v., and a second two-electron reduction is observed at -2.5 v., ascribed to further reduction of the ketone produced by cleavage of hydroxide (333).

Cpe of several phenacyl compounds, such as ω-hydroxyacetophenone (**278**, R = H), phenacyl acetate (**278**, R = $COCH_3$), and phenacyl phenyl ether (**278**, R = ϕ) produces acetophenone. In all

$$\underset{\mathbf{278}}{\phi COCH_2OR} \xrightarrow{e^-, (H^+)} RO^-(H^+) + \phi COCH_2^{\cdot} \xrightarrow[\text{fast}]{e^-, H^+} \phi COCH_3$$

these cases cleavage of the carbon–oxygen bond is similar to cleavage of a carbon–halogen bond (334). The half-wave potential for reduction of ω-hydroxyacetophenone depends upon pH, as expected if a proton assists removal of hydroxide, as it does with removal of fluoride (298). Cpe of esters of *p*-methoxymandelonitrile (**279**, Ar = *p*-$MeOC_6H_4$) produces *p*-methoxyphenylacetonitrile (**280**) (335). Cpe

$$\underset{\mathbf{279}}{ArCH(CN)OCOR} \xrightarrow{2e^-, H^+} \underset{\mathbf{280}}{ArCH_2CN} + RCO_2^-$$

of 2,4-dichlorophenoxyacetamidine (**281**, Ar = 2,4-$Cl_2C_6H_3$) produces acetamidine and 2,4-dichlorophenol; $E_{1/2}$ is nearly independent of pH (201).

$$\underset{\mathbf{281}}{ArOCH_2C(NH_2){=}NH} \xrightarrow{2e^-, H^+} ArO^- + CH_3C(NH_2){=}NH$$

In DMF containing phenol, xanthone (**94**) shows two two-electron waves; the first represents reduction to xanthydrol (**282**), and the second represents reduction to xanthene (**283**), isolated from cpe (336). In DMF, anthrahydroquinone diethyl ether and anthrahydroquinone

$$\mathbf{94} \xrightarrow[\text{DMF}]{2e^-, 2\phi OH} \mathbf{282}\ (\text{H, OH}) \xrightarrow[\text{higher } a_{e^-}]{2e^-, \phi OH} \mathbf{283}\ (H_2) + OH^-$$

OR OR + $\xrightarrow[\text{DMF}]{4e^-,\ 2H^+}$ $2RO^-$ + $\xrightarrow{2e^-,\ 2H^+}$

diacetate show several waves, with a total height corresponding to four-electron reduction, but 9,10-dihydroanthracene was isolated from cpe (336). These reductions represent polarographic cleavage of benzylic carbon–oxygen bonds.

3. Carbon–Sulfur Bonds

Phenacyl phenyl thioether (**284**, R = ϕ) undergoes two-electron reduction to thiophenol and acetophenone. Phenacyl thiocyanate (**284**, R = CN) undergoes two-electron reduction to acetophenone and thiocyanate ion; $E_{1/2}$ is independent of pH above pH 3 (334). Cpe of

$$\underset{\mathbf{284}}{\phi COCH_2SR} \xrightarrow{e^-} RS^- + \phi COCH_2^{\cdot} \xrightarrow[\text{fast}]{e^-,\ H^+} \phi COCH_3$$

isonicotinic thioamide (**285**) in 1*M* HCl hydrogenates the carbon–sulfur double bond to produce **286**, which is stable below 5°C. At more negative potentials, cpe indicates that the carbon–sulfur bond is split (199).

$$\underset{\mathbf{285}}{\text{S=C(NH}_2\text{)–C}_5\text{H}_4\text{N}} \xrightarrow{2e^-,\ 2H^+} \underset{\mathbf{286}}{\text{HS–CH(NH}_2\text{)–C}_5\text{H}_4\text{N}} \xrightarrow{2e^-,\ 2H^+} \underset{\mathbf{189}}{\text{NH}_2\text{CH}_2\text{–C}_5\text{H}_4\text{N}} + H_2S$$

Phenacyldialkyl sulfonium salts (**287**, R = Me, *n*-Bu) undergo two-electron reduction to the dialkyl sulfide and acetophenone, whose reduction wave is observed at more negative potentials. Below pH 7, $E_{1/2}$ is independent of pH, but above pH 7, $E_{1/2}$ depends upon pH because the substrate is converted to its nonreducible conjugate base (337).

$$\phi COCH{=}SR_2 \overset{H^+}{\rightleftharpoons} \underset{\mathbf{287}}{\phi COCH_2SR_2^+} \xrightarrow{e^-} R_2S + \phi COCH_2^{\cdot} \xrightarrow[\text{fast}]{e^-,\ H^+} \phi COCH_3$$

Trialkylsulfonium salts undergo irreversible two-electron reduction, with $E_{1/2}$ independent of pH (338).

According to cpc, *p*-thiocyano-*N*,*N*-dimethylaniline (**288**, Ar = *p*-$Me_2NC_6H_4$) undergoes two-electron reduction; cyanide ion and *p*-*N*,*N*-dimethylaminothiophenol were detected after cpe (339).

$$\underset{\mathbf{288}}{\mathrm{ArSCN}} + 2e^- \rightarrow \mathrm{ArS}^- + \mathrm{CN}^-$$

4. Carbon–Nitrogen Bonds

The polarography of tetraalkylammonium ions has been discussed by Southworth, Osteryoung, Fleischer, and Nachod, who have in addition reviewed the evidence for the formation of ammonium amalgams, analogous to alkali metal amalgams. The reduction of such ions usually represents hydrogen discharge, catalyzed by the formation of the amalgam:

$$\mathrm{R_4N^+} \overset{e^-}{\rightleftharpoons} \mathrm{R_4N\cdot(Hg)} \xrightarrow{\mathrm{H_2O}} \mathrm{R_4N^+} + \mathrm{OH^-} + {}^1/_2\mathrm{H_2}$$

However, when one of the alkyl groups is readily cleaved, reduction produces a tertiary amine plus that radical, which either dimerizes or adds a second electron and a proton. Such behavior occurs with benzyldimethylanilinium ion (**301**) (340).

In acid, preparative electrolysis (not cpe) of several cyclic α-amino ketones leads to reductive cleavage of the carbon–nitrogen bond. Thus **289** is reduced at a lead cathode to **290** (341).

$$\mathbf{289} \xrightarrow{4e^-,\ 4H^+} \mathbf{290}$$

289 **290**

Various quaternary ethyleneimonium salts (e.g. **291**) undergo two-electron reduction, with opening of the three-membered ring suggested. Surprisingly, $E_{1/2}$ depends upon pH (342). Quinones with

$$\underset{\mathbf{291}}{\text{(aziridinium)}\ \overset{+}{\mathrm{N}}(\mathrm{Et})\mathrm{CH_2CH_2Cl}} \xrightarrow{2e^-,\ H^+} \mathrm{H{-}CH_2CH_2{-}N(Et)CH_2CH_2Cl}$$

ethyleneimine substituents show a normal, reversible reduction wave for reduction to the hydroquinone (**292**), but an irreversible wave is observed at more negative potentials. This second wave is a kinetic wave, limited by the rate of protonation of the nitrogen. The ethyleneimine (**292**) is not itself reducible, but protonation of the nitrogen makes possible a much better leaving group (343):

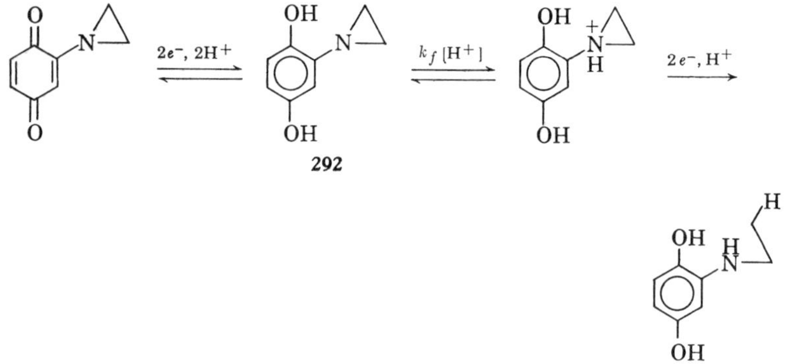

$$\xrightleftharpoons{2e^-,\ 2H^+} \quad \mathbf{292} \quad \xrightleftharpoons{k_f[H^+]} \quad \xrightarrow{2e^-,\ H^+}$$

In acid, aminomethylpyridines (e.g. **189**) show a two-electron reduction wave, to form ammonia and the corresponding picoline (**293**) (239).

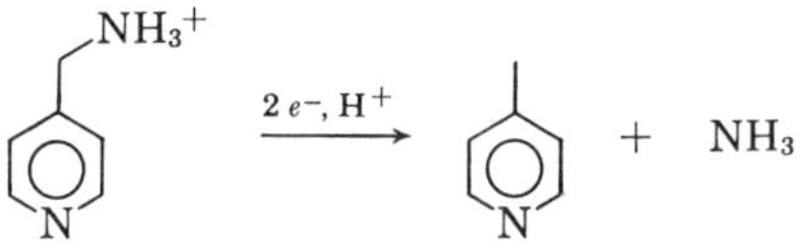

$$\mathbf{189}\ (NH_3^+) \xrightarrow{2e^-,\ H^+} \mathbf{293} + NH_3$$

Above pH 2.5, diazoacetophenone (**294**) shows two waves, of heights corresponding to six- and two-electron processes, respectively. The second represents reduction of protonated ω-aminoacetophenone (**295**) to ammonia and acetophenone, whose reduction is observed at more negative potentials (77,344). Below pH 2.5, diazoacetophenone is protonated to form phenacyl diazonium ion (**296**), and the height of the first wave drops to that of a two-electron reduction; here the reduction process is displacement of the excellent leaving group, nitrogen, in a fashion similar to the reductive cleavage of phenacyl halides (77).

Phenacyltrialkylammonium (**297**) salts show two waves, the first corresponding to two-electron reductive cleavage of the carbon–nitrogen bond, to form the trialkyl amine and acetophenone, whose $E_{1/2}$ is

$$\phi COCH{=}\overset{+}{N}{=}\overset{-}{N} \xrightarrow{6e^-,\,7H^+} NH_3 + \phi COCH_2NH_3^+ \xrightarrow[\text{higher } a_{e^-}]{2e^-,\,H^+} \phi COCH_3 + NH_3$$

294 295

H^+ ⇅

$$\phi COCH_2N_2^+ \xrightarrow{2e^-,\,H^+} \phi COCH_3 + N_2$$

296

the same as the $E_{1/2}$ of the second wave, and which was identified after cpe on the first plateau. Below pH 9, $E_{1/2}$ is surprisingly dependent

$$\phi COCH_2NR_3^+ \xrightarrow{2e^-,\,H^+} R_3N + \phi COCH_3$$

297

upon pH. Phenacylpyridinium salts (**298**) and β-triethylammonioacrolein salts (**299**) behave similarly. But ω-trialkylammoniopropiophenones (**300**) are not capable of cleavage with phenacyl partici-

$\phi COCH_2NC_5H_5^+$ (298) $R_3N^+{-}CH{=}CHCH{=}O$ (299) $\phi COCH_2CH_2NR_3^+$ (300)

pation, and only the reduction wave due to the carbonyl group is observed (337).

Cpc of the benzyl halide analog, benzyldimethylanilinium (**301**) bromide, in which the leaving group is *N*,*N*-dimethylaniline, shows that between 1.4 and 2.0 electrons are consumed per mole of substrate. When $n = 2$, only toluene was observed as a product of cpe, but when n was less than 2.0, both toluene and bibenzyl were detected. The potential-determining step involves one electron, to form benzyl radical (**302**), and the observed value of n, and the product composition, depend upon the competition between dimerization of the benzyl radical and its further reduction. Increasing the concentration of substrate favors the bimolecular one-electron process relative to the uni-

$$\phi CH_2NMe_2\phi^+ \xrightarrow{e^-} \phi NMe_2 + \phi CH_2\cdot \xrightarrow{e^-} \phi CH_2^- \xrightarrow[H_2O]{} \phi CH_3$$

301 302

$$\downarrow$$

$$^1/_2\phi CH_2CH_2\phi$$

molecular two-electron process. Also, decreasing the temperature from 25° to −35° slows down the transfer of the second electron, which has an appreciable activation energy, and thereby favors the bimolecular process, since dimerization of two radicals has a much smaller activation energy (345).

The polarographic cleavage of carbon–nitrogen bonds of tetra-substituted ammonium ions shows many similarities to the reaction of these ions with sodium in liquid ammonia. The active reagent in the latter case is the solvated electron (346). Studies of the reactions of organic compounds with the solvated electron can be expected to help in the understanding of polarographic reaction mechanisms, especially since it is then possible to study reactions of organic compounds with electrons under the conditions of homogeneous kinetics.

5. Oxygen–Oxygen Bonds

The bicyclic peroxide, ascaridole (**303**) shows a reduction wave, with $E_{1/2}$ independent of pH (347). Cumene hydroperoxide (**304**) behaves similarly (348). Above pH 12, reduction of alkyl hydroperoxides

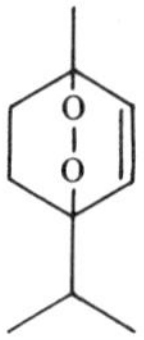

303 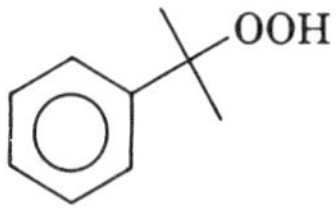 **304**

becomes more difficult, presumably because of ionization to a non-reducible anion (349). According to i_d, $n = 2$, but according to the log plot slope, the reduction is irreversible (350). The mechanism may be written simply as

$$\mathrm{ROO^-} \overset{\mathrm{H^+}}{\rightleftharpoons} \mathrm{ROOH} \xrightarrow{2e^-} \mathrm{RO^-} + \mathrm{OH^-}$$

although it is not clear why protons are incapable of assisting the cleavage of the oxygen–oxygen bond (cf. ω-hydroxyacetophenone). The ease of reducibility of various peroxides is in the order $RCO_3H \sim (RCO_2)_2 > ROOH > RCO_3\text{-}t\text{-Bu} > t\text{-BuOOH} > ROOR > (t\text{-BuO})_2$ (351,352).

6. Miscellaneous Bonds

Cleavage of nitrogen–nitrogen and nitrogen–oxygen bonds has been discussed in the section on polarography of hydrazone and oxime derivatives of carbonyl compounds. To summarize, the first step in the reduction of oximes is the cleavage of the nitrogen–oxygen bond, to form an imine; the reduction of hydrazones sometimes involves reductive cleavage of the nitrogen–nitrogen bond, but sometimes it is the carbon–nitrogen double bond which is electrolytically hydrogenated. We have also mentioned the electrolytic cleavage, in acid, of the nitrogen–oxygen bond of *N*-arylhydroxylamines and of the nitrogen–nitrogen bond of hydrazobenzene, and the facile cleavage of the nitrogen–oxygen bond of nitrate esters.

In addition, hydroxyurea undergoes irreversible two-electron reduction to urea (353). In acid, isonicotinoyl hydrazide (**305**) undergoes

$$HONHCONH_2 \xrightarrow{2e^-,\, 2H^+} NH_2CONH_2 + H_2O$$

two-electron reduction, with cleavage of the nitrogen–nitrogen bond, to form ammonia and isonicotinamide (**306**) (354).

$$\underset{\mathbf{305}}{\text{4-}C_5H_4N\text{-}CONHNH_2} \xrightarrow{2e^-,\, 2H^+} \underset{\mathbf{306}}{\text{4-}C_5H_4N\text{-}CONH_2} + NH_3$$

Chloramine T (*N*-chloro-*p*-toluenesulfonamide) is easily reduced to *p*-toluenesulfonamide; the cleavage of the nitrogen–chlorine bond is in-

$$p\text{-}CH_3C_6H_4SO_2NHCl \xrightarrow{e^-} Cl^- + p\text{-}CH_3C_6H_4SO_2NH\cdot \xrightarrow[\text{fast}]{e^-,\, H^+} p\text{-}CH_3C_6H_4SO_2NH_2$$

dependent of pH (355). *N*-Haloiminothiadiazoles (**307**) undergo two electron reduction, with cleavage of the nitrogen–halogen bond (356). *N*-Bromosuccinimide gives a two-electron reduction wave (357).

$$\underset{\mathbf{307}}{\text{(}\phi\text{-thiadiazoline, N-CH}_3\text{)=NX}} \xrightarrow{2e^-,\, H^+} \text{(}\phi\text{-thiadiazoline, N-CH}_3\text{)=NH} + X^-$$

E. HETEROCYCLIC COMPOUNDS

The polarography of some nitrogen-containing heterocycles has been reviewed by Volke (358). In summary, the polarographic behavior of aromatic heterocycles parallels that of aromatic hydrocarbons, with differences in behavior due to the different electronegativity of the heteroatom, and, where applicable, the charge type. Thus, although naphthalene and anthracene are reducible, indole and carbazole show no reduction wave in the accessible potential range. Although benzene is not polarographically reducible, *N*-alkylpyridinium salts do undergo polarographic reduction. However, the product of one-electron reduction is an uncharged radical, rather than a radical anion, and does not add a proton to give a still more readily reduced species.

1. *Pyridine Derivatives*

Pyridine and its derivatives produce catalytic hydrogen waves. The mechanism involves pyridinium ion, PyH^+, as the effective catalyst, which adds an electron to give the radical, $PyH\cdot$. This radical then evolves one-half mole of hydrogen gas and regenerates pyridine, which is rapidly protonated. The species being reduced is hydrogen ion, but pyridine provides a mechanism for decreasing the magnitude of the overvoltage required for hydrogen discharge on mercury. Since only pyridinium ion is effective as a catalyst, the wave height depends upon pH, and falls off near pK_a. The catalytic nature of this wave may be detected by the abnormally high limiting current in acid (359).

In DMF, pyridine and quinoline show a true reduction wave, not due to hydrogen discharge. According to i_d, the reduction involves two electrons (337).

N-Alkylpyridinium ions show two one-electron waves in aqueous media. $E_{1/2}$ of the first wave is independent of pH, and the log plot slope suggests that it corresponds to a reversible transfer of one electron to form an uncharged radical. This radical can either dimerize or add another electron, at more negative potentials. $E_{1/2}$ of this second wave depends upon pH, and suggests reduction to an *N*-alkyl dihydropyridine (360). Coulometry (not cpc) of *N*-methylnicotinamide (**308**, R = Me) indicates that n is near 1.6; n decreases at higher concentrations of substrate, because of dimerization of the intermediate radical (361). In dioxane, the log plot slope for *N*-methylpyridinium ion indicates that the transfer of the first electron is not reversible, although it is reversible in aqueous media (362).

$$\underset{\mathbf{308}}{C_5H_4N^+(R)CONH_2} \underset{}{\overset{e^-}{\rightleftharpoons}} C_5H_4\dot{N}(R)CONH_2 \xrightarrow{e^-,\,H^+} H_2C_5H_4N(R)CONH_2$$

$$\downarrow$$

dimer

Also, the wave shape for *N*-methylpyridinium ion does not follow eq. (19) exactly; the dependence of $E_{1/2}$ on concentration is slightly less than that expected for reversible electron transfer followed by irreversible dimerization. Apparently the electron transfer is not completely reversible (363).

Diphosphopyridine nucleotide (DPN^+, NAD^+) (**308**), a biologically important *N*-alkyl nicotinamide, shows a reduction wave whose $E_{1/2}$ is independent of pH between pH 4 and 8, but which depends upon pH above pH 8 (364). Cpc indicates two-electron reduction to DPNH, whose spectrum was observed after cpe. The two-electron reduction is polarographically irreversible, since $E_{1/2}$ is much more negative than the potentiometrically determined standard potential (365).

Above pH 8, where *N*-(diethylmalonato)pyridinium ion, $(EtOCO)_2CHNC_5H_5^+$, is present as the uncharged $(EtOCO)_2\overset{-}{C}—\overset{+}{N}C_5H_5$, a one-electron reduction wave is observed. $E_{1/2}$ is independent of pH, and the wave is ascribed to the formation of a radical anion (345). *N*-Cyclopentadienylpyridinium ion shows a one-electron reduction wave in acid. In neutral solution, the substrate is present as a neutral species (**309**) which gives two waves, the first a kinetic wave limited by the rate of protonation of this species. The second wave represents reduction of the neutral species to a radical anion (**310**), which dimerizes. The neutral species also gives a one-electron anodic wave, corresponding to oxidation to a radical cation (**311**) (366).

$$\underset{\mathbf{311}}{C_5H_5N^+—C_5H_4\cdot} \xleftarrow{-e^-} \underset{\mathbf{309}}{C_5H_5N^+—C_5H_4^-} \xrightarrow{e^-} \underset{\mathbf{310}}{C_5H_5\dot{N}—C_5H_4^-}$$

p,p′-Bipyridine shows two waves, the first apparently corresponding to reversible one-electron transfer; the blue color of a reduced form was observed at the electrode (367).

N,N′-Dialkylbipyridinium ions (**312**, R = Me, ϕCH_2) show two one-electron waves. $E_{1/2}$ of the first wave is independent of pH, and the log plot slope indicates that the reduction is reversible. Chemical reduction by one-electron reductants produces violet solutions of the radical cation (**313**) which shows anodic and cathodic waves at the same potentials as the two waves of **312**. $E_{1/2}$ of the second wave of **312** is independent of pH above pH 5, but depends upon pH below pH 5. Cpe on the second plateau produces a solution showing no anodic wave, indicating that the presumed two-electron product (**314**) reacts further (368). Similar results are obtained for **312** (R =

R—N⁺(pyridinium)—(pyridinium)N⁺—R (**312**) $\underset{}{\overset{e^-}{\rightleftharpoons}}$ R—N⁺(pyridinium)—(pyridinyl radical)N—R (**313**) $\xrightarrow[\text{higher } a_{e^-}]{e^-}$ R—N(dihydropyridylidene)=(dihydropyridylidene)N—R (**314**)

Et), and cpe on the first plateau produces **313** (R = Et). *N*-Ethyl-isonicotinonitrile (**315**, R = Et, X = CN) undergoes one-electron reduction, with $E_{1/2}$ independent of pH, but cpe produces **313** (R = Et), according to the mechanism

2 R—N⁺(pyridinium)—X (**315**) $\overset{2e^-}{\rightleftharpoons}$ 2 R—N•(pyridinyl)—X $\xrightarrow{\text{slow}}$

$2X^-$ + R—N⁺(pyridinium)—(pyridinium)N⁺—R (**312**) $\underset{\text{fast}}{\overset{e^-}{\rightleftharpoons}}$ R—N⁺(pyridinium)—(pyridinyl radical)N—R (**313**)

Methyl *N*-ethyl-isonicotinate (**315**, R = Et, X = COOMe) and *N*-thyl-isonicotinamide (**315**, R = Et, X = $CONH_2$) show pH-dependent, irreversible waves. In acetonitrile all three isonicotinic derivatives (**315**) show reversible one-electron waves (369).

Pyridine *N*-oxide shows an irreversible two-electron wave in acid; $E_{1/2}$ depends upon pH. Above pH 5, the limiting current decreases. Pyridine *N*-oxide is not reducible, but is conjugate acid, *N*-hydroxy-pyridinium ion (**316**), is reduced to pyridine. Near pH 5, the wave height is independent of *h*, indicating a kinetic current limited by the rate of protonation of the substrate (370,371). Quinoline *N*-oxide behaves similarly (371).

$$C_5H_5N{\rightarrow}O \overset{k[H^+]}{\rightleftharpoons} \underset{\mathbf{316}}{C_5H_5N^+{-}OH} \xrightarrow{2e,\,-H^+} C_5H_5N + H_2O$$

Pyrilium ions (e.g., **317**) undergo one-electron reduction, according to i_d (372). Below pH 5, $E_{1/2}$ is independent of pH, and cpc indicates one-electron reduction, reversible according to the log plot slope. Cpe produces a dimeric dipyran (**318**) which can be re-oxidized to the

317 $\xrightleftharpoons{e^-}$ → 1/2 ( )$_2$ 318

pyrilium ion at a platinum electrode (373).

Smith and Elving have studied the polarographic behavior of various pyrimidines and purines (374). Pyrimidine (**319**) gives several waves. In acid the mechanism is

319 $\xrightarrow{e^-, H^+}$ HN (radical) $\xrightarrow[\text{higher } a_{e^-}]{e^-, H^+}$ HN (H_2) $\xrightarrow[\text{wave 3}]{2e^-, 2H^+}$ tetrahydro pyrimidine

↓ dimer

This mechanism is supported by cpc, indicating the number of electrons per wave, and the UV spectra of the solutions after cpe. Near pH 4, the first two waves merge to give a single two-electron wave. Cytosine (**320**) undergoes three-electron reduction. The first step is a two-electron reduction to **321**, which eliminates ammonia, detected chemically, to form **322**, which rapidly undergoes further one-electron reduction, as verified by cpc of authentic **322**. 4-Amino-2,6-dimethylpyrimidine (**323**) undergoes four-electron reduction; the process is analogous to the reduction of cytosine, except that the two-electron

320 $\xrightarrow{2e^-, 2H^+}$ 321 → NH_3 + 322 $\xrightarrow[\text{fast}]{e^-, H^+}$ 1/2 ( )$_2$

product (**324**) undergoes further two-electron reduction, rather than only one-electron reduction (374).

323 $\xrightarrow{2e^-, 2H^+}$ 324 $\xrightarrow{-NH_3}$ $\xrightarrow[\text{fast}]{2e^-, 2H^+}$

In acid, alloxan (**325**) undergoes reversible two-electron reduction to dialuric acid (**326**). However, the reduction wave is a kinetic wave, limited by the rate of dehydration of the ketone hydrate. Nevertheless, $E_{1/2}$ of the cathodic wave of alloxan is the same as $E_{1/2}$ of the anodic wave of dialuric acid, and also the same as the potentiometrically determined standard potential. The one-electron reduction state, alloxantin (**327**) may be an intermediate in the electrode reaction (375).

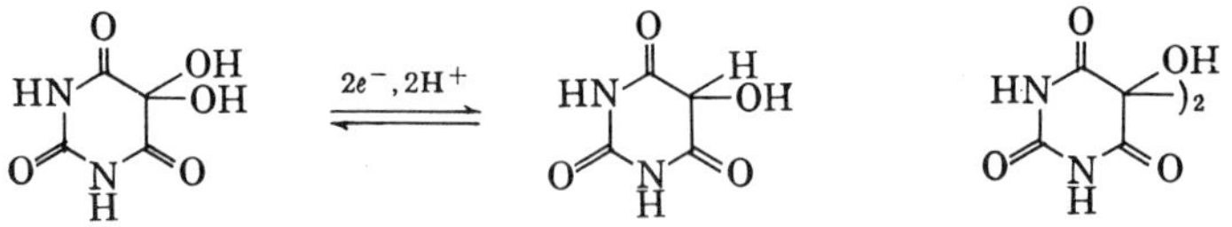

325 ⇌ ($2e^-$, $2H^+$) 326 ; 327

In nitromethane, chronopotentiometry indicates one-electron reduction for the dithiolium cation, **328** (376).

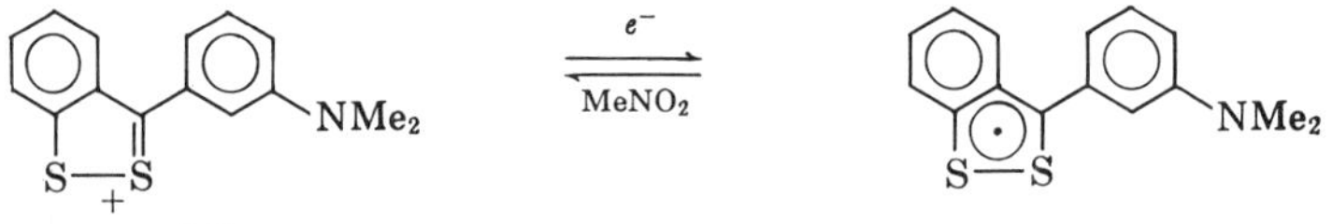

328 ⇌ (e^-, $MeNO_2$)

Anodic polarography of furan in methanol containing ammonium bromide produces a mixture of *cis*- and *trans*-2,5-dimethoxy-2,5-dihydrofuran (**329**) (377). The oxidation does not proceed via electrolytically generated bromine (378), but presumably involves radical cations of furan as intermediates.

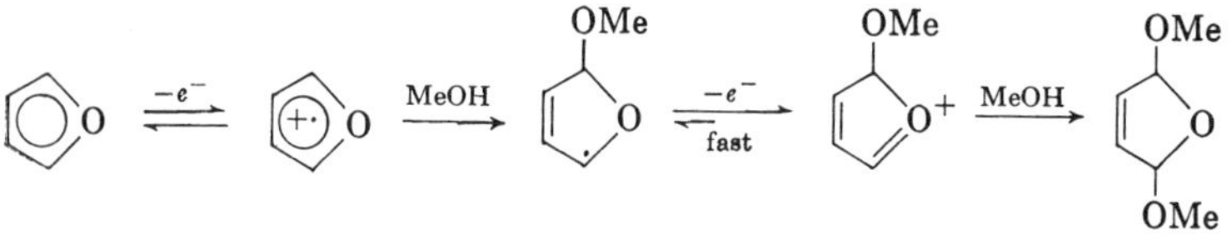

($-e^-$) ⇌ → (MeOH) → ($-e^-$, fast) → (MeOH) → 329

2. Polycyclic Heterocyclic Compounds

One of the first examples of the application of cpe to organic synthesis was the preparation of 9-(*o*-iodophenyl)-9,10-dihydroacridine (**330**) from 9-(*o*-iodophenyl)acridine (**331**). The latter compound shows two two-electron waves in alkali; the first corresponds to hydrogenation of the acridine moiety, and the second represents reductive cleavage of iodine. The versatility of preparative cpe was demonstrated by the selective reduction of **331** to **330**, leaving the iodine substituent intact. All chemical reducing agents tried either effected no reduction or reduced both the acridine and the iodine, to give 9-phenyl-9,10-dihydroacridine (**332**). In acid, cpe of 9-phenylacridine (**333**, R = ϕ) produces a red precipitate, presumably a radical resulting from addition of one electron to the conjugate acid. On dissolving

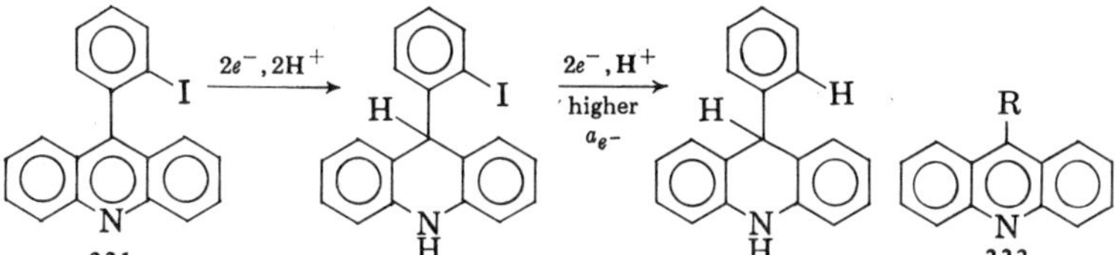

$$\mathbf{331} \xrightarrow{2e^-,\,2H^+} \mathbf{330} \xrightarrow[\text{higher } a_{e^-}]{2e^-,\,H^+} \mathbf{332} \qquad \mathbf{333}$$

the precipitate in alcohol, it disproportionates to 9-phenylacridine (**333**, R = ϕ) and 9-phenyl-9,10-dihydroacridine (**332**) (379).

Acridine itself (**333**, R = H) is reduced in two one-electron waves; $E_{1/2}$ depends upon pH, with a change of slope appearing at pK_a (380). The reductions are irreversible, according to the log plot slope, and complications arise from adsorption of the intermediate radical (381).

Quinaldinic acid (**334**) shows one pH-dependent two-electron wave in acid, and also a two-electron wave in base. The product is presumably the dihydro compound (**335**) (382).

Quinoxaline (**336**) undergoes two-electron reduction, according to i_d. *N*-Methylquinoxalinium ion undergoes two-electron reduction in acid. Quinoxaline-*N*,*N*′-dioxide shows a four-electron wave followed by a two-electron wave at the same potential as the wave of quinoxaline (383).

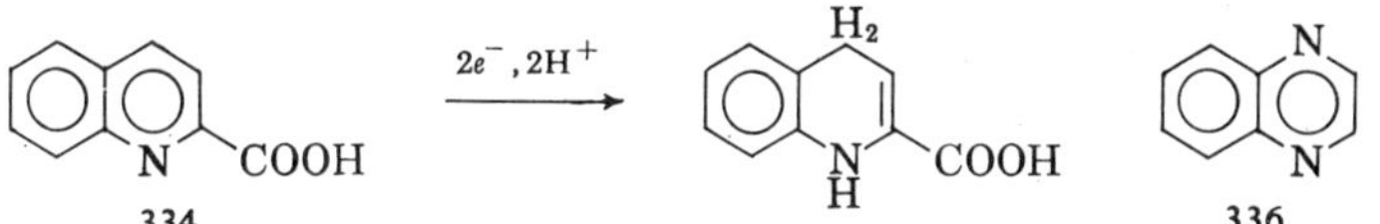

$$\mathbf{334} \xrightarrow{2e^-,\,2H^+} \mathbf{335} \qquad \mathbf{336}$$

Cpe of phenazine (**337**) in acetonitrile produces a radical anion, whose ESR spectrum was observed. Cpe of the conjugate di-acid (**338**) produces the radical cation (**339**), which could also be prepared by anodic cpe of 9,10-dihydrophenazine (**340**) (384).

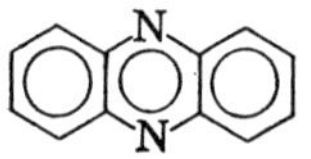

337

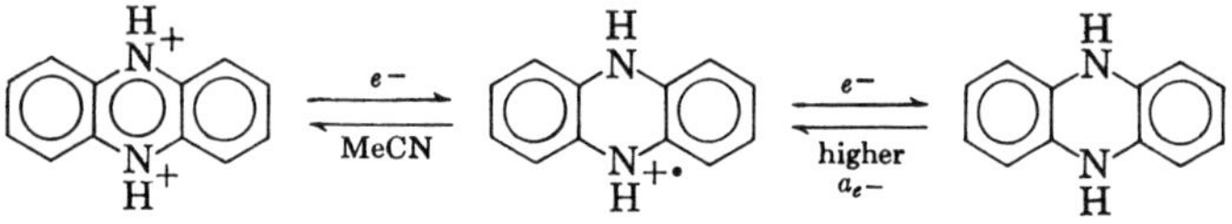

338 339 340

In acid, purine (**341**, X = H) shows two two-electron waves. The first represents hydrogenation of the 1,6 carbon–nitrogen bond; the resulting dihydropurine (**342**, X = H) can be re-oxidized to purine by oxygen. The second wave represents reduction to a tetrahydropurine (**343**, X = H), which can be opened to 4-amino-5-aminomethylimidazole (**344**) plus formaldehyde. Adenine (**341**, X = NH_2) undergoes four-electron reduction to a tetrahydro derivative (**343**, X =

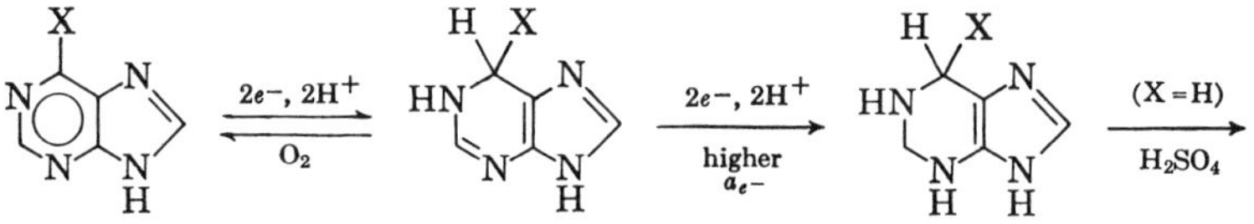

341 342 343

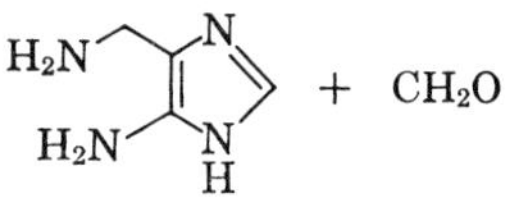

344

NH_2). However, cpc indicates six-electron reduction, since during the time of electrolysis, the tetrahydro derivative (**343**, X = NH_2) eliminates ammonia to give a dihydropurine (isomeric to **342**, X = H) which is readily reducible to the same tetrahydropurine (**343**, X = H) produced on the second plateau of the polarogram of purine (374).

Acridizinium ion (**345**) shows two one-electron waves in acid. The first wave represents reversible transfer of one electron. Below pH 7, these waves are independent of pH, but in base, these two waves disappear and are replaced by the waves of **346** (385).

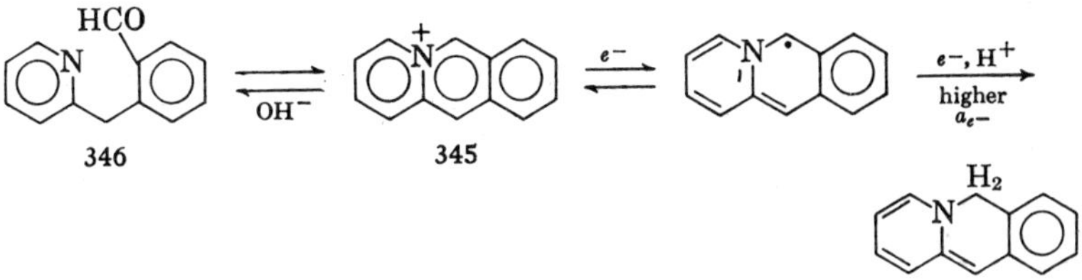

HCO N OH⁻ 346 N⁺ 345 e⁻ N e⁻, H⁺ higher a_{e^-} H_2 N

The isobenzpyrilium ion, **347** shows a one-electron reduction wave in acid (386).

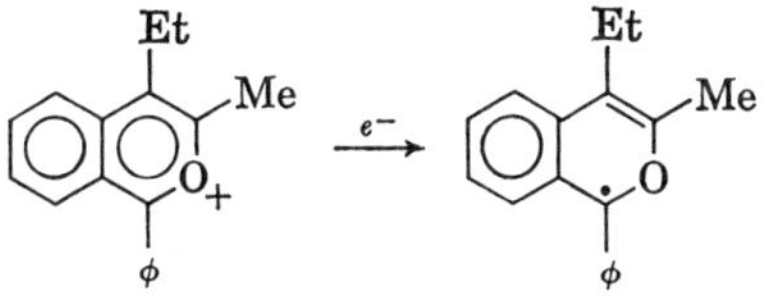

Et Me O+ φ 347 e⁻ Et Me O φ

3. Heterocycles with Two Different Heteroatoms

Above pH 5, sydnones (**348**) undergo irreversible four-electron reduction to **349**; $E_{1/2}$ is independent of pH. In acid, $E_{1/2}$ depends upon pH, and six-electron reduction proceeds to the amino acid (**350**). The log plot slope suggests that the potential-determining step involves reversible transfer of a single electron (387).

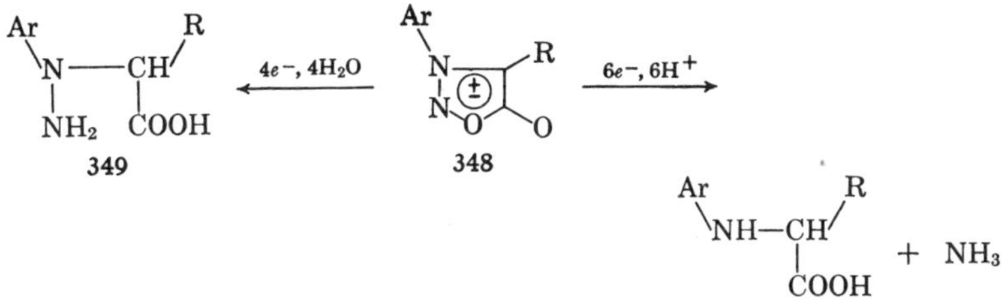

Ar R N—CH NH₂ COOH 349 $4e^-, 4H_2O$ Ar N R N O O (±) 348 $6e^-, 6H^+$ Ar R NH—CH COOH 350 + NH_3

The phenothiazine (**351**, R = alkyl, X = Cl) undergoes two-electron oxidation to the sulfoxide (**352**, R = alkyl, X = Cl) (388). In acetonitrile, phenothiazine (**351**, R = H, X = H) shows two one-electron anodic waves. Cpe on the first plateau produces an orange radical cation (**353**, R = H, X = H), whose ESR spectrum was observed, and cpe on the second plateau produces the red dication (**354**, R = H, X = H) (389), which can also be produced from phenothiazine *S*-oxide

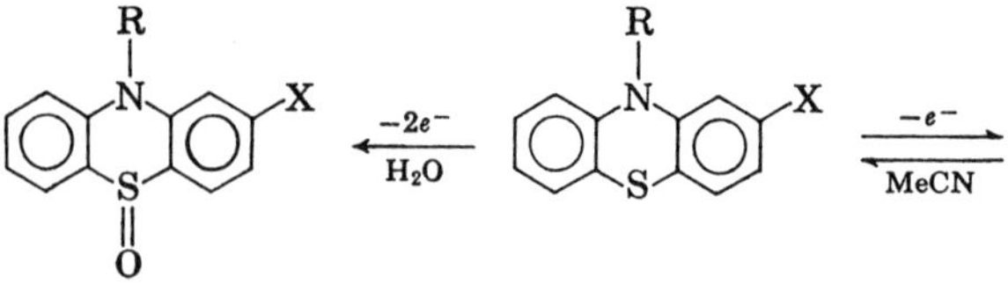

352 $\xleftarrow[H_2O]{-2e^-}$ 351 $\underset{MeCN}{\overset{-e^-}{\rightleftharpoons}}$

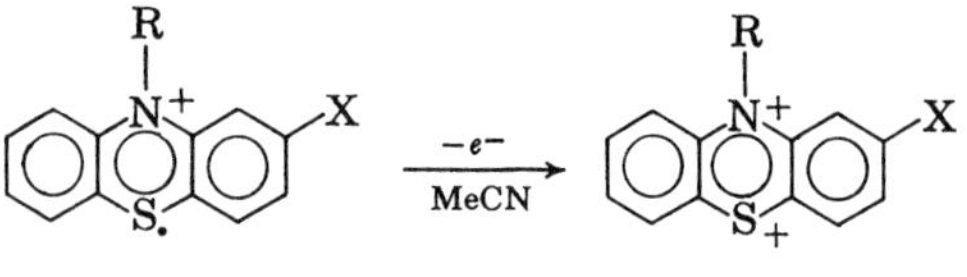

353 $\xrightarrow[MeCN]{-e^-}$ 354

(**352**, R = H, X = H) by the action of perchloric acid. The first wave is reversible, but the second is not. Addition of water or alkali renders the radical cation unstable to disproportionation, and the two waves merge to form a single two-electron wave (390).

F. SULFUR COMPOUNDS

An extended discussion of the polarography of sulfur compounds is irrelevant to an account of the mechanisms of organic polarography, since so much of the polarographic behavior of such compounds is a reflection of the properties of the mercury–sulfur bond. Therefore, we shall pay scant attention to this topic.

The system, cystine (**355**, R = $HOCOCH(NH_2)CH_2$)–cysteine (**356**, R = $HOCOCH(NH_2)CH_2$), is not reversible at a platinum electrode, solely because of the slowness of the electrode reaction

$$\underset{\mathbf{355}}{RSSR} + 2e^- + 2H^+ \underset{\text{slow}}{\rightleftharpoons} \underset{\mathbf{356}}{2RSH}$$

Between pH 3 and 9.2, cystine shows two pH-dependent waves at the dme, with the height of the total wave corresponding to two-electron reduction to cysteine. The second wave corresponds to the reduction of cystine to cysteine, which requires an overvoltage. The first wave corresponds to the reversible reduction of cystine to cysteine, and occurs at the standard potential of the system, with no overvoltage. This part of the wave is due to rapid reduction of a portion of the cystine which diffuses to the electrode surface, and proceeds rapidly as a result of a catalytic effect of the mercury, which forms an adsorbed layer of mercuric cysteinate, according to the mechanism (391)

$$\text{RSSR} + \text{Hg} \rightleftharpoons (\text{RS})_2\text{Hg} \xrightarrow[\text{fast}]{2e^-,\,2\text{H}^+} 2\text{RSH} + \text{Hg}$$

Thus, the polarographic behavior of mercaptan-disulfide systems is not a phenomenon of organic chemistry, but rather a phenomenon of the chemistry of mercury mercaptides.

Single bonds to sulfur can often be cleaved polarographically; some cases have already been discussed. Benzenesulfonyl chloride (**357**, X = Cl) shows a pH-independent two-electron wave, representing reduction to chloride plus benzenesulfinate, which is not further reduci-

$$\underset{\textbf{357}}{\text{ArSO}_2\text{X}} \xrightarrow{2e^-} \text{ArSO}_2^- + \text{X}^-$$

ble. Esters of arylsulfonic acids (**357**, X = OR) behave similarly (392), as do thiol esters (**357**, X = SR) of *p*-toluenesulfonic acid, for which cpc indicates two-electron reduction, and cpe produces *p*-toluenesulfinic acid (393).

Sulfoxides undergo two-electron reduction to the sulfide (392), verified by cpc and cpe (394). Anodic cpe of sulfides at a platinum electrode produces the corresponding sulfoxide (cf. phenothiazine, **351**, above). Both the oxidation and the reduction are irreversible (395,396).

$$\text{R}_2\text{S{=}O} + 2e^- + 2\text{H}^+ \underset{\text{slow}}{\rightleftharpoons} \text{R}_2\text{S} + \text{H}_2\text{O}$$

Diphenyl sulfone undergoes irreversible two-electron reduction to benzene and benzenesulfinic acid (397), isolated from cpe and verified by cpc (394). The reduction is analogous to the reductive cleavage

$$\phi_2\text{SO}_2 \xrightarrow{e^-} \phi\text{SO}_2^- + \phi\cdot \xrightarrow[\text{fast}]{e^-,\,\text{H}^+} \phi\text{H}$$

of halobenzenes. Benzothiophene dioxide (**358**) and methyl vinyl sulfone, $\phi\text{SO}_2\text{CH{=}CH}_2$, both show a pH-independent two-electron wave, although the reduction process is considered to be electrolytic hydrogenation of the carbon–carbon double bond, as in the polarography of α,β-unsaturated ketones (398).

[Scheme: **358** (benzothiophene dioxide, SO_2) $\underset{}{\overset{e^-}{\rightleftharpoons}}$ radical anion ($\bar{S}O_2$) $\xrightarrow{e^-,\,2H^+}$ dihydrobenzothiophene dioxide (SO_2)]

358

The alkyl thiosulfate, cysteine sulfite (**359**, R = $HOCOCH(NH_2)$-CH_2) undergoes two-electron reductive cleavage of the sulfur–sulfur bond, to form cysteine plus sulfite ion (399).

$$\underset{\mathbf{359}}{RSSO_3^-} \xrightarrow{2e^-} RS^- + SO_3^=$$

Hydroxymethylsulfinate anion (**360**) shows a pH-dependent two-electron anodic wave, to form formaldehyde bisulfite (**361**), which is rapidly hydrolyzed (400).

$$\underset{\mathbf{360}}{HOCH_2SO_2^-} \xrightarrow{2OH^-,\ -2e^-} \underset{\mathbf{361}}{HOCH_2SO_3^-} \xrightarrow[\text{fast}]{OH^-} CH_2O + SO_3^=$$

G. ORGANOMETALLICS

Organomercuric halides (RHgX, R = alkyl, phenyl) show two waves. $E_{1/2}$ of the first wave is independent of pH, and approximately independent of the nature of the R group, but does depend upon the nature of X. The log plot slope of the first wave suggests reversible transfer of a single electron, but $E_{1/2}$ is not independent of concentration. Cpe of phenylmercuric chloride on the first plateau produces diphenyl mercury, which is not further reducible. Therefore, the second wave must be due to reduction of a species which is capable of producing diphenyl mercury. $E_{1/2}$ of the second wave is independent of the nature of X, but does depend upon pH and the nature of R. A mechanism consistent with these observations is (401,402,282)

$$RHgX + e^- \rightleftharpoons X^- + RHg\cdot \xrightarrow[\text{higher } a_{e^-}]{e^-,\ H^+} RH + Hg$$
$$RHg\cdot \downarrow\ {}^1/_2R_2Hg + {}^1/_2Hg$$

although diphenyldimercury, $\phi HgHg\phi$, may be a further intermediate in the formation of diphenyl mercury (401).

Triethyl lead chloride undergoes one-electron reduction to chloride ion plus one-half mole of hexaethyl dilead (403). $E_{1/2}$ is independent

$$Et_3PbCl \xrightarrow{e^-} Cl^- + {}^1/_2Et_3PbPbEt_3$$

of pH, but depends upon concentration; the one-electron nature of the reduction was verified by cpc (404). Triethyl tin chloride shows two one-electron waves, according to cpc. The first wave is analogous to the wave of triethyl lead chloride, and the second wave represents further reduction to ethane plus diethyl tin, which disproportionates to tin metal plus tetraethyl tin (405). Diethyl tin dichloride, Et_2SnCl_2,

$$Et_3SnCl \xrightarrow{e^-} Cl^- + Et_3Sn\cdot \xrightarrow[\text{higher } a_{e^-}]{e^-,\ H^+} EtH + Et_2Sn \rightarrow {}^1/_2\,Sn + {}^1/_2\,Et_4Sn$$

$$Et_3Sn\cdot \downarrow \; {}^1/_2 Et_3SnSnEt_3$$

shows two one-electron waves, corresponding to stepwise reductive cleavage of chloride (406). Diethyl thallium bromide, Et_2TlBr, shows three one-electron waves, according to cpc. The first is attributed to reductive cleavage of chloride, followed by dimerization to tetraethyl dithallium (407).

In acetonitrile, anodic cpc of tetraphenyl borate (**362**) indicates between 1- and 1.8-electron oxidation, depending upon concentration. Two-electron oxidation would lead to biphenyl, isolated from cpe, plus diphenylorbinium ion (**363**). However, diphenylborinium ion, or protons formed by its hydrolysis, attack tetraphenyl borate, and decrease the effective concentration of the latter by converting it to triphenyl boron (**364**). During workup after cpe, triphenyl boron was oxidized to phenyl diphenylborinate (**365**), which was isolated (408).

$$\underset{\mathbf{362}}{\phi_4B^-} \xrightarrow[\text{MeCN}]{-2e^-} \phi_2 + \underset{\mathbf{363}}{\phi_2B^+} \xrightarrow{\phi_4B^-} \underset{\mathbf{364}}{2\phi_3B} \xrightarrow{O_2} \underset{\mathbf{365}}{\phi_2BO\phi}$$

Ferrocene, Cp_2Fe, undergoes reversible one-electron oxidation to ferricinium ion, Cp_2Fe^+. Ruthenocene behaves similarly. Cobalticinium ion undergoes one-electron reduction to cobaltocene, and rhodicinium ion behaves similarly. In 90% ethanol, the system, nickelocene–nickelocinium ion, is reversible. Dicyclopentadienyltitanium(IV) and dicyclopentadienylvanadium(IV) undergo reversible one-electron reduction to the +3 oxidation state (409). Osmocene also gives a reversible one-electron oxidation wave (410).

Diobenzenechromium–dibenzenechromium cation forms a reversible one-electron redox system. Dibenzenemolybdenum undergoes irreversible oxidation to an oxidation state higher than $(C_6H_6)_2Mo^+$ (411).

VI. Structure and Reactivity

A. POLAR EFFECTS

Zuman (412) has made most extensive studies of the effect of substituents on polarographic half-wave potentials. Most of the data may be correlated by a modified Hammett equation, where $E_{1/2}{}^{\mathrm{X}}$ and $E_{1/2}{}^{\mathrm{O}}$ are the half-wave potentials of substituted and standard com-

$$\Delta E_{1/2} = E_{1/2}{}^{\mathrm{X}} - E_{1/2}{}^{\mathrm{O}} = \rho_\pi \sigma_{\mathrm{X}} = \frac{RT}{\alpha nF} \ln \frac{k_{f,h}{}^0(\mathrm{X})}{k_{f,h}{}^0(\mathrm{O})}$$

pounds, respectively, ρ_π is a polarographic reaction constant, and σ_{X} is a substituent constant. The relation between $\Delta E_{1/2}$ and heterogeneous rate constants of substituted and standard compounds follows directly from eq. (14). It has been common practice in polarography to express reaction constants, ρ_π, in units of volts (per sigma unit). However, physical organic chemists use dimensionless reaction constants, ρ. The polarographic reaction constant may be converted to dimensionless form, reflecting rate constant ratios, by multiplying by $\alpha n_a F/RT$. Thus, (for αn_a chosen arbitrarily as 0.5) a value for ρ_π of +0.25 v. corresponds to a value for ρ of +2.1.

It is clear from the definition of ρ_π that its sign has the same significance as the sign of ρ in homogeneous reaction kinetics. A positive value for ρ_π means that a substrate with an electron-withdrawing substituent is reduced at less negative potentials than is the standard substrate, or that a substrate with such a substituent reacts more rapidly.

For benzene derivatives, application of the modified Hammett equation almost invariably leads to positive values of ρ_π, ranging from +0.02 v. to +0.37 v. The activating ability of electron-withdrawing substituents may be traced to two effects. The first is the essentially nucleophilic nature of the polarographic reduction process. Insofar as electrons are transferred in the potential-determining step, the transition state is more electron rich than is the reactant, and electron-withdrawing substituents will facilitate such a process (412). The second effect is ascribable to the nature of the interaction of substituent with reaction site in the reactant molecule itself. Polarographically reducible groups—carbonyl, nitro, halogen, azo—are electron-withdrawing in their own right, and interact with an electron-withdrawing substituent in such a way as to destabilize the reactant. On the other hand, electron-donating substituents interact with the reducible group in such a way as to stabilize the reactant and render it less reactive.

The uniformly positive value of the reaction constant, ρ_π, in polarographic reductions makes this quantity practically useless as a guide in elucidating reaction mechanisms. Clearly, the observation of a positive value for ρ_π is no more than corroboration of what was already undeniable, namely, that the electron is a nucleophilic reagent.

Only in exceptional cases can mechanistic inferences be drawn from ρ_π–σ relationships. Such cases include those in which ρ_π is negative, and those which yield poor ρ_π–σ correlations. The only well-substantiated case of a polarographic reaction series with a negative ρ_π is the reduction of substituted nitrobenzenes in concentrated sulfuric acid (412). In sulfuric acid, the reacting species is probably the conjugate acid (**366**) of the nitrobenzene, and electron-donating substituents permit a greater concentration of this form. If only a fraction of an electron is transferred at the transition state, the transition state will bear a fraction of a positive charge, relative to the reactant, and electron-donating substituents will facilitate the reduction.

NO_2 (X-substituted benzene) $\underset{H_2SO_4}{\overset{H^+}{\rightleftharpoons}}$ NO_2H^+ (X-substituted benzene) **366** $\xrightarrow{\alpha e^-}$ $[NO_2H^{(1-\alpha)+}$ (X-substituted benzene)$]^\ddagger$ $\rightarrow$ product

Another case is reduction of substituted benzyl bromides, which give a parabolic plot of $E_{1/2}$ vs. σ (413), and benzyl chlorides, for which ρ_π is apparently positive, but the correlation is very poor. The positive character of ρ_π is still ascribed to the nucleophilic character of the electrode, but the poor fit is ascribed to a stabilizing effect of all substituents on a radical-like transition state (**367**). This stabilizing effect is not simply related to the polar effect, and the *p*-methylthio group deviates quite markedly from a linear correlation. This poor correlation offers further support for the radical-like mechanism for polarography of organic halides (414). Similarly poor ρ_π–σ correlations might be expected for other reductions proceeding via radical-like transition states, such as that for ketone reduction (**368**), except that for such systems, the polar effect of substituents on the stability of the reactant (**369**) is much greater, and overshadows the smaller effect of substituents on radical stabilities, so that a good correlation is obtained, nevertheless.

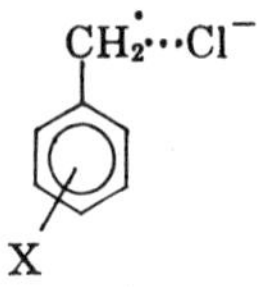

$CH_2^{\cdot}\cdots Cl^-$ R O$^-\cdots H^+$ O

X X X

367 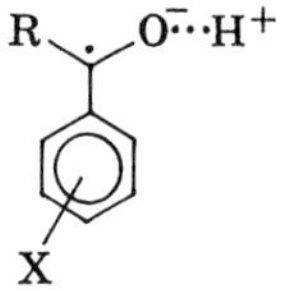 **368** 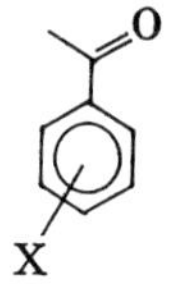 **369**

Zuman (412) has also applied to polarographic reductions Taft's extension (415) to polar substituent constants, $\sigma_X{}^*$. For correlations involving mostly alkyl groups, $\rho_\pi{}^*$ is sometimes positive, but occasionally it is negative, as, for example, in the reduction of substituted bromoacetic acids and their esters (416). The positive values of $\rho_\pi{}^*$ are to be expected for polarographic reductions, but no satisfactory explanation for negative values of $\rho_\pi{}^*$ is available. Perhaps negative values arise because steric effects do not permit meaningful $\rho_\pi{}^*$–σ^* correlations to be made. It has not yet been firmly established whether steric substituent constants (415) are applicable to reactions involving so bulky a reactant as the electrode surface.

Analogous correlations are observed between polarographic half-wave potentials and nuclear magnetic resonance chemical shifts (417), between $E_{1/2}$ and carbonyl stretching frequencies for dialkyl ketones and substituted benzophenones (418), and between $E_{1/2}$ and bond dissociation energies of alkyl iodides (419). Such correlations result because both quantities involved correlate according to σ or σ^*.

Although polar effects on half-wave potentials are not very helpful in elucidating polarographic reaction mechanisms, a recognition of the general facilitating effect of electron-withdrawing substituents is necessary for the interpretation of polarograms. We have already had occasion to invoke such effects, as in the discussion of the doubling of the waves in the system pyruvic acid–pyruvate anion (electron-withdrawing effect of the COOH group), the stepwise cleavage of halogen from *gem* polyhalides, and the effects of nitro and amino groups on half-wave potentials of substituted nitrobenzenes. Similarly, the differences in behavior between heterocyclics and aromatic hydrocarbons may be viewed as substituent effects.

It is more important to recognize the possibility of substituent effects on the mechanism of the reduction. Obviously, the effect of a substituent on $E_{1/2}$ of a given reaction cannot be determined if the substituent undergoes reduction more readily than does the reaction site under investigation. Thus, for example, the effect of a nitroso group on the reduction of acetophenone cannot be measured, since the nitroso group is reduced at less negative potentials than is the carbonyl.

We have already covered several reactions in which a substituent not only affects the half-wave potential, but also permits the occurrence of further reactions which are impossible in the unsubstituted compound. Examples of such behavior are the effect of an *ortho* or *para* amino or hydroxyl substituent on reduction of nitrobenzene, nitrosobenzene, and azobenzene, the reduction of *gem* dinitro compounds in a different fashion from the reduction of nitroalkanes, and the reduction of *o,o'*-dinitrosobiphenyl (**54**).

Finally, a warning must be injected. In order that a comparison of the effects of various substituents on $E_{1/2}$ be meaningful, it must be established that all compounds react by the same mechanism, and with the same step rate-limiting. Thus it is necessary to show that αn_a and i_d are constant in the reaction series, and that the effect of pH is the same for all compounds studied (420,412).

B. MOLECULAR ORBITAL CORRELATIONS

Correlations of half-wave potentials with MO calculations have been reviewed by Streitwieser (421) and by Elving and Pullman (16).

For many sets of reactions, half-wave potentials have been found to correlate with MO calculations. For example, the half-wave potentials of a series of alternant hydrocarbons correlate with the Hückel MO energy of the lowest vacant MO (LVMO), but not with the differences in delocalization energy between substrate and its dihydro product. This is evidence that the potential-determining step involves addition of the first electron, and not the protonation steps, which are subsequent (105). Anodic half-wave potentials correlate with the energy of the highest occupied MO (HOMO) (422).

In view of such correlations, it is not surprising that half-wave potentials correlate well with the frequency of the longest wave-length strong absorption band of alternant hydrocarbons. The absorption is due to excitation of an electron from the HOMO to the LVMO, and by the pairing theorem for alternant hydrocarbons, the energies of the HOMO and the LVMO are linearly related. Further consistency with the MO calculations is shown by the fact that the slope of the correlation line is 2.0, as required by the pairing theorem (423). Nonalternant hydrocarbons lie on a separate correlation line, and nonalternant hydrocarbons with two odd-membered rings seem to define a third correlation line (424).

Likewise, polarographic half-wave potentials of aromatic hydrocarbons correlate with photo-ionization potentials (425).

Half-wave potentials for ketones (426), quinones (427), chloromethyl aromatics (414), aromatic nitro compounds (428), and various substituted ethylenes (429) have also been correlated with MO calculations.

As with substituent effects, correlations of half-wave potentials with MO calculations are of marginal utility for elucidating polarographic reaction mechanisms. Such correlations are a necessary consequence of the fact that it is easier to add an electron to a larger "box"; the observation that derivatives of larger polycyclic systems are reduced at less negative potentials than those of smaller ones, or that α,β-unsaturated compounds are more readily reduced than are saturated analogs, does not ordinarily cast doubt on the validity of a possible mechanism.

On the other hand, the existence of such correlations permits the estimation of theoretical parameters by a simple method. Polarographic half-wave potentials are certainly easier to determine than ionization potentials or electron affinities. Furthermore, polarographic data serve to test theoretical models, especially MO theory. We have already seen that the simple Hückel MO treatment fails to correlate half-wave potentials of the conjugate acids of aromatic hydrocarbons, but that an SCF treatment provides a better model (124). Also, the unique polarographic behavior of cyclooctatetraene is a reflection of the relative instability of the radical anion, and the stability of the dianion (112,115). Such results suggest the possibility that anodic electrolysis of cyclooctatetraene in a nonnucleophilic solvent might produce a stable dication. We have also referred to a few examples of the use of polarography and cpe as a means of preparing stable radical anions, to be studied by ESR and spectroscopy; such results provide tests of theoretical calculations.

C. STERIC EFFECTS

Steric effects on polarographic half-wave potentials have been reviewed by Zuman (430).

Steric effects usually show up as a retarding influence on electrochemical reactions. We have already discussed the interpretations of steric effects on half-wave potentials of alkyl halides. It should be clear that steric effects and conformational effects offer great promise for the elucidation of polarographic reaction mechanisms.

Other examples include steric hindrance of *ortho* methyl groups in the polarographic reduction of nitromesitylene (**370**) and nitrodurene (**371**) (431), methylated acetophenones and benzophenones (432), and methylated acetyl- and benzoyl-azulenes (433). Steric hindrance has

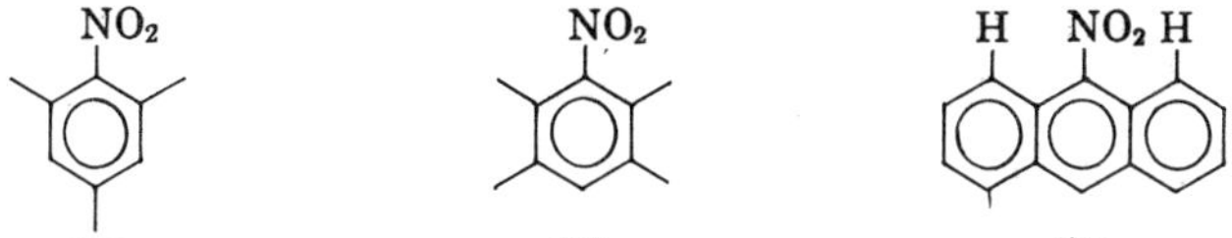

NO_2 — **370**

NO_2 — **371**

H NO_2 H — **372**

also been observed in polarography of Girard derivatives (**203**) of branched ketones (434), aliphatic nitro compounds (435), aryl alkyl sydnones (**348**) (387), 3-aryl-β-hydroxy coumarins (**120**, R = *ortho* substituted phenyl) (436), and a series of substituted nitrobenzenes and nitroanilines (437). Only a slight effect is observed in 9-nitroanthracene (**372**); apparently *peri* hydrogens do not interfere to as great an extent as do *ortho* methyl groups (438).

Zuman (412) has treated steric effects on polarographic half-wave potentials by Taft's equation (415) where $E_X{}^S$ is a steric substituent

$$\Delta E_{1/2} = \rho_\pi{}^*\sigma_X{}^* + \delta_\pi E_X{}^S$$

constant, and δ_π is a steric reaction constant. Polarographic half-wave potentials for alkyl hydroperoxides, dialkyl disulfides, and alkyl bromides have been analyzed accordingly. Seven alkyl bromides, including neopentyl bromide, follow the correlation, but isopropyl bromide, and especially *t*-butyl bromide, lie far from the correlation line, in the direction expected if the mechanism of reduction is becoming more S_N1-like.

Conformational and ring-size effects are also observed, as we have seen in the reduction of the benzene hexachlorides, *meso* and *dl vic* dibromides, α-halocyclohexanones, and cycloalkyl bromides. Other examples include Girard derivatives of cyclic ketones, among which the cyclohexanone derivative is most readily reduced, and the cyclooctanone derivative least readily reduced (439), and benzocycloalkenones, among which α-tetralone and 1-indanone are most readily reduced (440).

Steric effects traceable to hydrogen bonding usually facilitate polarographic reduction. Thus, the polar effect of a *p*-hydroxyl group causes *p*-nitrophenol to be reduced at more negative potentials than is *m*-nitrophenol. But in acid, *o*-nitrophenol (**373**) is reduced still more readily, because the effect of hydrogen bonding is greater than the polar effect (88). The hydrogen bond serves to polarize the nitro group and facilitate addition of an electron. The effect of this proton is equivalent to that of protons from the solution, and represents a type of general acid catalysis. *o*-Nitroanisole, which is incapable of internal hydrogen bonding, behaves normally (441). Similar results are observed with *o*-nitrobenzamidine (**374**) (442) and salicylaldehyde (**375**), although this isomer is not reduced more readily than is *m*-hydroxybenzaldehyde (443).

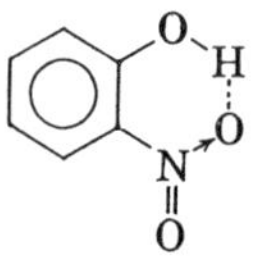

373 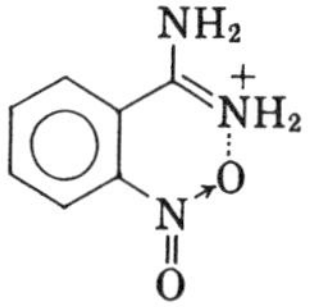 **374** 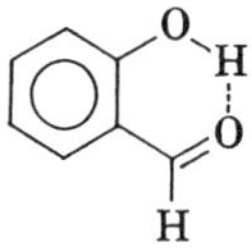 **375**

One exception to this role is the reduction of tropolone (**132**), which is reduced less readily than either β-tropolone (**135**) or tropolone methyl ether (**134**). Hydrogen bonding stabilizes tropolone and renders it less readily reducible (180). Apparently the proton used in hydrogen bonding does not facilitate addition of an electron to the aromatic system. Similar effects are operative in reduction of hydrogen-bonded osazones (e.g. **213**) (247).

VII. Applications

In any branch of organic chemistry, an understanding of the mechanisms of the reactions involved is an invaluable aid to the utilization of such reactions. The first step in applying organic electrode processes is a knowledge of what functional groups undergo reaction with electrons. The second step is a knowledge of how readily an electrochemical reaction takes place, i.e., what the half-wave potential is. We have discussed a number of reducible functional groups; more extensive tables of such groups are available (444,445). We have put little emphasis on actual values of half-wave potentials for different groups, since only relative reducibilities of similar functional groups are pertinent to a discussion of mechanism. However, half-wave potentials of many organic compounds have been compiled by Zuman (446) and by Gardner and Lyons (13) or are readily available through recourse to monographs (1,3) and bibliographies (9,447).

The third step in utilizing an organic electrode process is a knowledge of the products of the electrochemical reaction, and how the nature of the products is influenced by reaction conditions—electrode potential, concentration, and pH. The final step is an understanding of the mechanism of the reaction, especially the intermediates formed during the course of the reaction, but also the effect of substituents and the possibilities for side reactions.

The principal areas of application of electrode reactions to organic and physical organic chemistry are organic synthesis, structure proof, and quantitative analysis, especially as applied to reaction kinetics. We will mention a few examples, in order to suggest the versatility of the polarographic method.

A. ORGANIC SYNTHESIS

Cpe offers excellent possibilities for selective oxidations and reductions. Allen (7) has described a large number of organic electrolyses; we will cite only a few.

We have already described the electrolytic synthesis of 9-(*o*-iodophenyl)-9,10-dihydroacridine (**330**); this synthesis could not be accomplished by the usual chemical methods (379). Other examples have included the reduction of an *N*-alkyltetrachlorophthalimide (**165**) to the cyclic *N*-alkyl tetrachlorophthalaldehydamide (**166**) (222), the preparation of both isomers of *p,p*′-bis-dimethylaminohydrobenzoin (**376**, R = Me, R′ = H) (139), the synthesis of medium-ring azacyclanols (e.g. **290**) (341), the preparation of 3,4-diphenyl-$\Delta^{3,4}$-dehydroadipic acid (**159**, R = ϕ) (216), of the 1-aryl-1,2-dihydronaphthalene (**93**) (135), of steroidal pinacols (156,157), and of the stable free radical, di-*t*-butyl nitroxyl (**58**) (102).

Also, cpe of *p*-aminoacetophenone in acid produces the corresponding pinacol (**376**, R = H, R′ = Me), which could not be obtained by the usual chemical reduction methods (448). The unsymmetrical pinacol (**377**) could be obtained in 37% yield by cpe of a mixture of the two acetophenones (449). Cpe of various *para*-substituted nitro-

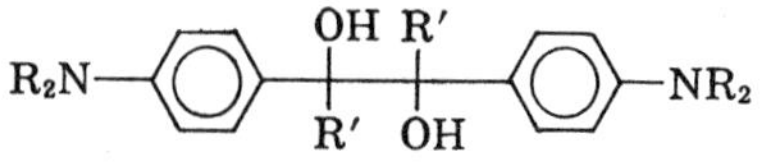

376

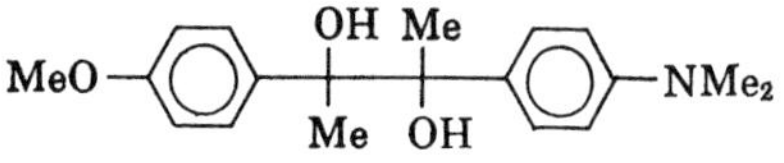

377

benzenes (**378**, X = OH, ϕCOO, ϕNH, MeCONH, ϕCONH) in sulfuric acid produces the corresponding aniline (**379**). Cpe of *m*-halonitrobenzenes (**380**, X = Br, Cl) in sulfuric acid produces either the aniline (**381**) or the *p*-hydroxyaniline (**382**); rearrangement of the intermediate arylhydroxylamine may or may not compete with further reduction, depending upon conditions (450).

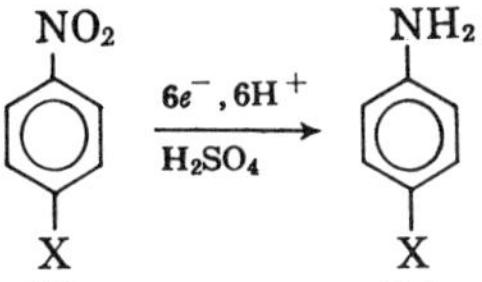

$$\text{378} \xrightarrow[\text{H}_2\text{SO}_4]{6e^-,\,6\text{H}^+} \text{379} \qquad \text{380} \xrightarrow[\text{H}_2\text{SO}_4]{6\text{ or }4e^-} \text{381 or 382}$$

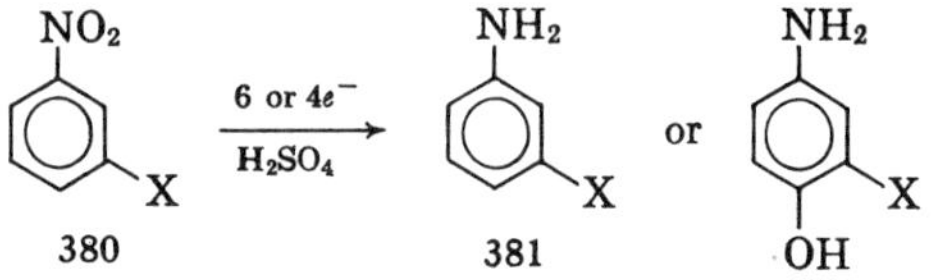

We have not mentioned the Kolbe synthesis, whereby electrolytic oxidation of a carboxylate salt produces a hydrocarbon, since this synthesis is not carried out under polarographic conditions. However, polarographic experiments and cpe have shown that several acetoxylations of aromatic hydrocarbons do not proceed via the mechanism of the Kolbe reaction, but rather via hydrocarbon radical cations (133). A recent review of the Kolbe synthesis is available (451).

One technique, which has already proved to be of great value, is cpe in aprotic media, to produce stable radicals for further study, as by ESR. We have mentioned a few examples above (42,53,109,252). Adams has reviewed the electrochemical generation of radicals (452).

B. STRUCTURE DETERMINATION

Since structure determination is not properly a part of mechanistic organic chemistry, no examples of application of polarography will be included here. It should suffice to point out that the characteristic reduction waves of organic functional groups may be used to ascertain the presence or absence of such groups. Most of the applications have been to natural products, and the interested reader is referred to monographs (1,3) for examples.

C. KINETICS

Widespread use has been made of polarography as a tool for quantitative analysis. Many examples are given in monographs (1–3), or are referred to in bibliographies (9,447). We will restrict ourselves to citing a few examples of the application of polarographic analysis to following reaction kinetics.

The use of polarographic analysis in following reaction kinetics has been reviewed by Semerano (453), Page (11), and Zuman (454). More recent examples include the reaction of substituted benzyl bromides with methoxide and thiophenoxide (455), the decomposition of *t*-butyl pertosylate (456), the Meerwein–Ponndorf–Verley reduction of benzophenone by aluminum isopropoxide (457), the reaction of bromoacetic acid and bromoacetamide with arylsulfinates (458), and the reaction of benzaldehyde with dimedone (459). The trimerization of "formiform" (triformylmethane) to 1,3,5-triformylbenzene was followed polarographically. During the course of the reaction, a new wave, at potentials more positive than the $E_{1/2}$ of either reactant or product, was observed, and ascribed to the buildup of some intermediate (460). The cleavage of tropanones (**383**, X = Me_2N, MeS) with base, to form a mixture of cycloheptadienones, was followed polarographically. At pH 10, the quaternary tropanone (**383**, X = Me_2N) was cleaved directly to the final product (**384**). But the polarographic studies indicated that the sulfur analog (**383**, X = MeS) reacts in two separate steps, with the first proceeding readily at pH 4, and the second proceeding readily only at pH above 10. These results were used to determine the optimum conditions for isolating the intermediate (**385**, X = MeS), which was indeed prepared by carrying out the hydrolysis between pH 5 and 9 (461).

X^+ =O —OH⁻→ X =O —OH⁻→ =O

383 **385** **384**

In order to apply the polarographic method of analysis to following reaction kinetics, it is necessary to determine the half-wave potentials of reactants and products, and to ascertain that none of the other species present interferes with the analysis. In the above examples, the polarographically active reactant is named first, to indicate how the polarographic method was applied. But a physical organic chemist, armed with a list of reducible functional groups and their half-wave potentials, should have little difficulty deciding whether polarography is applicable to his reaction kinetics. The advantages of this method are that it can be quite selective, that it requires only small amounts of material, and that a recording polarograph can be used as a continuous analyzer, without necessarily removing aliquots from the reaction mixture. Furthermore, polarographic analysis is most suitable for millimolar concentrations, so that a large excess of a non-reducible reactant may be present, in order to achieve pseudo-first-order kinetics.

VIII. Conclusions

It is apparent that the electron is quite a versatile reactant, whose reactions may be studied by all the techniques of the physical organic chemist. In recent years, both analytical and organic chemists have been making great progress in elucidating the mechanisms of such reactions. However, there still remain many reactions whose reduction mechanism is obscure, and many compounds whose polarographic behavior has not even been investigated. I hope that this article will stimulate interest in polarography among more organic chemists, and show what can be done in the field. Although a polarograph is a more complicated and more expensive "dispenser" than an ordinary reagent bottle, the reagent itself is one of the cheapest and most readily available.

As yet, mechanistic interpretations of organic polarography have largely been limited to specifying the products of the electrode reaction and cataloguing the intermediates. Organic polarography, combined with cpe in aprotic media, has been an especially productive technique, providing a means for generating and studying many otherwise unstable intermediates, and this technique will be used even more frequently in the future. But the next stage of development will encompass a description of the transition states. However, progress in this direction must await a better understanding, on the part of the organic chemists, of the mechanisms of electron transfer reactions, the nature of the electrode surface, the effects of the electric field on the surrounding layer, and the structure of the solvent.

Acknowledgments

I am very grateful to Dr. S. Wawzonek for the help his bibliographies (9) provided.

References

1. Kolthoff, I. M., and J. J. Lingane, *Polarography*, Interscience, New York, 1952.
2. Zuman, P., *Organic Polarographic Analysis*, Macmillan, New York, 1964.
3. Březina, M., and P. Zuman, *Polarography in Medicine, Biochemistry, and Pharmacy*, translated by S. Wawzonek, Interscience, New York, 1958.
4. Meites, L., *Polarographic Techniques*, 2nd ed., Interscience, New York, 1965.
5. Delahay, P., *New Instrumental Methods in Electrochemistry*, Interscience, New York, 1954.
6. Charlot, G., J. Badoz-Lambling, and B. Trémillon, *Electrochemical Reactions*, Elsevier, Amsterdam, 1962.
7. Allen, M. J., *Organic Electrode Processes*, Reinhold, New York, 1958.
8. Müller, O. H., in *Physical Methods of Organic Chemistry*, 3rd ed., Vol. I, Part IV, A. Weissberger, ed., Interscience, New York, 1960, pp. 3155–3279.
9. Wawzonek, S., *Anal. Chem.*, *21*, 61 (1949); *22*, 30 (1950); *26*, 65 (1954); *28*, 638 (1956); *30*, 661 (1958); *32*, 1445 (1960); *34*, 182R (1962); *36*, 220R (1964).
10. Laitinen, H. A., *Ann. Rev. Phys. Chem.*, *1*, 309 (1950).
11. Page, J. E., *Quart. Rev.*, *6*, 262 (1952).
12. Tanford, C., and S. Wawzonek, *Ann. Rev. Phys. Chem.*, *3*, 247 (1952).
13. Gardner, H. J., and L. E. Lyons, *Rev. Pure Appl. Chem.*, *3*, 134 (1953).
14. Elving, P. J., *Record Chem. Progr.* (*Kresge-Hooker Sci. Lib.*), *14*, 99 (1953).
15. Nünberg, H. W., *Angew. Chem.*, *72*, 433 (1960).
16. Elving, P. J., and B. Pullman, in *Advances in Chemical Physics*, Vol. I, I. Prigogine, ed., Interscience, New York, 1961, p. 1.
17. Zuman, P., and S. Wawzonek, in *Progress in Polarography*, Vol. I, P. Zuman and I. M. Kolthoff, eds., Interscience, New York, 1962, p. 303.
18. Elving, P. J., *Pure Appl. Chem.*, *7*, 423 (1963).
19. Koutecký, J., *Collection Czech. Chem. Commun.*, *18*, 597 (1953).
20. Brdička, R., *Advan. Polarography*, *2*, 655 (1960), This series, in three volumes, is the Proc. Second Intern. Congr. Polarography, held at Cambridge, 1959.
21. Müller, O. H., ref. 8, p. 3204.
22. Müller, O. H., ref. 8, p. 3218ff.
23. Ref. 1, p. 250ff.
24. Müller, O. H., ref. 8, p. 3221ff.; O. H. Müller, *Proc. First Intern. Congr. Polarog. Prague*, *1*, 159 (1951).
25. Delahay, P., ref. 5, p. 83ff.
26. Kivalo, P., *Acta Chem. Scand.*, *9*, 221 (1955).
27. Ref. 6, p. 36.
28. Koutecký, J., and V. Hanuš, *Collection Czech. Chem. Commun.*, *20*, 124 (1955).
29. Meites, L., in *Physical Methods of Organic Chemistry*, 3rd ed., Vol. I, Part IV, A. Weissberger, ed., Interscience, New York, 1960, pp. 3281–3333.
30. Ref. 1, p. 47ff.
31. Geddes, A. L., and R. B. Pontius, in *Physical Methods of Organic Chemistry*,

3rd ed., Vol. I, Part II, A. Weissberger, ed., Interscience, New York, 1960, pp. 895–1005.
32. Pearson, J., *Trans. Faraday Soc.*, *44*, 683 (1948).
33. Breiter, M., M. Kleinerman, and P. Delahay, *J. Am. Chem. Soc.*, *80*, 5111 (1958); P. Delahay, *Proc. First Intern. Congr. Polarog. Prague*, *1*, 65 (1951).
34. Hammond, G. S., *J. Am. Chem. Soc.*, *77*, 334 (1955).
35. Prelog, V., O. Häfliger, and K. Wiesner, *Helv. Chim. Acta*, *31*, 877 (1948).
36. Müller, O. H., *Ann. N. Y. Acad. Sci.*, *40*, 91 (1940); ref. 8; ref. 1, p. 253ff.
37. Edsberg, R. L., D. Eichlin, and J. J. Garis, *Anal. Chem.*, *25*, 798 (1953).
38. Wawzonek, S., R. Berkey, E. W. Blaha, and M. E. Runner, *J. Electrochem. Soc.*, *103*, 456 (1956).
39. Given, P. H., M. E. Peover, and J. Schoen, *J. Chem. Soc.*, *1958*, 2674.
40. Peover, M. E., and J. D. Davies, *J. Electroanal. Chem.*, *6*, 46 (1963).
41. Austen, D. E. G., P. H. Given, D. J. E. Ingram, and M. E. Peover, *Nature*, *182*, 1784 (1958).
42. Dehl, R., and G. K. Fraenkel, *J. Chem. Phys.*, *39*, 1793 (1963).
43. Given, P. H., and M. E. Peover, *Collection Czech. Chem. Commun.*, *25*, 3195 (1960); *J. Chem. Soc.*, *1960*, 385.
44. Gill, R., and H. I. Stonehill, *J. Chem. Soc.*, *1952*, 1857.
45. Berg, H., *Naturwiss.*, *48*, 714 (1961).
46. Smith, L. I., I. M. Kolthoff, S. Wawzonek, and P. M. Ruoff, *J. Am. Chem. Soc.*, *63*, 1018 (1941).
47. Doskočil, J., *Collection Czech. Chem. Commun.*, *15*, 599 (1950).
48. Wheeler, C. M., Jr. and R. P. Vigneault, *J. Am. Chem. Soc.*, *74*, 5232 (1952).
49. Sartori, G., and C. Cattaneo, *Gazz. Chim. Ital*, *72*, 525 (1942).
50. Knobloch, E., *Collection Czech. Chem. Commun.*, *14*, 508 (1949).
51. Brdička, R., and E. Knobloch, *Z. Elektrochem.*, *47*, 721 (1941).
52. Piette, L. H., P. Ludwig, and R. N. Adams, *Anal. Chem.*, *34*, 916 (1962).
53. Kuwata, K., and D. H. Geske, *J. Am. Chem. Soc.*, *86*, 2101 (1964).
54. Souchay, P., and F. Tatibouët, *J. Chim. Phys.*, *49*, C108 (1952).
55. Arcamone, F., C. Prévost, and P. Souchay, *Bull. Soc. Chim. France*, *1953*, 891.
56. Brdička, R., and P. Zuman, *Collection Czech. Chem. Commun.*, *15*, 766 (1950).
57. Holubek, J., and J. Volke, *Advan. Polarog.*, *3*, 847 (1960).
58. Ono, S., M. Takagi, and T. Wasa, *J. Am. Chem. Soc.*, *75*, 4369 (1953); *Bull. Chem. Soc. Japan*, *31*, 356 (1958).
59. Leonard, N. J., H. A. Laitinen, and E. H. Mottus, *J. Am. Chem. Soc.*, *75*, 3300 (1953); *76*, 4737 (1954).
60. Modiano, J., *Ann. Chim.* (*Paris*), *10*, 541 (1955).
61. Pasternak, R., *Helv. Chim. Acta*, *31*, 753 (1948).
62. Arai, T., *J. Electrochem. Soc. Japan* (*Overseas Ed.*), *30*, E-46 (1962).
63. Berg, H., *Z. Chem.*, *2*, 237 (1962).
64. Philp, R. H., Jr., R. L. Flurry, and R. A. Day, Jr., *J. Electrochem. Soc.*, *111*, 328 (1964); R. H. Philp, Jr., T. Layloff, and R. N. Adams, *ibid.*, *111*, 1189 (1964).
65. Holleck, L., and O. Lehmann, *Monatsh.*, *92*, 499 (1961); L. Holleck, O. Lehmann, and A. Mannl, *Naturwiss.*, *47*, 108 (1960).

66. Wawzonek, S., and J. D. Frederickson, *J. Am. Chem. Soc.*, *77*, 3985 (1955).
67. Nygård, B., *Arkiv Kemi*, *20*, 163 (1963).
68. Rüetschi, P., and G. Trümpler, *Helv. Chim. Acta*, *35*, 1021, 1486, 1957 (1952).
69. Gilbert, G. A., and E. K. Rideal, *Trans. Faraday Soc.*, *47*, 396 (1951); G. Costa, P. Rozzo, and P. Batti, *Ann. Chim. (Rome)*, *45*, 387 (1955); G. Costa and A. Puxeddu, *Ric. Sci.*, *27*, S94 (1957); through *Angew. Chem.*, *72*, 433 (1960); H. A. Laitinen and T. J. Kneip, *J. Am. Chem. Soc.*, *78*, 736 (1956).
70. McKeown, G. G., and J. L. Thomson, *Can. J. Chem.*, *32*, 1025 (1954).
71. Florence, T. M., and G. H. Aylward, *Austral. J. Chem.*, *15*, 65, 416 (1962).
72. Costa, G., *Gazz. Chim. Ital.*, *83*, 875 (1953).
73. Atkinson, E. R., H. H. Warren, P. I. Abell, and R. E. Wing, *J. Am. Chem. Soc.*, *72*, 915 (1950).
74. Elofson, R. M., *Can. J. Chem.*, *36*, 1207 (1958).
75. Kochi, J. K., *J. Am. Chem. Soc.*, *77*, 3208 (1955).
76. Rüetschi, P., and G. Trümpler, *Helv. Chim. Acta*, *36*, 1649 (1953).
77. Foffani, A., L. Salvagnini, and C. Pecile, *Ann. Chim. (Rome)*, *49*, 1677 (1959).
78. Calzolari, C., and A. D. Furlani, *Gazz. Chim. Ital.*, *87*, 862 (1957).
79. Solon, E., and A. J. Bard, *J. Am. Chem. Soc.*, *86*, 1926 (1964).
80. Smith, J. W., and J. G. Waller, *Trans. Faraday Soc.*, *46*, 290 (1950).
81. Holleck, L., and H. J. Exner, *Naturwiss.*, *39*, 159 (1952).
82. Kemula, W., and R. Sioda, *Bull. Acad. Polon. Sci.*, *10*, 107 (1962); through *Bull. Soc. Chim. France*, *1963*, 2359.
83. Le Guillanton, G., *Bull. Soc. Chim. France*, *1963*, 2359.
84. Schindler, R., W. Lüttke, and L. Holleck, *Ber.*, *90*, 157 (1957).
85. Stočesova, D., *Collection Czech. Chem. Commun.*, *14*, 615 (1949).
86. Bergman, I., and J. C. James, *Trans. Faraday Soc.*, *48*, 956 (1952); *50*, 60 (1954).
87. Astle, M. J., and W. V. McConnell, *J. Am. Chem. Soc.*, *65*, 35 (1943).
88. Pearson, J., *Trans. Faraday Soc.*, *44*, 692 (1948); *45*, 199 (1949).
89. Shreve, O. D., and E. C. Markham, *J. Am. Chem. Soc.*, *71*, 2993 (1949).
90. Mark, H. B., Jr., E. M. Smith, and C. N. Reilley, *J. Electroanal. Chem.*, *3*, 98 (1962).
91. Stone, K. G., and N. H. Furman, *J. Am. Chem. Soc.*, *70*, 3062 (1948).
92. Testa, A. C., and W. H. Reinmuth, *J. Am. Chem. Soc.*, *83*, 784 (1961).
93. Alberts, G. S., and I. Shain, *Anal. Chem.*, *35*, 1859 (1963).
94. Elofson, R. M., and J. G. Atkinson, *Can. J. Chem.*, *34*, 4 (1956).
95. Ross, S. D., G. J. Kahan, and W. A. Leach, *J. Am. Chem. Soc.*, *74*, 4122 (1952).
96. Martin, R. B., and M. O. Tashdjian, *J. Phys. Chem.*, *60*, 1028 (1956).
97. Lund, H., *Acta Chem. Scand.*, *11*, 990 (1957).
98. Holleck, L., and R. Schindler, *Z. Elektrochem.*, *62*, 942 (1958).
99. Kolthoff, I. M., and A. Liberti, *J. Am. Chem. Soc.*, *70*, 1885 (1948).
100. Elving, P. J., and E. C. Olson, *J. Am. Chem. Soc.*, *79*, 2697 (1957).
101. Armand, J., *Compt. Rend.*, *258*, 207 (1964).
102. Hoffmann, A. K., and A. T. Henderson, *J. Am. Chem. Soc.*, *83*, 4671 (1961).
103. Wawzonek, S., and H. A. Laitinen, *J. Am. Chem. Soc.*, *64*, 2365 (1942).

104. Hoijtink, G. J., J. Van Schooten, E. De Boer, and W. I. Aalbersberg, *Rec. Trav. Chim.*, *73*, 355 (1954).
105. Maccoll, A., *Nature*, *163*, 178 (1949); A. Pullman, B. Pullman, and G. Berthier, *Bull. Soc. Chim. France*, *1950*, 591; G. J. Hoijtink, *Rec. Trav. Chim.*, *77*, 555 (1958); A. Streitwieser, Jr., and I. Schwager, *J. Phys. Chem.*, *66*, 2316 (1962).
106. Hoijtink, G. J., and J. Van Schooten, *Rec. Trav. Chim.*, *71*, 1089 (1952); G. J. Hoijtink, E. De Boer, P. H. Van Der Meij, and W. P. Weijland, *ibid.*, *74*, 277 (1955); G. J. Hoijtink, *ibid.*, *76*, 885 (1957).
107. Wawzonek, S., E. W. Blaha, R. Berkey, and M. E. Runner, *J. Electrochem. Soc.*, *102*, 235 (1955).
108. Wawzonek, S., and D. Wearring, *J. Am. Chem. Soc.*, *81*, 2067 (1959).
109. Pointeau, R., and J. Favede, *Fifth Intern. Symp. Free Radicals, Uppsala, 1961*, 52-1.
110. Chopard-dit-Jean, L. H., and E. Heilbronner, *Helv. Chim. Acta*, *36*, 144 (1953).
111. Wawzonek, S., and J. W. Fan, *J. Am. Chem. Soc.*, *68*, 2541 (1946).
112. Elofson, R. M., *Anal. Chem.*, *21*, 917 (1949).
113. Glover, J. H., and H. W. Hodgson, *Analyst*, *77*, 473 (1952).
114. Craig, L. E., R. M. Elofson, and I. J. Ressa, *J. Am. Chem. Soc.*, *75*, 480 (1953).
115. Katz, T. J., *J. Am. Chem. Soc.*, *82*, 3784 (1960); T. J. Katz, W. H. Reinmuth, and D. E. Smith, *J. Am. Chem. Soc.*, *84*, 802 (1962).
116. Laitinen, H. A., and S. Wawzonek, *J. Am. Chem. Soc.*, *64*, 1765 (1942).
117. von Stackelberg, M., and W. Stracke, *Z. Elektrochem.*, *53*, 118 (1949).
118. Volke, J., and J. Holubek, *Collection Czech. Chem. Commun.*, *27*, 1777 (1962).
119. Hoijtink, G. J., *Rec. Trav. Chim.*, *73*, 895 (1954).
120. Thiec, J., and J. Wiemann, *Bull. Soc. Chim. France*, *1956*, 177.
121. Kemula, W., and J. Kornacki, *Roczniki Chem.*, *36*, 1835, 1857 (1962); through *Chem. Abstr.*, *59*, 218bd.
122. Grodzka, P. G., and P. J. Elving, *J. Electrochem. Soc.*, *110*, 225, 231 (1963).
123. Aten, A. C., and G. J. Hoijtink, *Z. Physik. Chem.* (*Frankfurt*), *21*, 192 (1959).
124. Aalbersberg, W. I., and E. L. Mackor, *Trans. Faraday Soc.*, *56*, 1351 (1960).
125. Wawzonek, S., R. Berkey, and D. Thomson, *J. Electrochem. Soc.*, *103*, 513 (1956).
126. Elving, P. J., and J. M. Markowitz, *J. Phys. Chem.*, *65*, 686 (1961).
127. Zhdanov, S. I., and A. N. Frumkin, *Proc. Acad. Sci. USSR* (*English Transl.*), *122*, 659 (1958).
128. Breslow, R., W. Bahary, and W. Reinmuth, *J. Am. Chem. Soc.*, *83*, 1763 (1961).
129. Lund, H., *Acta Chem. Scand.*, *11*, 1323 (1957).
130. Friend, K. E., and W. E. Ohnesorge, *J. Org. Chem.*, *28*, 2435 (1963).
131. Geske, D. H., *J. Am. Chem. Soc.*, *81*, 4145 (1959).
132. Mango, F. D., and W. A. Bonner, *J. Org. Chem.*, *29*, 1367 (1964).
133. Eberson, L., and K. Nyberg, *Acta Chem. Scand.*, *18*, 1568 (1964); S. D. Ross, M. Finkelstein, and R. C. Petersen, *J. Am. Chem. Soc.*, *86*, 4139 (1964).

134. Inoue, T., K. Koyama, T. Matsuoka, and S. Tsutsumi, *Tetrahedron Letters, 1963*, 1409.
135. O'Connor, J. J., and I. A. Pearl, *J. Electrochem. Soc.*, *111*, 335 (1964).
136. Ashworth, M., *Collection Czech. Chem. Commun.*, *13*, 229 (1948).
137. Swann, S., Jr., et al. *Trans. Electrochem. Soc.*, *85*, 231 (1944).
138. Gardner, H. J., *Chem. Ind. (London)*, *1951*, 819.
139. Allen, M. J., *J. Chem. Soc.*, *1951*, 1598.
140. Arai, T., *J. Electrochem. Soc. Japan (Overseas Ed.)*, *30*, E-100 (1962).
141. Elving, P. J., and J. T. Leone, *J. Am. Chem. Soc.*, *80*, 1021 (1958).
142. Whitman, W. E., and L. A. Wiles, *J. Chem. Soc.*, *1956*, 3016.
143. Suzuki, M., and P. J. Elving, *J. Phys. Chem.*, *65*, 391 (1961).
144. Valenta, P., *Advan. Polarog.*, *3*, 1004 (1960).
145. Wawzonek, S., and J. H. Fossum, *Trans. Electrochem. Soc.*, *96*, 234 (1949).
146. Wawzonek, S., H. A. Laitinen, and S. J. Kwiatkowski, *J. Am. Chem. Soc.*, *66*, 827 (1944).
147. Wawzonek, S., and A. Gundersen, *J. Electrochem. Soc.*, *107*, 537 (1960).
148. Cisak, A., and P. J. Elving, *Rev. Polarog. (Kyoto)*, *11*, 21 (1963); through *Pure Appl. Chem.*, *7*, 423 (1963).
149. Lund, H., *Acta Chem. Scand.*, *11*, 491 (1957).
150. Kemula, W., Z. R. Grabowski, and M. K. Kalinowski, *Collection Czech. Chem. Commun.*, *25*, 3306 (1960).
151. Grabowski, Z. R., and M. K. Kalinowski, *Fifth Intern. Symp. Free Radicals, Uppsala, 1961*, 221.
152. Moshier, R. W., *Ind. Eng. Chem. Anal. Ed.*, *15*, 107 (1943).
153. Adkins, H., and F. W. Cox, *J. Am. Chem. Soc.*, *60*, 1151 (1938).
154. Fields, M., and E. R. Blout, *J. Am. Chem. Soc.*, *70*, 930 (1948).
155. Korshunov, I. A., and Yu. V. Vodzinskiĭ, *Zh. Fiz. Khim.*, *27*, 1152 (1952), through *Chem. Abstr.*, *48*, 5674c.
156. Kabasakalian, P., and J. McGlotten, *J. Am. Chem. Soc.*, *78*, 5032 (1956).
157. Lund, H., *Acta Chem. Scand.*, *11*, 283 (1957).
158. Zuman, P., *J. Electrochem. Soc.*, *105*, 758 (1958).
159. Šantavý, F., *Collection Czech. Chem. Commun.*, *12*, 422 (1947).
160. Párkányi, C., and R. Zahradník, *Collection Czech. Chem. Commun.*, *27*, 1355 (1962).
161. Knobloch, E., *Advan. Polarog.*, *3*, 875 (1960).
162. Geissman, T. A., and S. L. Friess, *J. Am. Chem. Soc.*, *71*, 3893 (1949).
163. Holleck, L., and D. Marquarding, *Naturwiss.*, *49*, 468 (1962).
164. Lavrushin, V. F., V. D. Bezuglyi, and G. G. Belous, *J. Gen. Chem. USSR (English Transl.)*, *33*, 1667 (1963).
165. Wawzonek, S., and J. H. Fossum, *Proc. First Intern. Congr. Polarog. Prague*, *1*, 548 (1951).
166. Wawzonek, S., R. C. Reck, W. W. Vaught, Jr., and J. W. Fan, *J. Am. Chem. Soc.*, *67*, 1300 (1945).
167. Harle, A. J., and L. E. Lyons, *J. Chem. Soc.*, *1950*, 1575.
168. Čapka, O., *Collection Czech. Chem. Commun.*, *15*, 965 (1950).
169. Patzak, R., and L. Neugebauer, *Monatsh.*, *82*, 662 (1951); *83*, 776 (1952).

170. Hartnell, E. D., and C. E. Bricker, *J. Am. Chem. Soc.*, *70*, 3385 (1948).
171. Zuman, P., and J. Michl, *Nature*, *192*, 655 (1961).
172. McMullen, W. H., III, B. S. Thesis, Polytechnic Institute of Brooklyn, 1956; through *Physical Methods of Organic Chemistry*, 3rd ed., Vol. I, Part IV, A. Weissberger, ed., Interscience, New York, 1960, p. 3317.
173. Given, P. H., and M. E. Peover, *J. Chem. Soc.*, *1960*, 465.
174. Zhdanov, S. I., and M. I. Polievktov, *J. Gen. Chem. USSR* (*English Transl.*) *31*, 3607(1961).
175. Prévost, C. A., P. Souchay, and J. Chauvelier, *Bull. Soc. Chim. France*, *1951*, 714.
176. Beckmann, P., *Australian J. Chem.*, *14*, 229 (1961).
177. Bartek, J., T. Mukai, T. Nozoe, and F. Šantavý, *Collection Czech. Chem. Commun.*, *19*, 885 (1954).
178. James, J. C., and J. C. Speakman, *Trans. Faraday Soc.*, *48*, 474 (1952).
179. Neish, W. J. P., and O. H. Müller, *Rec. Trav. Chim.*, *72*, 301 (1953).
180. Jambor, B., J. Bartek, and F. Šantavý, *Collection Czech. Chem. Commun.*, *20*, 1244 (1955).
181. Semerano, G., and A. Chisini, *Gazz. Chim. Ital.*, *66*, 504 (1936).
182. Doskočil, J., and M. Vondráček, *Chem. Listy*, *46*, 564 (1952); through *Chem. Abstr.*, *47*, 4947i.
183. Krestýnová-Tělupilová, O., V. Mačak, and F. Šantavý, *Collection Czech. Chem. Commun.*, *19*, 234 (1954).
184. Elving, P. J., and P. G. Grodzka, *Anal. Chem.*, *33*, 2 (1961).
185. Veselý, K., and R. Brdička, *Collection Czech. Chem. Commun.*, *12*, 313 (1947).
186. Bieber, R., and G. Trümpler, *Helv. Chim. Acta*, *30*, 706 (1947).
187. Bieber, R., and G. Trümpler, *Helv. Chim. Acta*, *30*, 1109 (1947).
188. Valenta, P., *Collection Czech. Chem. Commun.*, *25*, 853 (1960).
189. Bieber, R., and G. Trümpler, *Helv. Chim. Acta*, *30*, 2000 (1947); *31*, 5 (1948).
190. Elving, P. J., and C. E. Bennett, *J. Am. Chem. Soc.*, *76*, 1412 (1954).
191. Kirrman, A., J.-M. Savéant, and N. Moe, *Compt. Rend.*, *253*, 1106 (1961).
192. Wiesner, K., *Collection Czech. Chem. Commun.*, *12*, 64 (1947).
193. Brdička, R., *Collection Czech. Chem. Commun.*, *12*, 212 (1947).
194. Becker, M., and H. Strehlow, *Z. Elektrochem.*, *64*, 129, 813 (1960).
195. Los, J. M., and N. J. Gaspar, *Z. Elektrochem.*, *64*, 41 (1960).
196. Heyrovský, M., *Collection Czech. Chem. Commun.*, *28*, 26 (1963).
197. Przhiyalgovskaya, N. M., L. N. Lavrishcheva, G. T. Mondodoev, and V. N. Belov, *J. Gen. Chem. USSR* (*English Transl.*), *31*, 2163 (1961).
198. Lund, H., *Acta Chem. Scand.*, *17*, 972 (1963).
199. Lund, H., *Acta Chem. Scand.*, *17*, 2325 (1963).
200. Lund, H., *Collection Czech. Chem. Commun.*, *25*, 3313 (1960).
201. Kane, P. O., *Z. Anal. Chem.*, *173*, 50 (1960).
202. Arthur, P., and H. Lyons, *Anal. Chem.*, *24*, 1422 (1952).
203. Wawzonek, S., S. C. Wang, and P. Lyons, *J. Org. Chem.*, *15*, 593 (1950).
204. Dineen, E., T. C. Schwan, and C. L. Wilson, *Trans. Electrochem. Soc.*, *96*, 226 (1949).

205. Parravano, G., *J. Am. Chem. Soc.*, *73*, 628 (1951).
206. Kern, W., and H. Quast, *Makromol. Chem.*, *10*, 202 (1953).
207. Bargain, M., *Compt. Rend.*, *254*, 130 (1962); *255*, 1948 (1962); *256*, 1990 (1963).
208. Crain, A. V. R., B. S. Thesis, Polytechnic Institute of Brooklyn, 1957 through *J. Am. Chem. Soc.*, *83*, 58 (1961).
209. Missan, S. R., E. I. Becker, and L. Meites, *J. Am. Chem. Soc.*, *83*, 58 (1961).
210. Elving, P. J., and C. Teitelbaum, *J. Am. Chem. Soc.*, *71*, 3916 (1949).
211. Wessely, F., and J. Wratil, *Mikrochem. Mikrochim. Acta*, *33*, 248 (1947).
212. Rosenthal, I., J. R. Hayes, A. J. Martin, and P. J. Elving, *J. Am. Chem. Soc.*, *80*, 3050 (1958).
213. Miller, D. M., *Can. J. Chem.*, *33*, 1806 (1955).
214. Seo, E. T., and T. Kuwana, *J. Electroanal. Chem.*, *6*, 417 (1963).
215. Maturová, M., A. Němečková, and F. Šantavý, *Collection Czech. Chem. Commun.*, *27*, 1021 (1962); A. Němečková, M. Maturová, M. Pergál, and F. Šantavý, *Collection Czech. Chem. Commun.*, *26*, 2749 (1961).
216. Fierz-David, H. E., L. Blangey, and M. Uhlig, *Helv. Chim. Acta*, *32*, 1414 (1949).
217. Fencl, Z., *Collection Czech. Chem. Commun.*, *19*, 1339 (1954).
218. Whitnack, G. C., J. Reinhart, and E. St. C. Gantz, *Anal. Chem.*, *27*, 359 (1955).
219. Ryvolová, A., and V. Hanuš, *Collection Czech. Chem. Commun.*, *21*, 853 (1956).
220. Peover, M. E., *Trans. Faraday Soc.*, *58*, 2370 (1962).
221. Tirouflet, J., R. Robin, and M. Guyard, *Bull. Soc. Chim. France*, *1956*, 571; J. Tirouflet, and R. Dabard, *Compt. Rend.*, *242*, 2839 (1956).
222. Allen, M. J., and J. Ocampo, *J. Electrochem. Soc.*, *103*, 452 (1956).
223. Abrahamson, E. A., *J. Am. Chem. Soc.*, *81*, 3692 (1959).
224. Okubo, T., and S. Tsutsumi, *Technol. Rept. Osaka Univ.*, *13*, 495 (1963); through *Chem. Abstr.*, *61*, 6637e.
225. Rieger, P. H., I. Bernal, and G. K. Fraenkel, *J. Am. Chem. Soc.*, *83*, 3918 (1961).
226. Acker, D. S., and W. R. Hertler, *J. Am. Chem. Soc.*, *84*, 3370 (1962).
227. Melby, L. R., R. J. Horder, W. R. Hertler, W. Mahler, R. E. Berson, and W. E. Mochel, *J. Am. Chem. Soc.*, *84*, 3374 (1962).
228. Peover, M. E., *Proc. Chem. Soc.*, *1963*, 167; *Trans. Faraday Soc.*, *60*, 417 (1964).
229. Sevast'yanova, I. G., and A. P. Tomilov, *J. Gen. Chem. USSR (English Transl.)*, *33*, 2741 (1963).
230. Rieger, P. H., I. Bernal, W. H. Reinmuth, and G. K. Fraenkel, *J. Am. Chem. Soc.*, *85*, 683 (1963).
231. Laviron, E., *Compt. Rend.*, *250*, 3671 (1960).
232. Volke, J., and J. Holubek, *Collection Czech. Chem. Commun.*, *28*, 1597 (1963).
233. Schmidt, H., and J. Noack, *Z. anorg. allgem. Chem.*, *296*, 262 (1958).
234. Zuman, P., *Nature*, *165*, 485 (1950).
235. Lund, H., *Acta Chem. Scand.*, *13*, 249 (1959).
236. Coufalík, E., and F. Šantavý, *Collection Czech. Chem. Commun.*, *19*, 457 (1954).

237. Souchay, P., and S. Ser, *J. Chim. Phys.*, *49*, C172 (1952).
238. Gardner, H. J., and W. P. Georgans, *J. Chem. Soc.*, *1956*, 4180.
239. Volke, J., R. Kubíček, and F. Šantavý, *Collection Czech. Chem. Commun.*, *25*, 871 (1960).
240. Lund, H., *Acta Chem. Scand.*, *18*, 563 (1964).
241. Spritzer, M., and L. Meites, *Anal. Chim. Acta*, *26*, 58 (1962); R. I. Gelb and L. Meites, *J. Phys. Chem.*, *68*, 2599 (1964).
242. Sartori, G., and A. Gaudiano, *Gazz. Chim. Ital.*, *78*, 77 (1948).
243. Prelog, V., and O. Häfliger, *Helv. Chim. Acta*, *32*, 2088 (1949).
244. Tirouflet, J., *Advan. Polarog.*, *2*, 740 (1960).
245. Kitaev, Yu. P., and A. E. Arbusov, *Bull. Acad. Sci. USSR, Div. Chem. Sci.* (*English Transl.*), *1957*, 1068; *1960*, 1306; A. E. Arbusov and Yu. P. Kitaev, *Proc. Acad. Sci. USSR* (*English Transl.*), *113*, 243 (1957); Yu. P. Kitaev, G. K. Budnikov, and A. E. Arbusov, *Bull. Acad. Sci. USSR, Div. Chem. Sci.* (*English Transl.*), *1961*, 1132, 1652; Yu. P. Kitaev and T. V. Troepol'skaya, *Bull. Acad. Sci. USSR, Chem. Sci.* (*English Transl.*), *1963*, 408, 418.
246. O'Connor, R., *J. Org. Chem.*, *26*, 4375 (1961).
247. Jámbor, B., and L. Mester, *Acta Chim. Acad. Sci. Hung.*, *9*, 485 (1956).
248. Seagers, W. J., and P. J. Elving, *Proc. First Intern. Congr. Polarog. Prague*, *1*, 281 (1951).
249. Suzuki, M., *J. Electrochem. Soc. Japan*, *22*, 112 (1954); through *Chem. Abstr.*, *48*, 13472f.
250. Kastening, B., *Naturwiss.*, *47*, 443 (1960).
251. Holleck, L., and H. J. Exner, *Z. Elektrochem.*, *56*, 46 (1952).
252. Geske, D. H., and A. H. Maki, *J. Am. Chem. Soc.*, *82*, 2671 (1960).
253. Kemula, W., and R. Sioda, *J. Electroanal. Chem.*, *7*, 233 (1964).
254. Levin, É. S., *Proc Acad. Sci. USSR, Phys. Chem. Sect.* (*English Transl.*) *151*, 734 (1963).
255. Holleck, L., and H. J. Exner, *Z. Elektrochem.*, *56*, 677 (1952).
256. Holleck, L., and H. Schmidt, *Z. Elektrochem.*, *59*, 56 (1955).
257. Schmidt, H., and L. Holleck, *Z. Elektrochem.*, *59*, 531 (1955).
258. Holleck, L., and G. Perret, *Z. Elektrochem.*, *59*, 114 (1955).
259. Hartley, A. M., and R. E. Visco, *Anal. Chem.*, *35*, 1871 (1963).
260. Bellavita, V., N. Fedi, and N. Cagnoli, *Ric. Sci.*, *25*, 504 (1955).
261. Runner, M. E., and E. C. Wagner, *J. Am. Chem. Soc.*, *74*, 2529 (1952).
262. Meites, L., and T. Meites, *Anal. Chem.*, *28*, 103 (1956).
263. Astle, M. J., and S. P. Stephenson, *J. Am. Chem. Soc.*, *65*, 2399 (1943).
264. Petrů, F., *Collection Czech. Chem. Commun.*, *12*, 620 (1947).
265. Radin, N., and T. De Vries, *Anal. Chem.*, *24*, 971 (1952).
266. De Vries, T., and R. W. Ivett, *Ind. Eng. Chem. Anal. Ed.*, *13*, 339 (1941).
267. Seagers, W. J., and P. J. Elving, *J. Am. Chem. Soc.*, *72*, 5183 (1950).
268. Deswarte, S., and J. Armand, *Compt. Rend.*, *258*, 3865 (1964).
269. Masui, M., and H. Sayo, *Pharm. Bull.* (*Tokyo*), *4*, 332 (1956); M. Masui, H. Sayo, and Y. Nomura, *ibid.*, 337; through *Chem. Abstr.*, *51*, 9370ab.
270. Masui, M., and H. Sayo, *Yakugaku Zasshi*, *78*, 703 (1958); through *Chem. Abstr.*, *52*, 16089g.

271. Stock, J. T., *J. Chem. Soc., 1957*, 4532.
272. Masui, M., and H. Sayo, *J. Chem. Soc., 1961*, 4773.
273. Mausi, M., and H. Sayo, *J. Chem. Soc., 1961*, 5325.
274. Masui, M., and H. Sayo, *J. Chem. Soc., 1962*, 1733.
275. Kaufman, F., H. J. Cook, and S. M. Davis, *J. Am. Chem. Soc., 74*, 4997 (1952).
276. Whitnack, G. C., J. M. Nielsen, and E. St. C. Gantz, *J. Am. Chem. Soc., 76*, 4711 (1954).
277. Pasternak, R., and H. von Halban, *Helv. Chim. Acta, 29*, 190 (1946).
278. Elving, P. J., and C.-S. Tang, *J. Am. Chem. Soc., 72*, 3244 (1950).
279. Elving, P. J., I. Rosenthal, J. R. Hayes, and A. J. Martin, *Anal. Chem., 33*, 330 (1961).
280. Marple, L. W., L. E. I. Hummelstedt, and L. B. Rogers, *J. Electrochem. Soc., 107*, 437 (1960).
281. Wawzonek, S., R. C. Duty, and J. H. Wagenknecht, *J. Electrochem. Soc., 111*, 74 (1964).
282. Hush, N. S., and K. B. Oldham, *J. Electroanal. Chem., 6*, 34 (1963).
283. Rosenthal, I., C. H. Albright, and P. J. Elving, *J. Electrochem. Soc., 99*, 227 (1952).
284. Elving, P. J., J. M. Markowitz, and I. Rosenthal, *J. Electrochem. Soc., 101*, 195 (1954).
285. Lambert, F. L., and K. Kobayashi, *Chem. Ind. (London), 1958*, 949.
286. Elving, P. J., J. M. Markowitz, and I. Rosenthal, *Chem. Ind. (London), 1959*, 1192.
287. Lambert, F. L., and K. Kobayashi, *J. Am. Chem. Soc., 82*, 5324 (1960).
288. Sease, J. W., P. Chang, and J. L. Groth, *J. Am. Chem. Soc., 86*, 3154 (1964).
289. Lambert, F. L., A. H. Albert, and J. P. Hardy, *J. Am. Chem. Soc., 86*, 3155 (1964).
290. Saito, E., *Bull. Soc. Chim. France, 1948*, 404.
291. Gergely, E., and T. Iredale, *J. Chem. Soc., 1951*, 13, 3502.
292. Elving, P. J., and R. E. Van Atta, *Anal. Chem., 26*, 295 (1954); *27*, 1908 (1955).
293. Schwabe, K., and H. Berg, *Z. Elektrochem., 56*, 952 (1952).
294. Elving, P. J., and R. E. Van Atta, *J. Electrochem. Soc., 103*, 676 (1956).
295. Wilson, A. M., and N. L. Allinger, *J. Am. Chem. Soc., 83*, 1999 (1961).
296. Kabasakalian, P., and J. McGlotten, *Anal. Chem., 34*, 1440 (1962).
297. Elving, P. J., and J. T. Leone, *J. Am. Chem. Soc., 79*, 1546 (1957).
298. Elving, P. J., and J. T. Leone, *J. Am. Chem. Soc., 82*, 5076 (1960).
299. Armand, J., *Compt. Rend., 254*, 2777 (1962).
300. Kolthoff, I. M., T. S. Lee, D. Stočesova, and E. P. Parry, *Anal. Chem., 22*, 521 (1950).
301. Elving, P. J., I. Rosenthal, and M. K. Kramer, *J. Am. Chem. Soc., 73*, 1717 (1951).
302. Rosenthal, I., C.-S. Tang, and P. J. Elving, *J. Am. Chem. Soc., 74*, 6112 (1952).
303. Meites, T., and L. Meites, *Anal. Chem., 27*, 1531 (1955).
304. Elving, P. J., and C. E. Bennett, *J. Am. Chem. Soc., 76*, 4473 (1954).

305. Elving, P. J., and C. E. Bennett, *J. Electrochem. Soc.*, *101*, 520 (1954).
306. Wawzonek, S., and R. C. Duty, *J. Electrochem. Soc.*, *108*, 1135 (1961).
307. Lund, H., *Acta Chem. Scand.*, *13*, 192 (1959).
308. Cohen, A. I., B. T. Keeler, N. H. Coy, and H. L. Yale, *Anal. Chem.*, *34*, 216 (1962).
309. Schwabe, K., and H. Frind, *Z. physik. Chem.* (*Leipzig*), *196*, 342 (1951).
310. Kemula, W., and A. Cisak, *Roczniki Chem.*, *28*, 275 (1954), through *Chem. Abstr.* *48*, 11217d.
311. Nakazima, M., Y. Katumura, and T. Okubo, *Proc. First Intern. Congr. Polarog. Prague*, *1*, 173 (1951).
312. Neĭman, M. B., A. V. Ryabov, and E. M. Sheyanova, *Dokl. Akad. Nauk SSSR*, *68*, 1065 (1949); through *Chem. Abstr.*, *44*, 1360e.
313. Rosenthal, I., G. J. Frisone, and R. J. Lacoste, *Anal. Chem.*, *29*, 1639 (1957).
314. Feoktistov, L. G., A. P. Tomilov, and M. M. Gol'din, *Izv. Akad. Nauk SSSR, Otdel. Khim. Nauk*, *1963*, 1352; through *Chem. Abstr.*, *59*, 12624f.
315. Lund, H., *Acta Chem. Scand.*, *17*, 2139 (1963).
316. Elving, P. J., I. Rosenthal, and A. J. Martin, *J. Am. Chem. Soc.*, *77*, 5218 (1955).
317. McKeon, M. G., *J. Electroanal. Chem.*, *3*, 402 (1962).
318. Page, J. E., *Proc. First Intern. Congr. Polarog. Prague*, *1*, 193 (1951).
319. Elving, P. J., and C. L. Hilton, *J. Am. Chem. Soc.*, *74*, 3368 (1952).
320. Kitagawa, T., T. P. Layloff, and R. N. Adams, *Anal. Chem.*, *35*, 1086 (1963).
321. Březina, J., and J. Heyrovský, *Collection Czech. Chem. Commun.*, *8*, 114 (1936).
322. Mairanovskii, S. G., and L. D. Bergel'son, *Russ. J. Phys. Chem.* (*English Transl.*), *34*, 112 (1960).
323. Jura, W. H., and R. J. Gaul, *J. Am. Chem. Soc.*, *80*, 5402 (1958).
324. Wawzonek, S., and J. H. Wagenknecht, *J. Electrochem. Soc.*, *110*, 420 (1963).
325. Colichman, E. L., and H. P. Maffei, *J. Am. Chem. Soc.*, *74*, 2744 (1952); Colichman, E. L., and J. T. Matschiner, *J. Org. Chem.*, *18*, 1124 (1953).
326. Bachofner, H. E., F. M. Beringer, and L. Meites, *J. Am. Chem. Soc.*, *80*, 4269, 4274 (1958).
327. Fornasari, E., M. Scarpa, and P. Lanza, *Atti accad. Nazl. Lincei, Rend., Classe sci. fis. mat. e nat.*, *14*, 70, 272 (1953); through *Chem. Abstr.*, *47*, 9178hi.
328. Colichman, E. L., *Anal. Chem.*, *25*, 1124 (1953).
329. Lund, H., *Acta Chem. Scand.*, *14*, 359 (1960).
330. Šantavý, F., and M. Černoch, *Chem. Listy*, *46*, 81 (1951); through *Chem. Abstr.*, *46*, 11582b.
331. Ohloff, G., *Arch. Pharm.*, *286*, 242 (1953).
332. Fedoronko, M., *Chem. Zvesti*, *12*, 17 (1958); through *Progress in Polarography*, Vol. I, P. Zuman and I. M. Kolthoff, eds., Interscience, New York, 1962, p. 304.
333. Kabasakalian, P., and J. McGlotten, *Anal. Chem.*, *31*, 1091 (1959).
334. Lund, H., *Acta Chem. Scand.*, *14*, 1927 (1960).
335. Wawzonek, S., and J. D. Fredrickson, *J. Electrochem. Soc.*, *106*, 325 (1959).
336. Given, P. H., and M. E. Peover, *Nature*, *184*, 1064 (1959); P. H. Given, *J. Chem. Soc.*, *1958*, 2684.

337. Tang, S., and P. Zuman, *Collection Czech. Chem. Commun.*, *28*, 829, 1524 (1963); P. Zuman and V. Horák, *Advan. Polarog.*, *3*, 804 (1960); *Collection Czech. Chem. Commun.*, *26*, 176 (1961).
338. Colichman, E. L., and D. L. Love, *J. Org. Chem.*, *18*, 40 (1953).
339. Schwabe, K., and J. Voigt, *Z. Elektrochem.*, *56*, 44 (1952).
340. Southworth, B. C., R. Osteryoung, K. D. Fleischer, and F. C. Nachod, *Anal. Chem.*, *33*, 208 (1961).
341. Leonard, N. J., S. Swann, Jr., and co-workers, *J. Am. Chem. Soc.*, *74*, 2871, 4620, 6251 (1952).
342. Mantsavinos, R., and J. E. Christian, *Anal. Chem.*, *30*, 1071 (1958).
343. Wagner, H., and H. Berg, *J. Electroanal. Chem.*, *2*, 452 (1961).
344. Coombs, D. M., and L. L. Leveson, *Anal. Chim. Acta*, *30*, 209 (1964).
345. Mayell, J. S., and A. J. Bard, *J. Am. Chem. Soc.*, *85*, 421 (1963).
346. Grovenstein, E., Jr., and co-workers, *J. Am. Chem. Soc.*, *81*, 4842, 4850 (1959); *86*, 854 (1964).
347. Bitter, B., *Proc. First Intern. Congr. Polarog. Prague*, *1*, 771 (1951).
348. Willits, C. O., C. Ricciuti, H. B. Knight, and D. Swern, *Anal. Chem.*, *24*, 785 (1952).
349. Brüschweiler, H., and G. J. Minkoff, *Anal. Chim. Acta*, *12*, 186 (1955).
350. Skoog, D. A., and A. B. H. Lauwzecha, *Anal. Chem.*, *28*, 825 (1956).
351. Kuta, E. J., and F. W. Quackenbush, *Anal. Chem.*, *32*, 1069 (1960).
352. Swern, D., and L. S. Silbert, *Anal. Chem.*, *35*, 880 (1963).
353. Kofod, H., *Acta Chem. Scand.*, *9*, 455 (1955).
354. Lund, H., *Acta Chem. Scand.*, *17*, 1077 (1963).
355. Heller, K., and E. N. Jenkins, *Nature*, *158*, 706 (1946).
356. von Sturm, F., and W. Hans, *Angew. Chem.*, *67*, 743 (1955).
357. Zuman, P., ref. 3, p. 223.
358. Volke, J., *Acta Chim. Acad. Sci. Hung.*, *9*, 223 (1956).
359. Knobloch, E., *Collection Czech. Chem. Commun.*, *12*, 407 (1947).
360. Tompkins, P. C., and C. L. A. Schmidt, *Univ. Calif. Pub. Physiol.*, *8*, 237, 247 (1944); through *Polarography*, I. M. Kolthoff and J. J. Lingane, Interscience, New York, 1952, p. 815ff.
361. Leach, S. J., J. H. Baxendale, and M. G. Evans, *Australian J. Chem.*, *6*, 395 (1953).
362. Colichman, E. L., and P. A. O'Donovan, *J. Am. Chem. Soc.*, *76*, 3588 (1954).
363. Mairanovskii, S. G., *Proceedings of the Fourth Congress Electrochemistry, Moscow (1956)*, Consultants Bureau, New York, Vol. 1, 1961, 289.
364. Carruthers, C., and V. Suntzeff, *Arch. Biochem. Biophys.*, *45*, 140 (1953); C. Carruthers and J. Tech, *ibid.*, *56*, 441 (1955).
365. Ke, B., *Biochim. Biophys. Acta*, *20*, 547 (1956).
366. Zhdanov, S. I., and L. S. Mirkin, *Collection Czech. Chem. Commun.*, *26*, 370 (1961).
367. Zahlan, A. B., and R. H. Linnell, *J. Am. Chem. Soc.*, *77*, 6207 (1955).
368. Elofson, R. M., and R. L. Edsberg, *Can. J. Chem.*, *35*, 646 (1957).
369. Schwarz, W. M., E. M. Kosower, and I. Shain, *J. Am. Chem. Soc.*, *83*, 3164 (1961).
370. Foffani, A., and E. Fornasari, *Gazz. Chim. Ital.*, *83*, 1051 (1953).

371. Horn, G., *Acta Chim. Acad. Sci. Hung.*, *27*, 123 (1961).
372. Feldman, M., and S. Winstein, *Tetrahedron Letters*, *1962*, 853.
373. Balaban, A. T., C. Bratu, and C. N. Rentea, *Tetrahedron*, *20*, 265 (1964).
374. Smith, D. L., and P. J. Elving, *J. Am. Chem. Soc.*, *84*, 1412, 2741 (1962).
375. Struck, W. A., and P. J. Elving, *J. Am. Chem. Soc.*, *86*, 1229 (1964).
376. Voorhies, J. D., and E. J. Schurdak, *Anal. Chem.*, *34*, 939 (1962).
377. Clauson-Kaas, N., F. Limborg, and K. Glens, *Acta Chem. Scand.*, *6*, 531 (1952).
378. Hayashi and Wilson, through ref. 7, p. 164.
379. Lingane, J. J., C. G. Swain, and M. Fields, *J. Am. Chem. Soc.*, *65*, 1348 (1943).
380. Breyer, B., G. S. Buchanan, and H. Duewell, *J. Chem. Soc.*, *1944*, 360.
381. Kaye, R. C., and H. I. Stonehill, *J. Chem. Soc.*, *1951*, 27.
382. Stock, J. T., *J. Chem. Soc.*, *1949*, 763; *Proc. First Intern. Congr. Polarog. Prague*, *1*, 371 (1951).
383. Sartori, G., and C. Furlani, *Ann. Chim. (Rome)*, *45*, 251 (1955).
384. Hausser, K. H., A. Häbisch, and V. Franzen, *Z. Naturforsch.*, *16A*, 836 (1961).
385. Frost, J. G., and J. H. Saylor, *Rec. Trav. Chim.*, *82*, 828 (1963).
386. Vajda, M., *Advan. Polarog.*, *2*, 786 (1960).
387. Zuman, P., *Z. physik. Chem.*, July, *1958* (Sonderheft), 243; *Collection Czech. Chem. Commun.*, *25*, 3245 (1960).
388. Kabasakalian, P., and J. McGlotten, *Anal. Chem.*, *31*, 431 (1959).
389. Billon, J.-P., G. Cauquis, J. Combrisson, and A.-M. Li, *Bull. Soc. Chim. France*, *1960*, 2062.
390. Billon, J.-P., *Ann. Chim. (Paris)*, *7*, 183 (1962); *Bull. Soc. Chim. France*, *1961*, 1923.
391. Kolthoff, I. M., W. Stricks, and N. Tanaka, *J. Am. Chem. Soc.*, *77*, 4739 (1955).
392. Maĭranovskiĭ, S. G., and M. B. Neĭman, *Dokl. Akad. Nauk SSSR*, *79*, 85 (1951); *87*, 805 (1952); through *Chem. Abstr.*, *46*, 28e; *47*, 6271e.
393. Barnard, D., M. B. Evans, G. M. C. Higgins, and J. F. Smith, *Chem. Ind.*, *1961*, 20.
394. Bowers, R. C., and H. D. Russell, *Anal. Chem.*, *32*, 405 (1960).
395. Nicholson, M. M., *J. Am. Chem. Soc.*, *76*, 2539 (1954); *Anal. Chem.*, *27*, 1364 (1955).
396. Drushel, H. V., and J. F. Miller, *Anal. Chem.*, *29*, 1459 (1956).
397. Levin, E. S., and A. P. Shestov, *Dokl. Akad. Nauk SSSR*, *96*, 999 (1954); through *Chem. Abstr.*, *48*, 13471e.
398. Johnson, C. W., C. G. Overberger, and W. J. Seagers, *J. Am. Chem. Soc.*, *75*, 1495 (1953).
399. Kolthoff, I. M., and W. Stricks, *J. Am. Chem. Soc.*, *73*, 1728 (1951).
400. Kolthoff, I. M., and N. Tamberg, *J. Polarog. Soc.*, *1958*, No. 3, 54; through *Chem. Abstr.*, *53*, 13841f.
401. Benesch, R., and R. E. Benesch, *J. Am. Chem. Soc.*, *73*, 3391 (1951); *J. Phys. Chem.*, *56*, 648 (1952).
402. Leach, S. J., *Australian J. Chem.*, *13*, 520 (1960).

403. Riccoboni, L., *Gazz. Chim. Ital.*, *72*, 47 (1942).
404. Costa, G., *Ann. Chim. (Rome)*, *40*, 541 (1950).
405. Costa, G., *Gazz. Chim. Ital.*, *80*, 42 (1950).
406. Allen, R. B., *Dissertation Abstr.*, *20*, 897 (1959).
407. Costa, G., *Ann. Chim. (Rome)*, *40*, 559 (1950).
408. Geske, D. H., *J. Phys. Chem.*, *63*, 1062 (1959).
409. Wilkinson, G., and co-workers, *J. Am. Chem. Soc.*, *74*, 6149 (1952); *75*, 3586 (1953); *76*, 1970, 4281 (1954).
410. Tirouflet, J., E. Laviron, R. Dabard, and J. Komenda, *Bull. Soc. Chim. France, 1963*, 857.
411. Furlani, C., and E. O. Fischer, *Z. Elektrochem.*, *61*, 481 (1957); C. Furlani and G. Sartori, *Ric. Sci.*, *28*, 973 (1958).
412. Zuman, P., *Collection Czech. Chem. Commun.*, *19*, 599 (1954); *25*, 3225 (1960); *27*, 2035 (1962); *Advan. Polarog.*, *3*, 812 (1960).
413. Klopman, G., *Helv. Chim. Acta*, *44*, 1908 (1961).
414. Streitwieser, A., Jr., and C. Perrin, *J. Am. Chem. Soc.*, *86*, 4938 (1964).
415. Taft, R. W., Jr., in *Steric Effects in Organic Chemistry*, M. S. Newman, ed., Wiley, New York, 1956.
416. Elving, P. J., and J. M. Markowitz, *J. Org. Chem.*, *25*, 18 (1960).
417. Bennett, C. E., and P. J. Elving, *Collection Czech. Chem. Commun.*, *25*, 3213 (1960).
418. Fuson, N., M.-L. Josien, and E. M. Shelton, *J. Am. Chem. Soc.*, *76*, 2526 (1954).
419. Evans, M. G., and N. S. Hush, *J. Chim. Phys.*, *49*, C159 (1952).
420. Berzins, T., and P. Delahay, *J. Am. Chem. Soc.*, *75*, 5716 (1953).
421. Streitwieser, A., Jr., *Molecular Orbital Theory for Organic Chemists*, Wiley, New York, 1961, p. 173ff.
422. Pysh, E. S., and N. C. Yang, *J. Am. Chem. Soc.*, *85*, 2124 (1963).
423. Watson, A. T., and F. A. Matsen, *J. Chem. Phys.*, *18*, 1305 (1950).
424. Bergman, I., *Trans. Faraday Soc.*, *50*, 829 (1954); *52*, 690 (1956).
425. Neikam, W. C., G. R. Dimeler, and M. M. Desmond, *J. Electrochem. Soc.*, *111*, 1190 (1964).
426. Schmid, R. W., and E. Heilbronner, *Helv. Chim. Acta*, *37*, 1453 (1954).
427. Evans, M. G., J. Gergely, and J. de Heer, *Trans. Faraday Soc.*, *45*, 312 (1949).
428. Rieger, P. H., and G. K. Fraenkel, *J. Chem. Phys.*, *39*, 609 (1963).
429. Fueno, T., K. Asada, K. Morokuma, and J. Furukawa, *J. Polymer Sci.*, *40*, 511 (1959).
430. Zuman, P., *Acta Chim. Acad. Sci. Hung.*, *18*, 141 (1959).
431. Fields, M., C. Valle, Jr., and M. Kane, *J. Am. Chem. Soc.*, *71*, 421 (1949).
432. Prévost, C., P. Souchay, and C. Malen, *Bull. Soc. Chim. France, 1953*, 78.
433. Gerdil, R., and E. Heilbronner, *Helv. Chim. Acta*, *40*, 141 (1957).
434. Young, J. R., *J. Chem. Soc.*, *1955*, 1516.
435. Goward, G. W., C. E. Bricker, and W. C. Wildman, *J. Org. Chem.*, *20*, 378 (1955).
436. van Zanten, B., and W. T. Nauta, *Rec. Trav. Chim.*, *80*, 181 (1961).

437. Geske, D. H., J. L. Ragle, M. A. Bambenek, and A. L. Balch, *J. Am. Chem. Soc.*, *86*, 987 (1964).
438. Zahradník, R., and K. Boček, *Collection Czech. Chem. Commun.*, *26*, 1733 (1961).
439. Wolfe, J. K., E. B. Hershberg, and L. F. Fieser, *J. Biol. Chem.*, *136*, 653 (1940).
440. Huisgen, R., W. Rapp, I. Ugi, H. Walz, and E. Mergenthaler, *Ann.* *586*, 1 (1954).
441. Page, J. E., J. W. Smith, and J. G. Waller, *J. Phys. & Colloid Chem.*, *53*, 545 (1949).
442. Runner, M. E., M. L. Kilpatrick, and E. C. Wagner, *J. Am. Chem. Soc.*, *69*, 1406 (1947); M. E. Runner, *ibid.*, *74*, 3567 (1952).
443. Winkel, A., and G. Proske, *Ber.*, *69*, 1917 (1936).
444. Zuman, P., ref. 2, pp. 84–99.
445. Elving, P. J., in *Progress in Polarography*, Vol. II, P. Zuman and I. M. Kolthoff, eds., Interscience, New York, 1962, p. 625.
446. Zuman, P., *Collection Czech., Chem. Commun.*, *15*, 1107 (1950).
447. E. H. Sargent Co., *Bibliography of Polarographic Literature*, 1922–1955.
448. Allen, M. J., and A. H. Corwin, *J. Am. Chem. Soc.*, *72*, 114 (1950).
449. Levine, H. A., and M. J. Allen, *J. Chem. Soc.*, *1952*, 254.
450. Le Guyader, M., and Y. Mercier, *Compt. Rend.*, *257*, 1307 (1963); M. Le Guyader and M. Le Demezet, *ibid.*, *258*, 3046 (1964).
451. Weedon, B. C. L., in *Advances in Organic Chemistry*, Vol. I, R. A. Raphael, E. C. Taylor, and H. Wynberg, eds., Interscience, New York, 1960, p. 1.
452. Adams, R. N., *J. Electroanal. Chem.*, *8*, 151 (1964).
453. Semerano, G., *Proc. First Intern. Congr. Polarog. Prague*, *1*, 300 (1951).
454. Zuman, P., *Proc. Intern. Symp. Microchem., Birmingham (1958)*, Pergamon, New York, 1960, p. 294.
455. Klopman, G., and R. F. Hudson, *Helv. Chim. Acta*, *44*, 1914 (1961).
456. Bartlett, P. D., and T. G. Traylor, *J. Am. Chem. Soc.*, *83*, 856 (1961).
457. Moulton, W. N., R. E. Van Atta, and R. R. Ruch, *J. Org. Chem.*, *26*, 290 (1961).
458. Lindberg, B., *Acta Chem. Scand.*, *17*, 393 (1963).
459. Dawson, B. E., and T. Henshall, *J. Phys. Chem.*, *67*, 1187 (1963).
460. Žemlička, J., J. Krupička, and Z. Arnold, *Collection Czech. Chem. Commun.*, *27*, 2464 (1962).
461. Horák, V., and P. Zuman, *Tetrahedron Letters*, *1961*, 746.

Physical Organic Polarography

By P. Zuman

J. Heyrovský Institute of Polarography, Czechoslovak Academy of Sciences, Prague, Czechoslovakia

CONTENTS

I. Introduction

Polarography is an electrochemical method based on electrolysis with a dropping mercury electrode or another type of mercury electrode with a perpetually renewed surface. The current flowing during this electrolytic process is recorded as a function of the applied potential. In the presence of a substance that can undergo reduction or oxidation at the surface of the mercury electrode, this current–voltage curve, called a polarographic curve, shows an increasing current with increasing applied potential (Fig. 1). This stepwise increase of current is called the polarographic wave. In addition to information that can be obtained from polarographic curves, which is of interest for detailed electrochemical studies, there are two quantities of importance in the application of polarography to the solution of various chemical problems. These two quantities are: (*a*) the limiting current, i_{lim} corresponding to the increase in current due to the electrolytic process and measured as the wave height (as shown in Fig. 1), and (*b*) the half-wave potential $E_{1/2}$, i.e., the potential at that point of the polarographic curve where the current reaches half of its limiting value.

The height of the limiting current is usually limited by the rate of diffusion of the electroactive species to the surface of the electrode, but it can also be affected by adsorption, the rate of a chemical reaction taking place in the vicinity of the surface of the electrode, or catalytic reaction. The diffusion-controlled limiting current, usually called diffusion current, is in most cases linearly proportional to the concentration of the electroactive species in the electrolyzed solution. Hence, the height of the polarographic wave provides us with information concerning the quantitative composition of the solution.

The half-wave potential can be a measure of the free energy, activation or standard, necessary for carrying out the electrolytic process. It depends not only on the nature of the electroactive species but also on the composition of the solution in which the electrolysis is carried out. Keeping constant the composition of the solution containing a supporting electrolyte, often buffered, in addition to the electrolyzed substance, it is possible to compare half-wave potentials of various substances. When the mechanism of the electrode process is similar in all cases compared, the half-wave potential can be considered as a measure of the reactivity of the compound toward the electrode.

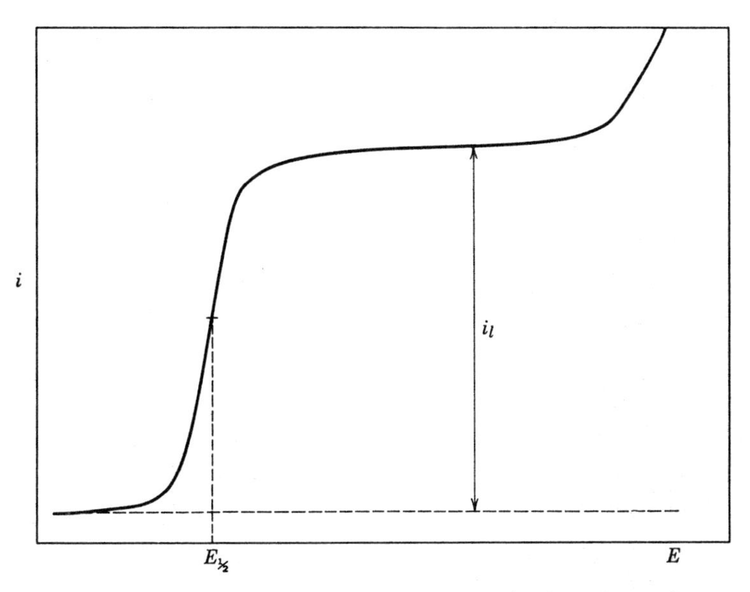

Fig. 1. The shape of a polarographic wave showing the dependence of current i on potential E; half-wave potential $E_{1/2}$ and limiting current i_l.

Hence half-wave potentials qualitatively characterize the electrolyzed compound or the composition of the electrolyzed solution, respectively.

It is useful to mention the analogy with absorptiometry. The limiting current corresponds to absorptivity at a given wavelength in its dependence on concentration and the wavelength of the absorption maximum corresponds to the half-wave potential since both depend on the nature of the substance studied and on the composition of the media in which the measurement is carried out. Because of a relatively smaller available scale, the resolving power of polarography can be considered smaller than that of absorptiometric methods; on the other hand, the strict additivity of limiting currents of various substances, which holds true for most cases in polarography, makes the evaluation of polarographic curves obtained in the electrolysis of mixtures simpler than that of absorption spectra of mixtures.

Polarography is firmly established in the solution of inorganic problems, mainly in analytical applications. The acceptance of polarography for solving theoretical problems of inorganic chemistry can perhaps be explained by the fact that the first important contributions showing the possibilities of such applications appeared during a renaissance of inorganic chemistry. That is, inorganic chemists were looking at this stage for new physical methods and were ready to accept them in their arsenal.

Application of polarography in organic chemistry shows another development. Even as an analytical method, polarographic determinations of organic substances have penetrated rather slowly into academic and industrial laboratories. Moreover, at the time when contributions showing the possibilities of polarography in solving problems of theoretical and physical organic chemistry were just published, organic chemists were concentrating on other physical methods, each of which was supposed during the early stages of its development to solve more or less all the problems of organic chemistry. Even when methods such as infrared spectroscopy, gas chromatography, nuclear magnetic resonance, mass spectrometry, or optical rotatory dispersion showed limitations in addition to their advantages, they remained in the focus of interest of organic chemists. Even though the application of polarography to the solution of theoretical problems of organic chemistry is definitely much more limited than that of the above-mentioned giants, there is still a good number of cases in which, at least in our opinion, polarography can be used with a comparable, and in some cases even greater, success. The possibilities that this technique offers in addition to its purely practical applications are not always recognized by the organic chemist. It is the aim of this paper to indicate some of these possibilities.

Because some aspects of applications of polarography to the study of organic compounds have already been discussed in one of the previous volumes of this series (1), we shall restrict our discussion to three points: mechanisms of organic electrode processes, correlations of half-wave potentials with structural parameters, and applications of polarography in the study of organic reactions.

II. Reversibility and Irreversibility of Organic Electrode Processes

The electrode processes can be distinguished as reversible and irreversible. A reversible system is one in which both the rate of the reduction and that of the oxidation process are fast. In irreversible processes, some of the steps of the total process are slow.

The term "fast" or "slow" is of course relative and depends on the method used for the measurement. In the present discussion, these terms will be used in connection with normal polarographic conditions, i.e., a drop-life of the order of 3 sec. and the rate of potential scanning of the order of 200–400 mV./min., providing practically constant potential during the life of a single drop. Under these conditions, it is clear that with a half-time of about 10^{-5} sec., any reaction or process can be considered instantaneous.

The half-wave potentials of reversible processes are given only by the difference in the standard free energy of the oxidized and reduced forms:

$$E_{1/2} = E^\circ - \frac{RT}{nF} \ln \left(\frac{D_{\text{Ox}}}{D_{\text{Red}}}\right)^{1/2} - \frac{RT}{nF} \ln \frac{f_{\text{Red}}}{f_{\text{Ox}}} \tag{1}$$

where R is the gas constant, T is the absolute temperature, n is the number of electrons transferred, F is the Faraday charge, D_{Ox}, and D_{Red} are the diffusion coefficients of the oxidized and reduced forms, respectively, and f_{Ox} and f_{Red} are the activity coefficients. At a given ionic strength at which the activity coefficients are constant and the diffusion coefficients of the oxidized and reduced forms are practically equal

$$E_{1/2} = E^\circ \tag{2a}$$

$$-E_{1/2} = \Delta G^\circ/nF \tag{2b}$$

Polarographic study of reversible systems allows us to make deductions concerning the state of the oxidized and reduced forms e.g., protonation, dimerization, etc., and about structural effects on the standard free energy difference of these two states. But, as for the measurement of equilibrium constants, no information concerning the course of the electrode process can be drawn from measurement of half-wave potentials of polarographically reversible systems.

The half-wave potential of an irreversible process depends, on the other hand, on the free activation energy of the potential-determining step of the sequence. For the simplest case of the so-called slow electrode process when

$$\mathrm{Ox} + ne \xrightarrow{k_e^\circ} \mathrm{Red}$$

where k_e° is the heterogeneous rate constant of the electrode process at the standard potential, the half-wave potential is given by expression (3):

$$E_{1/2} = E^\circ - (2.3RT/\alpha nF)\log 0.886\ k_e^\circ(t_1/D)^{1/2} \quad (3)$$

where α is the transfer coefficient and t_1 is the drop time.

There is a simple relation between the half-wave potential and the change in free activation energy $\Delta G^{\pm}$:

$$E_{1/2} \sim E^\circ - (\Delta G^{\pm}/\alpha n) \log 0.886\ (t_1/D)^{1/2} \quad (4)$$

Hence the half-wave potential of irreversible systems depends on the structure and energy distribution in the transition state of the electrode process, and these half-wave potentials can be treated for the elucidation of the mechanisms of electrode processes in a way similar to determining the rate constants of homogeneous chemical reactions.

Most of the organic electrode processes are irreversible; only quinoid systems, the nitrosobenzene–phenylhydroxylamine and azobenzene–hydrazobenzene systems, some radical formations, and a few others can be reversible under proper experimental conditions. Nevertheless in view of the above, it is essential prior to discussing the course and mechanism of the electrode process to prove its reversibility.*

Three conditions must be fulfilled to call a system polarographically reversible: (*1*) The half-wave potentials of the oxidized and reduced form of the studied system must be identical and practically th

* The character of polarographic limiting currents, i.e., the proof whether they are diffusion, adsorption, reaction, or catalytically controlled is to be discovered first, but this has been thoroughly discussed elsewhere (1–3). Diffusion currents are anticipated in the subsequent discussion when not explicitly stated otherwise.

same as measured by static, e.g., potentiometric, methods. (*2*) The shape of the polarographic curve must correspond to the equation derived for the particular type of electrode processes. (*3*) The change of the instantaneous current with time, the *i–t* curves, must possess the shape theoretically required for the given type of electrode process, not only in the region of the limiting current but also in the potential region corresponding to the rising portion of the polarographic curve and particularly at its lower bend.

Systems that do not fulfill one or more of these conditions are called polarographically irreversible.

Comparison of the half-wave potentials of the oxidized and reduced forms of the studied couple and potentiometric determination of the redox potential are the simplest methods of proof of reversibility. But often such proof is prevented by the fact that one form is not sufficiently stable to be prepared or kept or is not sufficiently stable to enable one to carry out the potentiometric titration. A common reason for such instability is the autoxidation of the reduced form. Hence, when only preparations of the oxidized form are available, the reduced form can be sometimes prepared by a chemical reaction within the polarographic vessel. The application of the familiar reducing agents, such as hydrazine, hydroxylamine, sulfite, or hydrosulfite, is in many cases prevented because these reagents undergo various side reactions with the studied system in addition to the reduction. Platinum or palladium sols saturated with hydrogen can result in catalytically increased anodic waves and their use is therefore limited. Sometimes it is possible, especially for systems reduced at positive potentials, to carry out the reduction by zinc in acid media. But the most successful and general reduction procedure seems to be the addition of platinum asbestos to the solution of the oxidized form of the studied system and subsequent introduction of hydrogen. The half-wave potential of the oxidized form is measured first after deaeration with nitrogen and compared with the half-wave potential of the reduced form measured after introduction of hydrogen into solution. When the half-wave potentials are equal, the system is considered reversible. To exclude the possibility of an irreversible cleavage, introducing oxygen or air into the reduced solution can be recommended. After deaeration with a stream of nitrogen, the wave should be identical with the wave of the oxidized form prior to reduction.

When the reduced form prepared in this way is not sufficiently stable in the given media to allow the recording of a polarographic curve, or if the stability of the oxidized form is too low, it is necessary to use some of the auxiliary methods. The choice depends on the stability of the unstable species.

When the reactive species is insufficiently stable when generated in the solution, i.e., when the half-time of its decay is of the order of minutes, but if it is relatively stable when formed at the surface of the mercury electrode, i.e., when its half-life is of the order of seconds, commutator or stripping methods can be used.

The commutator method (4,5) is based on polarization with a periodically changed rectangular voltage. In its most widely used modification, the voltage tapped off the potentiometer of a polarograph increases regularly and a fixed voltage taken from an auxiliary potentiometer is chosen to be in the region of the limiting current of the studied wave. The polarizing voltage applied to the cell is periodically switched between the regularly increasing voltage source and the auxiliary voltage having a constant value. If the product formed at the auxiliary potential, usually corresponding to the limiting current is capable of exhibiting a polarographic wave, this wave can be observed on the curve recorded with the commutator circuit. The half-wave potential of the wave appearing on the normal i–E polarographic curve is then compared with the half-wave potential of the electrolysis product obtained with the commutator. Identity of half-wave potentials is a proof of reversibility.

The stripping methods are most conveniently carried out using the hanging mercury drop electrode (6–8), but a mercury pool electrode can be used. When the oxidized form is stable and present in the examined solution, the constant surface electrode is first polarized continuously from positive to negative potentials. The voltage scanning is then interrupted for some time at a potential corresponding to the limiting current of the studied wave where the electrolysis product is formed. Then the direction of voltage scanning is reversed and the electrode is polarized from negative to more positive potentials. The "production period" depends on the size of the electrode and the concentration and adsorbability of the electrolysis product. In some cases this period can be very short and in fact the reversal of the direction of polarization can occur "immediately" after arriving at the negative potential corresponding to the limiting current.

The potentials at which the cathodic peak occurs in the forward run is then compared with the potential of the anodic peak recorded during the backward run. If the peak potentials differ only by some few tens of millivolts, the theoretical difference (7) depending on the number of transferred electrons, the system is considered to be polarographically reversible.

For still more short-lived electrolysis products, oscillographic methods, such as the single-sweep methods (9), in which the peak potentials of the cathodic and anodic peak are compared as in stripping methods, or oscillographic polarography (10) can be used. In the latter method, the potential of the indentations on the $dE/dt = f/E$ curves of the oxidized (in the cathode branch) and the reduced (in the anode branch) form are compared. Identical potentials of the cathodic and anodic indentations indicate reversibility. Nevertheless, because of the short time periods during which the measurement is performed in these methods and because of some additional complications due to the cyclic polarization and controlled current instead of potential, as in other discussed methods, electrolysis in oscillographic polarography, the meaning of the term "reversibility" in oscillographic methods and in classical polarography needs not be identical.

The comparison of polarographic half-wave potentials in the widest possible range of experimental conditions with standard redox potentials obtained under the same conditions is of primary importance for all cases in which the latter can be obtained by potentiometric or other measurements. An agreement of these two sets of data is a strong support for reversibility.

The correct shape of the polarographic wave is the second condition to be fulfilled by a wave of a reversible system. For a diffusion-controlled reversible system, not involving semiquinone (or generally radical) formation or dimerization of either oxidized or reduced form, the course of the polarographic wave follows (11) eq. (5):

$$E = E^\circ + \frac{RT}{nF} \ln \frac{\bar{\imath}_d - \bar{\imath}}{\bar{\imath} - \bar{I}_d} \left(\frac{D_{\text{Red}}}{D_{\text{Ox}}}\right)^{1/2} \tag{5}$$

where $\bar{\imath}_d$ is the mean diffusion cathodic current, $\bar{I}_d$ is the mean diffusion anodic current and $\bar{\imath}$ is the mean current at the potential E. This form of the equation assumes that both the oxidized and the reduced forms are present in the solution. We shall nevertheless restrict ourselves to the more often encountered case in which only the oxidized form is present and the equation possesses the form of eq. (6):

$$E = E_{1/2} + \frac{RT}{nF} \ln \frac{\bar{\imath}_d - \bar{\imath}}{\bar{\imath}} \tag{6}$$

To prove that the system is reversible and that it possesses the above characteristics, the mean diffusion limiting current $\bar{\imath}_d$ is measured first. Then the mean current, $\bar{\imath}_n$, is measured in the rising part of the polarographic wave at several potentials E_n. For each point in the i–E curve, the value of log $(\bar{\imath}_d - \bar{\imath}_n)/\bar{\imath}_n$ is computed and plotted against E_n. For a reversible system with the above-mentioned characteristics, a linear graph results with a slope of 2.3 RT/nF, i.e., $0.059/n$ at 20°C. For reversible systems involving semiquinone formation, dimerization, formation of insoluble or complex compounds with mercury, other equations have been derived in the same way as for reversible systems accompanied by an antecedent or successive chemical reaction. To verify the reaction scheme for a given system, these equations are treated in a similar manner, using the appropriate logarithmic terms, the so-called logarithmic analysis.

Logarithmic analysis, because easily performed directly from polarographic curves, is usually the first and sometimes the only proof of reversibility. It should not be forgotten that many irreversible systems give linear plots in logarithmic analysis corresponding to some unit value of n. A deduction of reversibility of the electrode process based only on logarithmic analysis is worthless unless supported by the results of some other measurements described in this chapter.

Another group of modern electrochemical methods that allow us to reach conclusions about reversibility are methods using superimposed alternating voltage. In ac polarography (12) the peak current for the small amplitudes of the superimposed alternating voltage is equal to

$$\bar{\imath}_S = knFD^{1/2}m^{2/3}t_1^{2/3}\omega^{1/2}c\, nFE_0/2RT \tag{7}$$

where k is a constant, m, the outflow velocity of the mercury, t_1, the drop time, c, the concentration of the electroactive substance, ω, the frequency, and E_0, the amplitude of the superimposed voltage. For an irreversible process, the peak current $\bar{i}_s$ is smaller by

$$\alpha[1 - 1.20\ (\omega t)^{-0.22}]$$

Hence to distinguish a reversible process from an irreversible one, the height of the peak current is compared first; the smaller peak current indicates an irreversible process. Moreover, for small amplitudes the peak current should be linearly proportional to the square root of the frequency $\omega^{1/2}$ or to the amplitude E_0.

Also when square-wave polarography (13) is used, considerably smaller or no current is observed for irreversible processes when compared with the sensitive signals for reversible ones.

Finally, the third important and general method for distinguishing the reversibility of the electrode process is the study of the changes of the instantaneous current with time (14). In this method, *i–t* curves are recorded showing the increase of current with time on a single drop using either a string galvanometer or a proper oscilloscope. To eliminate the effects of the transfer of concentration polarization, the current should be recorded on the "first drop," i.e., immediately after the voltage has been applied.

The current–time (*i–t*) curves are recorded in the form $i = kt^x$ and plotted in the logarithmic scale as $\log i = K + x \log t$. When comparing the curves obtained at various potentials in the region of the rising portion of the polarographic curve, reversible processes show for $t > 0.5$ sec. at all potentials a linear logarithmic plot with a mean value of the exponent $x = 0.192$. For irreversible processes, the logarithmic plots show departures from linearity. For a simple slow electrode process, the value of the exponent x of the linear part of such plots is the greater the more positive the potential at which the measurement is carried out. At the foot of the irreversible polarographic wave the exponent reaches the value $x = 0.66$.

III. Factors Affecting the Mechanism of the Electrode Process

After the irreversibility of the electrode process has been established by the methods discussed above, it is possible to gain some further information concerning the mechanism of the electrode process.

We shall first summarize the steps that are considered able to contribute to the irreversibility of the electrode process (Table I):

TABLE I
Scheme of the Electrode Process

Transported particle	
↓ k_1; Antecedent chemical reaction	(I)
Electroactive particle	
↓ Adsorption; orientation	(II)
[Electroactive particle]$_0$	
↓ n_1e; Interaction with the electrode	(III)
Transition state I	(III)
↓ Interaction with the electrode	(III)
[Primary unstable intermediate]$_0$	
↓ k_2; Consecutive chemical reaction	(IV)
[Stable intermediate]$_0$	
↓ n_2e; Interaction with the electrode	(V)
Transition state II	(V)
↓ Interaction with the electrode	(V)
[Product]$_0$	
↓ Desorption	(VI)
Product transported from the electrode	

[a] Subscript zero means at the electrode surface.

The scheme given in Table I is an oversimplified picture. The actual electrode process can contain either only a part of the depicted steps I–VI, or in other cases, one of the steps given in Table I consists of several processes or some of the steps are coupled together.

The first problem to be settled is whether the irreversibility results from a slow electrode process V or more frequently III or whether the reversible electrode process III is followed by a deactivating chemical process like IV which causes irreversibility of the electrode process. Because some of the organic systems are ascribed to the former and some to the latter type, these two mechanisms will be discussed separately. The criteria for distinguishing these two types have not been formulated clearly so far: for the time being there seem to be two possibilities.

1. As stated above, for a simple slow electrode process, i.e., those in which steps III and perhaps II affect the polarographic curve, the index x of the $i = kt^x$ dependence should increase regularly with increasingly positive potential in the region of the rising part of the polarographic curves. On the other hand, cases were observed in which the value of the coefficient x first increased and then decreased with increasingly positive potential. It has been suggested (14) that in such cases not a slow electrode process but a more complicated process is responsible for the irreversibility. The scarcity of data on i–t curves of organic systems unfortunately prevents the division of the systems according to i–t curves to be used for distinguishing the types of organic electrode processes.

2. It has been suggested that systems in which reversible step III is combined with an irreversible process like IV, produce a wave shape which more or less closely resembles a theoretical one for a reversible process and the half-wave potentials of which show a pH dependence similar to that obtained for reversible processes. Sometimes the logarithmic plot for an irreversible process is reasonably linear but with a slope corresponding to a number of transferred electrons n_1 smaller than the value of n determined from the limiting current. This is then sometimes interpreted as corresponding to a system that involves a reversible step III where there is an uptake of n_1 electrons, followed by one or more irreversible further electron uptakes n_2. . . The scope of validity of such deductions is yet to be determined. Nevertheless, it is possible to attribute with a certain degree of probability the irreversibility of waves with almost theoretical shapes to consecutive chemical reactions. On the other hand, it is possible to assume when adsorption phenomena as the main source of the deformations of the polarographic curves are excluded that drawn-out waves with nonlinear logarithmic analysis and small values of αn correspond to slow electrode process III.

To separate, identify, and explain factors given in Table I, both techniques used in the study of mechanisms of homogeneous reactions and specific electrochemical methods reflecting the heterogeneity of the electrode process must be used (Table II). The limiting current, wave shape, and half-wave potentials are measured. In analogy with the homogeneous kinetics it is studied how these parameters depend on the composition of the solution, in particular on the concentration of the studied substance, the acidity of the media, the ionic strength, and the kind and concentration of the solvent. Sometimes, but rather rarely, the effects of temperature, light, and pressure are studied as well. Further information, as in homogeneous kinetics, can be obtained from the changes of the above-mentioned polaro-

TABLE II
Techniques for Elucidating Mechanisms of Electrode Processes

- A. Effects of composition of the solution
  1. Acidity
  2. Concentration of the depolarizer
  3. Ionic strength
  4. Solvent
  5. Temperature, light, pressure
  6. Surface active agents
- B. Identification of products and intermediates
- C. Structural effects
  1. The same electroactive group on various molecular frames
  2. Comparison of various electroactive groups
  3. Polar, steric, and resonance substituent effects
  4. Isotopes
- D. Electrochemical techniques
  1. Capillary constants
  2. Electrode type: shape, properties, material
  3. Comparison with other electrochemical methods
  4. Specific salt effects
  5. Adsorption phenomena

graphic parameters with the structure of the studied compound. Various structurally similar electroactive groups attached to the same molecular frame can be compared. The effect of various types of molecular frames bearing the same electroactive group can be compared. The polar, steric, and resonance effects of substituents in a position on the molecular frame which is more or less remote from the electroactive grouping, can be studied. Products and intermediates of the electrode process are detected and identified. The use of isotopes, even if possible, has been relatively less often used.

The special electrochemical techniques used in the elucidation of electrode process mechanisms are the changes in the capillary characteristics, studies on effects of the electrode type (shape, properties, electrode material), comparison of results of the classical polarographic examination with the results of other electrochemical methods, study of specific salt effects, including specific effects of cations or anions that affect electrode processes but not homogeneous reactions, and studies of adsorption phenomena, including adsorption of the oxidized and reduced forms, added surface-active substances, and the solvent used.

The intensity with which this or that type of the above-mentioned questions is studied depends on the aim of the whole study. In electrochemical studies, interest is centered on the peculiarities that distinguish heterogeneous processes at the surface of the electrode from homogeneous processes. In physical organic polarography, on the other hand, we try to bring these irregularities under control and to look for the analogies between chemical reactions in which two particles usually participate and electrochemical reactions [in which the only particle present always has the electrode or the electron as its partner]. It is the aim of the present contribution to stress these latter aspects and the discussion of the diagnostic value of various factors for elucidating the mechanism of the electrode process will be carried out from this viewpoint.

IV. Effects of Acidity

For numerous polarographic curves of organic substances, it has been observed that either the half-wave potential, or the wave-height, or the shape of the wave, or all of them depend on acidity, most conveniently expressed as pH or corresponding acidity functions.

These changes can be due not only to acid–base equilibria participating in step I (Table I) preceding the electrode process proper, step III, but also to those participating in step IV preceding process V. In addition to acid–base equilibria affecting the electroactive species directly, changes in polarographic curves with pH can also result from effects of acidity on chemical reactions, such as ring-opening or dehydration, included in step I. As in homogeneous kinetics, the rate of these chemical reactions preceding the electrode process proper, III, can be affected by fastly established acid–base equilibria, rate of proton or hydroxyl ion transfer, or general acid–base catalysis.

Hence, step I can cause the observed pH-dependence either because of a change in the protonation of the electroactive species (8a) or because of a change in the rate of a chemical reaction transforming an electroinactive species into an electroactive species [e.g., (9–11)]:

$$A + H^+ \rightleftharpoons AH \qquad (8a)$$

$$AH \rightarrow [AH]_0 \qquad (8b)$$

$$[AH]_0 + n_1e \rightarrow \text{products} \qquad (8c)$$

or

$$B + H^+ \rightleftharpoons BH \quad (9a)$$

$$BH \xrightarrow{k} AH \quad (9b)$$

followed by (8b) and (8c), or

$$B + H^+ \xrightarrow{k} BH \quad (10a)$$

$$BH \xrightarrow{\text{fast}} AH \quad (10b)$$

followed by (8b) and (8c), or

$$B \xrightarrow[k_i]{\text{general catalysis}} AH \quad (11)$$

followed by (8b) and (8c).

These electrode processes in which III is an irreversible step can be affected only by an acid–base process preceding the slow step. The acid–base equilibria can hence affect only reactions (8)–(11). Explanations, found sometimes in the literature, according to which the observed pH dependence is attributed to an acid–base equilibria following a slow step (IV in Table I) are wrong.*

Those irreversible processes that involve a reversible electrochemical step can be affected also by acid–base processes following that reversible step. For example, when process III is reversible, any acid–base reactions involving process IV preceding the slowest chemical or electrochemical step can affect polarographic curves.

It is possible to consider the acid–base process as a change in the number of protons directly consumed in electrode process III. In some cases, it is difficult to exclude such a possibility, but usually this explanation offers no additional information when compared with the mechanism considering reaction I as pH-dependent. Because on the contrary, an explanation of the effects of pH on polarographic curves as due to the pH-dependence of reaction I offers a more thorough understanding and enables a comparison of various seemingly incoherent phenomena, it is preferred in further discussion here.

* When both processes III and V (Table I) are irreversible, an acid–base participation in step IV cannot affect the current due to process III, but can affect the part of the curve corresponding to process V. For process V, reaction IV is again an antecedent reaction.

It would also be possible to explain the pH-effects as resulting from a reaction of generated atomic hydrogen. To explain the various potentials at which the reduction of organic compounds occurs and the various pH-dependences observed for various compounds, it would be necessary to expect that the electrolytic formation of atomic hydrogen is a reversible process and that the potential-determining step is the reaction of the atomic hydrogen with the organic molecule:

$$H^+ + e \underset{}{\overset{\text{fast}}{\rightleftharpoons}} H \tag{12a}$$

$$H + B \underset{\text{slow}}{\longrightarrow} \text{products} \tag{12b}$$

This scheme, even when considered in the early periods of organic polarography, seems little probable in view of studies showing that formation of atomic hydrogen is a slow process, and will not be further considered here.

A. REACTIONS PRECEDING THE ELECTRODE PROCESS PROPER

Acid–base reactions preceding an irreversible electrode process will be discussed first and of these, those reactions (8a) in which the electroactive species AH is formed in the acid–base reaction. Some few simple examples of that type will be discussed here.

1. *Rapidly Established Acid–Base Equilibria*

The relative rate of establishing the acid–base equilibrium when compared with the drop-life and rate of the electrochemical reaction, is of primary importance for the classification of these reactions and in particular in the mode of their effect on polarographic curves. The equilibria that are established quickly, equally as fast, and slowly compared with the electrode process are discussed separately.

For rapidly established acid–base equilibria, one single wave is observed on polarographic curves, corresponding to process III (Table I). The height of this wave corresponds to the number of electrons transferred in process III and does not change with pH. The half-wave potentials for the scheme:

$$HA \overset{\text{fast}}{\rightleftharpoons} A^- + H^+ \tag{13a}$$

$$HA + n_1e \rightarrow \text{products} \tag{13b}$$

depend on pH according to eq. (14):*

$$E_{1/2} = \text{const} + \frac{RT}{\alpha nF} \ln \frac{[H^+]}{K + [H^+]} \quad (14)$$

For $[H^+] \gg K$ is

$$E_{1/2} = \text{const} \quad (15)$$

and the half-wave potential is pH independent. For $[H^+] \ll K$ is

$$E_{1/2} = \text{const} - \frac{2.3RT}{\alpha nF} \text{pH} \quad (16)$$

A more general derivation for the system:

$$\text{Ox} + m\text{H}^+ \rightleftharpoons \text{Ox H}_m \quad (17a)$$

$$\text{Ox H}_m + ne \rightarrow \text{Red} \quad (17b)$$

is given as follows. The current for the irreversible process is given by

$$\bar{\imath} = nF\bar{q}\, k_e^-[\text{OxH}_m]_0 \quad (18a)$$

The analytical concentration of the oxidized form S_{Ox} at the surface of the electrode is given by eq. (19a):

$$[S_{\text{Ox}}]_0 = [\text{OxH}_m]_0 + [\text{Ox}]_0 \quad (19a)$$

Using the equilibrium constant

$$K = \frac{[\text{Ox}][\text{H}^+]^m}{[\text{OxH}_m]} \quad (20)$$

it can be expressed as

$$[S_{\text{Ox}}]_0 = [\text{OxH}_m]_0 \left(1 + \frac{K}{[\text{H}^+]^m}\right) \quad (19b)$$

and the current

$$\bar{\imath} = nF\bar{q}k_e^- S_{\text{Ox}}(1 + K/[\text{H}^+]) \quad (18b)$$

After introducing

$$k_e^- = k_e^0 \exp\left[-\frac{\alpha nF}{RT}(E - E^0)\right] \quad (21)$$

the following equation is obtained:

$$E = E^0 + \frac{RT}{\alpha nF} \ln 0.886k_e^0 \left(\frac{t_1}{D}\right)^{1/2} + \frac{RT}{\alpha nF} \ln \frac{[\text{H}^+]^m}{K + [\text{H}^+]^m} - \frac{RT}{\alpha nF} \ln \frac{\bar{\imath}}{\bar{\imath}_d - \bar{\imath}} \quad (22)$$

and hence for $\bar{\imath} = \bar{\imath}_d/2$

$$E_{1/2} = \text{const} + \frac{RT}{\alpha n F} \ln \frac{[\text{H}^+]^m}{K + [\text{H}^+]^m} \tag{23}$$

and the half-wave potential is shifted towards more negative values. The two linear portions [plotted for (pK − 1) > pH > (pK + 1)] intersect at a pH value corresponding to pK. An example of the $E_{1/2}$–pH plot is given in Figure 2. The pH at the intersection of the two linear parts corresponds to the pK value of eq. (13a). A practical example of that type has been found in the case of phenacyl sulfonium salts (15). Because the cleavage of the C—S bond is an irreversible process, followed by proton transfers in consecutive steps, the half-wave potentials are expected to be pH-independent. The observed dependence of the type given in Figure 2 leads to a potentiometric

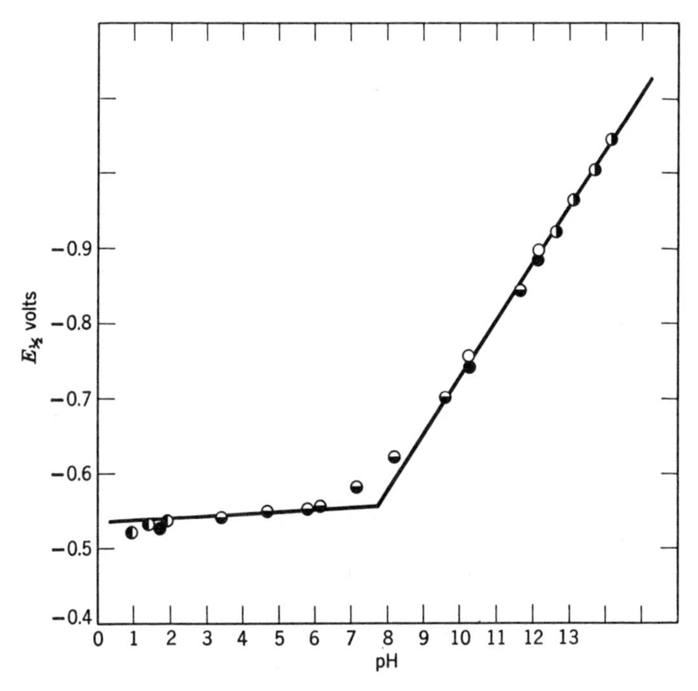

Fig. 2. Dependence of half-wave potentials $E_{1/2}$ of a system involving a rapidly established acid–base equilibria preceding the electrode process proper. Dependence of half-wave potentials of methyl butyl phenacyl sulfonium ion on pH (15).

examination that revealed that phenacyl sulfonium salts behave like moderately strong acids with pK values between 7 and 8. Potentiometrically determined pK values were in good agreement with polarographically found interactions on $E_{1/2}$–pH plots. Hence it was possible to deduce that the electrode process is:

$$C_6H_5COCH_2\overset{(+)}{S}R_2 \rightleftharpoons C_6H_5CO\overset{(-)}{C}H\overset{(+)}{S}R_2 + H_2 \qquad (24a)$$

$$C_6H_5COCH_2\overset{(+)}{S}R_2 + 2e \rightarrow C_6H_5COCH_2^{(-)} + SR_2 \qquad (24b)$$

followed by further reactions about which information can be obtained from the study of the more negative wave, corresponding to the reduction of the CO group (p. 119). The ylide form can be stabilized here by the enolate formation and by the *d*-orbital resonance. The other possible explanation based on formation of a pseudo-base of the type $C_6H_5COCH_2SR_2(OH)$ was excluded by comparison with the ultraviolet absorption spectra (16). Hence polarography enabled the detection of stable ylide formation.

There is another possibility, namely that the electroactive form will be the conjugate base form A^-. Since, as a rule, the more protonized forms are reduced at more positive potentials than the less protonized forms and since the data on oxidation processes are too scarce to allow discussion, this case has not been encountered in practice so far.

2. Acid–Base Equilibria Established with a Comparable Rate

For acid–base equilibria which involve interaction with the proton only, and which are established in a time period comparable with that needed for the electrode process, the following scheme is considered:

$$HA \underset{k_r}{\overset{\text{comparably slow}}{\rightleftharpoons}} A^- + H^+ \qquad (25a)$$

$$HA + n_1e \xrightarrow{E_1} \text{products}_1 \qquad (25b)$$

$$A^- + n_2e \xrightarrow{E_2} \text{products}_2 \qquad (25c)$$

In these cases two waves can appear on polarographic curves, that one corresponding to the reduction of the acid form HA at a potential E_1, the other one corresponding to the reduction of the conjugate base A^- usually at a more negative potential E_2. The more positive wave corresponds to an exchange of n_1 electrons, the more

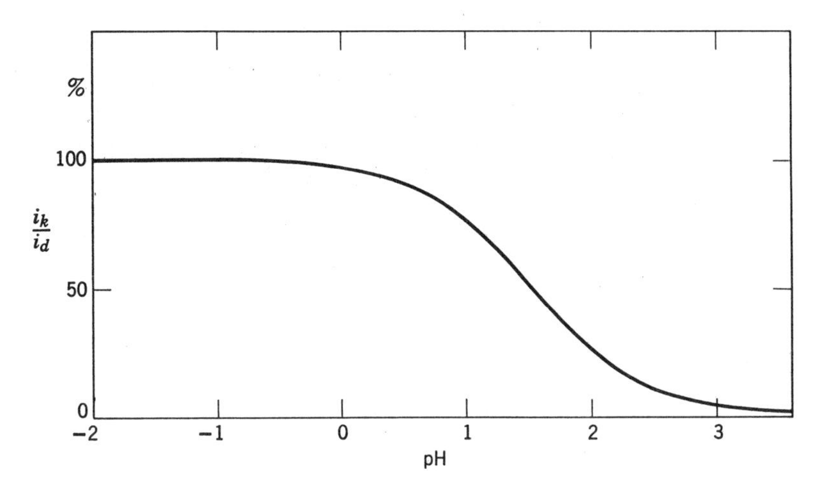

Fig. 3. pH Dependence of the kinetic current i_k expressed as a fraction of the total diffusion current i_d for a system $HA \rightleftharpoons A^- + H^+$; $HA + ne \rightarrow$ prod in which the rate of the establishment of the equilibrium is comparable with the rate of the electrode process. Relative pH scale.

negative wave of n_2-electrons. If $n_1 = n_2$ then the total wave height remains constant, but the ratio of the wave heights i_{HA}/i_A is pH-dependent. With increasing pH the height of the more positive wave i_{HA} decreases (17) in the form of a dissociation curve (Fig. 3) according to eq. (26):

$$\frac{i_{HA}}{i_{HA} + i_A} = \frac{0.886(k_r t_1/K_a)^{1/2}[H^+]}{1 + 0.886(k_r t_1/K_a)^{1/2}[H^+]} \tag{26}$$

where k_r is the rate constant of the recombination reaction, K_a is the dissociation constant of the acid HA, and other symbols are already known.

It should be stressed that the wave heights in these cases do not correspond to the actual concentrations of forms HA and A^- in the bulk of the solution. The height of the more positive wave is increased by the current corresponding to the amount of the acid form HA formed by recombination. The height of the more negative wave is decreased by the same amount. The wave heights are no more limited by diffusion but rather by the rate of the chemical process. The polarographically found dissociation curve is shifted towards higher pH-values when compared with the titrimetrically or spectro-

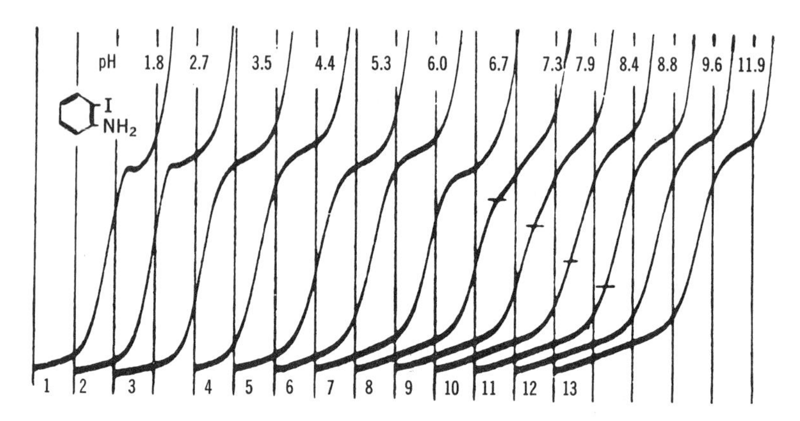

Fig. 4. pH Dependence of polarographic curves corresponding to a system $HA \rightleftharpoons A^- + H^+$; $HA + ne \rightarrow$ prod in which the rate of the establishment of the equilibrium is comparable with the rate of the electrode process. Dependence of *o*-iodoaniline waves on pH. $2 \times 10^{-4}M$ *o*-iodoaniline, Britton-Robinson buffers, pH given on the polarogram. Curves *1–3* starting at −0.6 v, *4–13* at −0.8 V. S.C.E., 200 mV./absc.; h = 80 cm.; m = 2.7 mg./sec.; t_1 = 0.85 sec. (capillary according to Smoler); full scale sensitivity 3.2 μamp.

photometrically determined dissociation curve. The polarographically found value of pK' i.e., pH at which $\bar{i} = \bar{i}_d/2$, is generally greater than pK_a.

Dependence of limiting currents on pH of the type shown in Figure 3 is quite common in organic polarography and allows us to deduce that the scheme (25) is principally valid. The most common types of such reactions, such as recombinations of α-ketoacids (18), pyridinecarboxylic acids (19,20), or carbonyl groups (21), are complicated by the hydration of the carbonyl group, the possibility of solvation of the carboxylic group, and tautomeric changes, as well as by the adsorption of the unprotonized and protonized forms and of buffer components. These latter factors enable the proton-transfer reaction to occur as a surface reaction, while the volume reaction takes place in the so-called reaction volume in the vicinity of the electrode. All these factors complicate the theoretical treatment, in particular the computation of the rate constants k_r. But these factors only slightly affect the shape of the observed pH dependence of the limiting currents we are interested in here. The adsorption phenomena can result in a decrease of the limiting current with increasingly negative potentials, but even then, the shape of the pH-dependence of the current measured at a chosen potential corresponds to the above equation.

Examples in which at least part of the complications mentioned above are excluded are the iodophenols and iodoanilines (Fig. 4). Hydration and tautomeric phenomena are absent in these cases, but if the reaction occurs as a volume or surface reaction, their absence is still to be proved.

Another example in which it has been proved (69) that the reaction occurs as a volume reaction, is the behavior of a Lewis acid, the tropylium ion. This seems to be the simplest case of acid–base reactions that has been studied and follows the scheme (27):

$$C_7H_7^+ + 2H_2O \rightleftharpoons C_7H_7(H)(OH) + H_3O^+ \qquad (27a)$$

$$C_7H_7^+ + e \longrightarrow \text{products} \qquad (27b)$$

A direct nucleophilic attack of hydroxyl ions on the tropylium ion would result in a different shaped dissociation curve than that shown in Figure 3. This possibility could thus be rejected.

Generally speaking, a pH-dependence of the limiting current corresponding to Figure 3 allows us to deduce that the conjugate acid form is reduced at more positive potentials than the conjugate base. Furthermore, it is possible to deduce from such a dependence that the acid–base equilibrium is established with a rate comparable with that of the electrode process. Nevertheless, it cannot be decided without a further, more detailed study whether the equilibrium corresponds only to step I or also to step II (Table I).

The pH dependence of the limiting current can sometimes have the shape of a dissociation curve twice as steep as that predicted by eq. (26). This happens for the first dissociation step of dibasic acids (22, 23), provided that $pK_2 > (pK_1 + 2)$ where K_1 and K_2 are the first and second potentiometric dissociation constants and $pK_1' > pK_2$ where pK_1' corresponds to the pH value at which the studied limiting current i_1 reaches half the value corresponding to the diffusion current. Examples of such behavior have been observed for maleic and fumaric acids (22), phthalic acid (24), and pyridoxine derivatives (25).

Not only the limiting currents but also the half-wave potentials of polarographic waves corresponding to the discussed system are changed with the change in pH. Nevertheless, much less attention has been paid to the study of these parameters. This is due mainly to the fact that in addition to the comparatively slow proton transfer that usually occurs on a group other than the electroactive center, e.g., on the carboxylic group in keto-acids in which the carbonyl group is reduced, the electrode process involves one or more rapidly established acid–base equilibria preceding the electrode process proper III (Table I). Hence the pH-dependence of the half-wave potential is caused not only by the antecedent chemical acid–base reaction but also by the consumption of protons in the reduction process itself.* For a theoretical treatment, a system such as iodophenol or iodoaniline in which the reduction process itself is known to be pH-independent would be suitable. No such system has been studied in sufficient detail. Nevertheless, it can be derived that the shape of the pH-dependence of the half-wave potentials should possess principally a form shown in Figure 5(*a*). The half-wave potentials of the acid form show two regions in which the half-wave potentials are pH-independent and a region in which the half-wave potentials are shifted towards more negative values with increasing pH. The intersection at lower pH values should correspond to the potentiometric pK_a value of the acid–base couple involved. The intersection at higher pH values corresponds to the pK' value, i.e., to the pH at which the limiting currents of the acid and base forms are identical or at which the limiting current of the acid form equals half of its diffusion limited value. The region corresponding to pK_a is often experimentally inaccessible, but for pK' the identity of values obtained from limiting currents and from half-wave potentials has often been demonstrated. The half-wave potentials of the conjugate base A are pH-independent [Fig. 5(*a*)] since no proton transfer precedes the irreversible pH-independent reduction process.

In some instances, the half-wave potentials of the acid form at $pH > pK'$ and those of the base form A differ too little to allow a separation of two waves. The observed dependence has the shape shown in Figure 5(*b*). This is probably the case for α-bromoalkanoic acids (26) where the separation of waves is rendered more difficult by the

* As discussed above, we understand that even these protons are transferred before the electrode process proper III (Table I).

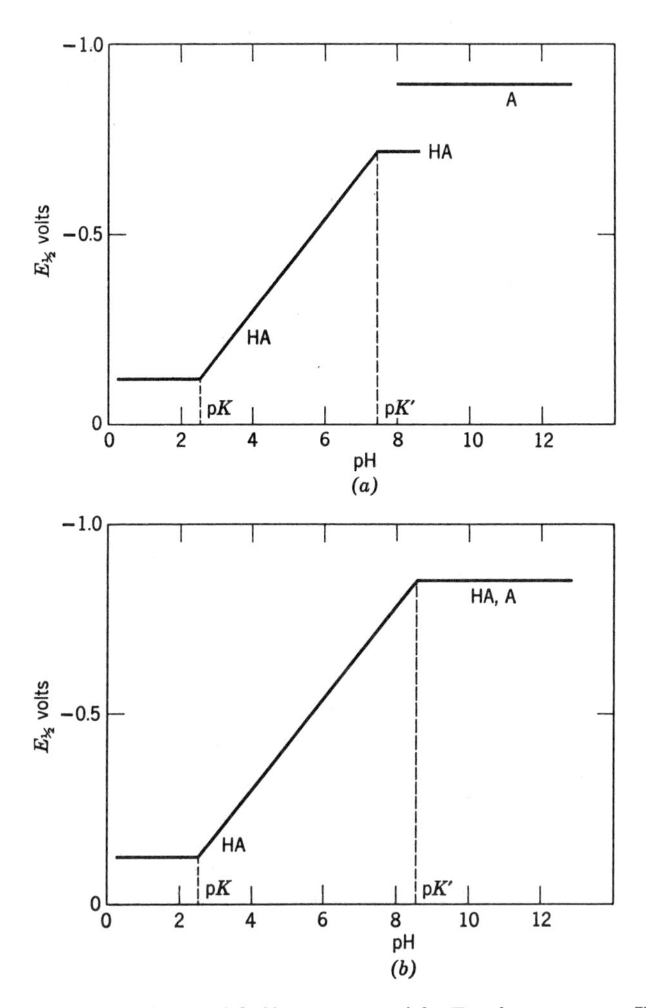

Fig. 5. pH Dependence of half-wave potentials $E_{1/2}$ for a system $HA \rightleftharpoons A^- + H^+$; $HA + ne \xrightarrow{E_1}$ prod; $A^- + n_2e \xrightarrow{E_1}$ prod for which the rate of the establishment of the equilibrium is comparable with the rate of the electrode process. (*a*) The half-wave potentials of the acid form $(E_{1/2})_1$ at pH > pK' differ from those of the conjugate base $(E_{1/2})_2$. (*b*) The half-wave potentials of the acid form $(E_{1/2})_1$ at pH > pK' differ so little from those of the conjugate base $(E_{1/2})_2$ that only one wave is actually observed. pK corresponds to the equilibrium acid dissociation constant and pK' to the "polarographic dissociation constant," i.e., pK' = pH at which $i_k = i_d/2$.

drown-out character of the irreversible waves. It is possible that distinguishing the waves could be accomplished by the use of logarithmic analyses at various pH values.

In the special type of acid–base reactions belonging to this group, the conjugate base not only reacts with the hydroxonium ion as the sole proton donor, but also with other proton donors as well, according to eqs. (28):

$$HA + B_1 \underset{}{\overset{\text{comparatively slow}}{\rightleftharpoons}} B_1H + A^- \tag{28a}$$

$$HA + B_2 \underset{}{\overset{\text{comparatively slow}}{\rightleftharpoons}} B_2H + A^- \tag{28b}$$

$$\vdots \qquad \qquad \vdots$$

$$HA + B_n \underset{}{\overset{\text{comparatively slow}}{\rightleftharpoons}} B_nH + A^- \tag{28c}$$

$$HA + n_1e \xrightarrow{E_1} \text{products}_1 \tag{28d}$$

$$A^- + n_2e \xrightarrow{E_2} \text{products}_2 \tag{28e}$$

Whereas in the simple type the height of the wave of the acid form i_{HA} is dependent only on the pH, in this reaction type, it also depends on buffer type and concentration. The shape of the pH dependence of the limiting current in simple buffers, i.e., containing only one weak acid or base, will be discussed first. Understandingly, the dissociation curve obtained in buffers containing varying amounts of both acids B_nH and base B_n, as prepared by mixing various volumes of equimolar solutions of acid and base, possesses a distorted shape. In the study of general acid-catalyzed homogeneous reactions, buffers are prepared by keeping the concentration of the acid component constant and changing the concentration of base component and hence the pH. When this procedure is applied in the study of polarographic curves corresponding to the above type, the shape of the pH dependence of the limiting current of the acid forms differs again in accordance with the theory (27) from the simple dissociation curve. Only when the base component concentration is kept constant and the pH changed by the change in the concentration of the acid component, does the observed dependence of $i_{HA}/(i_{HA} + i_A)$ on pH possess the form predicted by eq. (26). The inflection point of this curve, pK', is shifted to lower pH values with decreasing concentration of the base buffer component and depends on buffer type. The participation of various acids in the acid–base reaction can be detected from the increase of the current i_{HA} with increasing buffer concentration at a given pH and from differences in the current i_{HA} recorded in buffers made from various acids and bases showing the same pH value.

3. Slowly Established Acid–Base Equilibria

Finally, in some of the systems the establishment of the acid–base equilibria takes place very slowly when compared with the electrode process. The equilibrium remains practically undisturbed by the

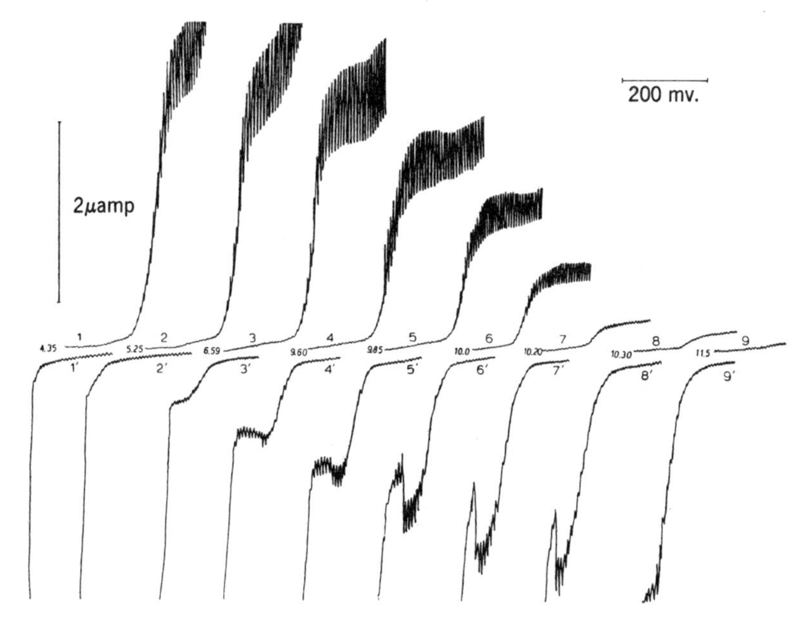

Fig. 6. pH Dependence of polarographic waves corresponding to a system $HA \rightleftharpoons A^- + H^+$; $HA + n_1e \rightarrow prod_1$; $A^- \rightarrow prod_2 + n_2e$, in which the equilibrium is slowly established. Waves of 3-thianaphthenone in Britton-Robinson buffers, pH: (*1*) 4.35; (*2*) 5.25; (*3*) 6.6; (*4*) 9.6; (*5*) 9.85; (*6*) 10.0; (*7*) 10.2; (*8*) 10.3; (*9*) 11.5. (1)–(9) reduction waves; (1′)–(9′) oxidation waves. Curves starting at: (*1, 2*) −1.0; (*3*) −1.1; (*4, 5*) −1.25; (*6*) −1.3; (*7*) −1.35; (*8*) −1.4; (*9*) −1.45. Backward recording of anodic waves started at: (*1′–3′*) −0.15; (*4′*) −0.25; (*6′*) −0.30; (*5′*)–(*9′*) −0.25 V. t_1 = 4.0 sec. (30).

electrolysis. Two waves can again be in principle observed on polarographic curves, but these waves are strictly proportional to the concentration of the two forms present in the solution. The corresponding scheme is:

$$\mathrm{HA} \underset{}{\overset{\text{slow}}{\rightleftharpoons}} \mathrm{A}^- + \mathrm{H}^+ \tag{29a}$$

$$\mathrm{HA} + n_1 e \xrightarrow{E_1} \text{products}_1 \tag{29b}$$

$$\mathrm{A}^- + n_2 e \xrightarrow{E_2} \text{products}_2 \tag{29c}$$

The pH dependence of wave i_{HA} and i_A follows the equation

$$i_{HA}/(i_{HA} + i_A) = [H^+]/(K_a + [H^+]) \tag{30}$$

This dependence is identical to those obtained with equilibrium methods, e.g., potentiometry or spectrophotometry, for identical values of K_a.

The number of slowly established acid–base equilibria is rather limited and hence the known examples of that type are restricted to C-acids such as some nitroparafins (28,29) and 3-thianaphthenone (30). In the latter case, the acid form undergoes reduction and shows a cathodic wave, whereas the conjugate base is oxidized and gives an anodic wave (Fig. 6). The scheme is as follows:

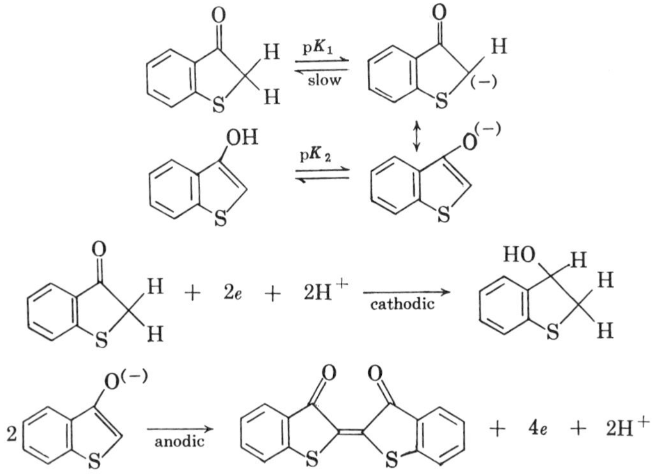

(31a) 3-thianaphthenone (C=O, CH_2) $\underset{\text{slow}}{\overset{pK_1}{\rightleftharpoons}}$ carbanion (H, (−)) ↕ enolate $O^{(-)}$

(31b) 3-hydroxythianaphthene (OH) $\overset{pK_2}{\rightleftharpoons}$ enolate $O^{(-)}$

(31c) 3-thianaphthenone + 2e + $2H^+$ $\xrightarrow{\text{cathodic}}$ 3-hydroxy-2,3-dihydrothianaphthene (HO, H, H, H)

(31d) 2 enolate $O^{(-)}$ $\xrightarrow{\text{anodic}}$ thioindigo (O, O, S, S) + 4e + $2H^+$

Information about pK_2 cannot be obtained from polarography and application of spectrophotometric methods is under consideration.

The shape of the pH dependence of the limiting current i_{HA} for scheme (25) and (29) differs slightly. Differentiation is possible on the basis of the character of the limiting current, kinetic in the former, diffusion-controlled* in the latter, and on the comparison of the obtained pK' value with that obtained with equilibrium methods, differing in the former, identical in the latter.

4. Chemical Reactions Affected by Antecedent Equilibria

The systems in which the pH dependence of polarographic curves results from the effects of acidity on the rate of chemical reactions in which the electroactive form is formed, are discussed next.

In the first type, acid–base equilibria affect the rate of the chemical reaction by changing the concentration of various ionic forms of the reactant differing in reactivity according to eqs. (9) and (10). Dehydration of pyridine aldehydes (32) is an example which can be depicted as in eqs. (32).

$$\text{Py}^{+}\text{H}-\text{CH(OH)}(\text{OH}_2^{(+)}) \underset{+H_2O}{\overset{v_3}{\rightleftharpoons}} \text{Py}^{(+)}\text{H}-\text{C}(=\text{OH}^{(+)})\text{H} \qquad (32a)$$

$$v_2 \downarrow\uparrow +H^+ \qquad\qquad -H^+ \downarrow\uparrow v_4$$

$$\underset{\text{(I)}}{\text{Py}^{(+)}\text{H}-\text{CH(OH)}_2} \underset{+H_2O}{\overset{v_a}{\rightleftharpoons}} \text{Py}^{(+)}\text{H}-\text{CHO} \qquad (32b)$$

$$v_1 \downarrow\uparrow +H^+, K_1 \qquad\qquad -H^+ \downarrow\uparrow v_5$$

$$\underset{\text{(II)}}{\text{Py}-\text{CH(OH)}_2} \underset{+H_2O}{\overset{v_{10}}{\rightleftharpoons}} \text{Py}-\text{CHO} \qquad (32c)$$

$$v_8 \downarrow\uparrow +H^+ \qquad\qquad -H^+ \downarrow\uparrow v_6$$

$$\text{Py}-\text{CHOH}(\text{O}^{(-)}) \underset{+H_2O}{\overset{v_7}{\rightleftharpoons}} \text{Py}-\overset{(-)}{\text{C}}=\text{O} \qquad (32d)$$

The observed pH dependence (Fig. 7) can be interpreted as follows: In region A the dehydration of the protonated form I occurs by

* To prove the character, it is important to study the behavior of the current when it is only about 10–15% of the original wave height.

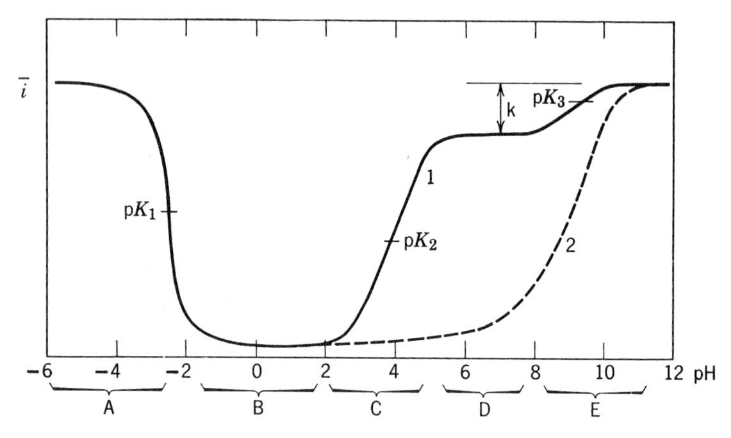

Fig. 7. pH Dependence of the limiting current $\bar{i}$ of a pyridine aldehyde(—) and *N*-methylpyridinium aldehyde (– –). pK_1, pK_2 and pK_3 correspond to eqs. (32) (schematically).

an acid-catalyzed* reaction following the path v_2, v_3, v_4; in region B the protonated form I is dehydrated by an uncatalyzed reaction v_a. In region D the unprotonized form II undergoes an uncatalyzed dehydration v_{10}, in region E form II is dehydrated by a base-catalyzed reaction following the path v_8, v_7, v_6. The change in the limiting current in region C is ascribed to the equilibrium between forms I and II with the equilibrium constant K_1. This explanation is based (31) on comparison with quaternized compounds and with spectra.

A similar type of pH dependence has been found for pyridoxal (25,32). In addition to the hydration of the aldehydic group, the decrease of current can be explained also by hemiacetal formation. The comparison with pyridoxal-5-phosphate is important for deciding between these two possibilities. Formation of hemiacetal is impossible with this ester and hence no decrease of limiting current would be expected for hemiacetal formation. In fact, a small decrease in the limiting current was observed (25) at pH 6–10 which is, moreover, dependent on the kind of buffer used. This observation suggests dehydration as the more probable current-limiting chemical reaction. The introduction of a phosphoric acid residue into position 5 either shifts the equilibrium towards the dehydrated form or increases the rate of dehydration.

* Only one simple mechanism for the acid and base catalyzed dehydrations is considered here.

Another example that can be included in this group is the dehydration of *N*-alkyl pyridinium aldehydes (31). The pH-dependence shown on curve 2 (Fig. 7) can be explained by:

$$\text{Py}^{(+)}(\text{CH}_3)\text{-CHOH(OH}_2^{(+)}) \underset{+\text{H}_2\text{O}}{\overset{v_2}{\rightleftharpoons}} \text{Py}^{(+)}(\text{CH}_3)\text{-C(=OH}^{(+)})\text{H} \tag{33a}$$

$$v_1 \Big\downarrow\Big\uparrow +\text{H}^+ \qquad\qquad +\text{H}^+ \Big\downarrow\Big\uparrow v_3$$

$$\text{Py}^{(+)}(\text{CH}_3)\text{-CH(OH)}_2 \underset{+\text{H}_2\text{O}}{\overset{v_7}{\rightleftharpoons}} \text{Py}^{(+)}(\text{CH}_3)\text{-CHO} \tag{33b}$$

$$v_6 \Big\downarrow\Big\uparrow +\text{H}^+ \qquad\qquad +\text{H}^+ \Big\downarrow\Big\uparrow v_4$$

$$\text{Py}^{(+)}(\text{CH}_3)\text{-CHCO(O}^{(-)}) \underset{+\text{H}_2\text{O}}{\overset{v_5}{\rightleftharpoons}} \text{Py}^{(+)}(\text{CH}_3)\text{-}\overset{(-)}{\text{C}}\text{=O} \tag{33c}$$

In region A, the acid-catalyzed reaction follows the path v_1, v_2, v_3, in regions B, C, and D, the uncatalyzed reaction follows the path v_7, and in region E, the base-catalyzed reaction follows the path v_6, v_5, v_4. Whether the limiting current depends solely on pH or also on buffer type and concentration remains to be proved. In the latter case this reaction would belong to the next group.

5. General Catalyzed Reactions

To the second group belong reactions in which the electroactive form is formed by a general acid or base-catalyzed reaction. Typical representative of this group are aliphatic aldehydes, such as formaldehyde (33), that follow eqs. (34):

$$\text{R—CH(OH)(OH)} + \text{B} \rightleftharpoons \text{R—CH(OH)(O}^{(-)}) + \text{BH}^+ \tag{34a}$$

$$\text{R—CH(OH)(O}^{(-)}) \rightleftharpoons \text{R—CH=O} + \text{OH}^- \tag{34b}$$

$$\text{R—CH=O} + 2e \rightarrow \text{products} \tag{34c}$$

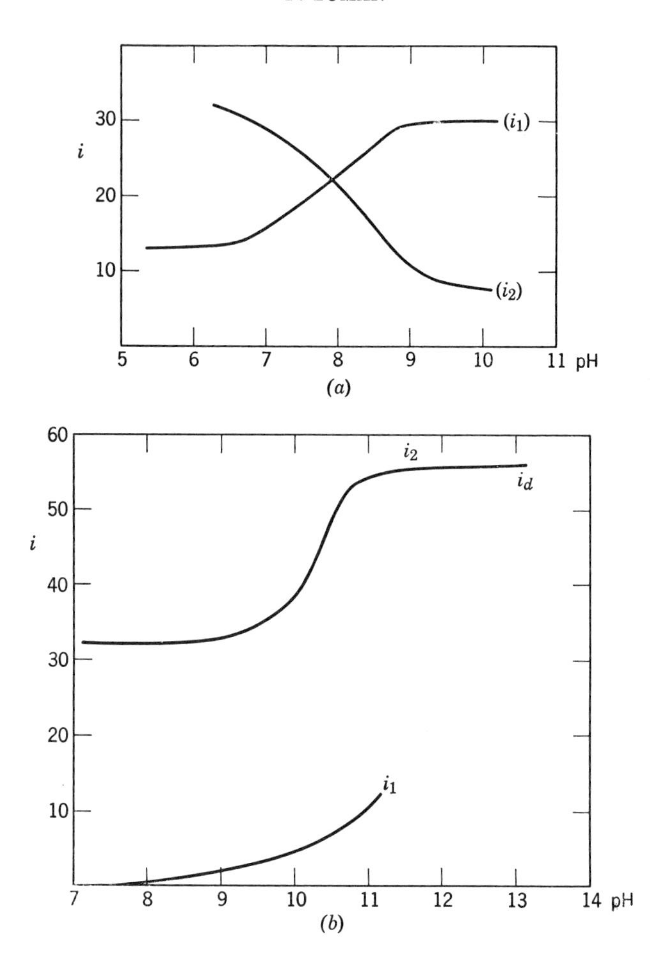

Fig. 8. pH Dependence of limiting currents i of (a) diacetyl and (b) methylglyoxal; i_1 is the more positive wave; i_2 is the more negative (according to Fedoroňko).

or any other equivalent mechanism suggested for general base-catalyzed additions of nucleophilic reagents to the carbonyl group (34). The pH dependence of the current possesses the form of a maximum, shown in the right-hand side of Figure 9. The pH dependence of half-wave potential for these systems has not been studied in detail so far.

A more complex scheme (35) is observed for the α-dicarbonyl compounds, such as diacetyl, R = CH_3, or methylglyoxal, R = H:

$$R{-}C(OH)_2{-}C(OH)_2{-}CH_3 \underset{+H_2O}{\overset{k_1}{\rightleftharpoons}} R{-}C(=O){-}C(OH)_2{-}CH_3 \qquad (35a)$$

$$RCOC(OH)_2CH_3 + 2e + 2H^+ \xrightarrow{E_2} \text{products} \qquad (35b)$$

$$R{-}C(=O){-}C(OH)_2{-}CH_3 \underset{+H_2O}{\overset{k_2}{\rightleftharpoons}} R{-}CO{-}CO{-}CH_3 \qquad (35c)$$

$$R{-}CO{-}CO{-}CH_3 + 2e + 2H^+ \xrightarrow{E_1} \text{products} \qquad (35d)$$

The course of pH dependence shows a different form according to the nature of the groups R [Fig. 8(a,b)]. For R = CH_3 at pH <6.5, the wave at more negative potentials E_2, corresponding to the reduction of the monohydrated form reaches its limiting value [Fig. 8(a)]. Either the equilibrium (35a) is shifted under these conditions towards the monohydrated form, or the reaction with the rate constant k_1 is sufficiently fast to transform all the dihydrated form into the monohydrated. The increase of the more positive wave at E_1, corresponding to the reduction of the unhydrated form, is due to the base-catalyzed reaction with the rate constant k_2. Because the value of current at E_1 does not decrease to zero values with decreasing pH, at pH <6, either some unhydrated form is present in equilibriums (35a) and (35c), or the rate constant k_2 also includes in addition to a pH-dependent component a pH-independent one, corresponding to a reaction with the solvent $(k_{H_2O})_2$.

The pH-dependence for R = H [Fig. 8(b)] shows that in this case the equilibrium (35a) is less shifted towards the monohydrated form than was observed for R = CH_3. The rate of dehydration with the constant k_1 is base-catalyzed as shown by the increase of the wave at E_2 at pH >9. The fact that this wave at E_2 does not fall to zero at pH <9 and that it is diffusion-controlled can be explained by the partial shift of the equilibrium (35a) toward the monohydrated form. Dehydration of the monohydrated form with the rate constant k_2 is also base-catalyzed, but the decrease of i_1 at pH < 9 shows that no considerable contribution of $(k_{H_2O})_2$ is involved.

6. *Complex Reactions*

An example of the third group is glyoxalic acid for which both acid–base equilibria and a general base-catalyzed reaction govern the formation of the electroactive form. The pH dependence of the limiting current (35) shown in Figure 9 can be explained by eqs. (36):

$$HC(OH_2^{(+)})(OH){-}COOH \underset{}{\overset{v_3}{\rightleftharpoons}} C(=OH^{(+)})(H){-}COOH \qquad (36a)$$

$$v_2 \upharpoonleft\downharpoonleft +OH^- \qquad\qquad +H^+ \upharpoonleft\downharpoonleft v_4$$

$$HC(OH)(OH){-}COOH \overset{v_7}{\rightleftharpoons} C(=O)(H){-}COOH \overset{E_1}{\rightarrow} \qquad (36b)$$

$$v_1 \upharpoonleft\downharpoonleft +OH^-;\ K_1 \qquad\qquad +H^+ \upharpoonleft\downharpoonleft v_5$$

$$HC(OH)(OH){-}COO^{(-)} \overset{v_8}{\rightleftharpoons} C(=O)(H){-}COO^- \overset{E_2}{\rightarrow} \qquad (36c)$$

$$v_{10} \upharpoonleft\downharpoonleft +OH^-,\ +B;\ K_2 \qquad\qquad +H^+ \upharpoonleft\downharpoonleft v_6$$

$$HC(OH)(O^{(-)}){-}COO^{(-)}\ (III) \overset{v_9}{\rightleftharpoons} \overset{(-)}{C}{=}O{-}COO^{(-)} \qquad (36d)$$

Region A corresponds to the acid-catalyzed dehydration of the free acid following path v_2, v_3, v_4; region B, to the uncatalyzed dehydration of the free acid v_7, and region D, to the uncatalyzed dehydration of the anion v_8. Region C corresponds to the establishment of the acid–base equilibrium with the constant K_1. The decrease of current in region F at pH >12 is interpreted (35) as due to the formation of anion III which is assumed not to undergo dehydration. Hence the general base catalysis causing the increase in current in region E cannot be explained by the path v_{10}, v_9, v_6, but other paths discussed (34) for general base-catalyzed reactions should be considered.

7. *Slow Chemical Reactions*

The rates of all chemical processes producing the electroactive form discussed so far are comparable to the rates of electrolysis, and hence

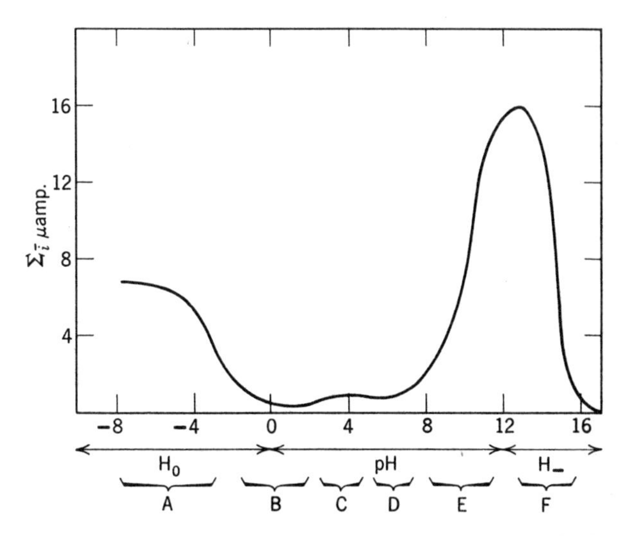

Fig. 9. Dependence of the total limiting current $\Sigma\bar{i}$ on values of H_0, pH, and H_- of glyoxalic acid. $4 \times 10^{-3}M$ glyoxalic acid in sulfuric acid, acetate, borate buffer, and sodium hydroxide solutions. Limiting current corrected for changes in viscosity (35).

the equilibria depicted were perturbed during electrolysis and kinetic currents were observed. The pH value can also affect the curves in the presence of slowly established equilibria, where diffusion-governed waves that are proportional to the concentration of electroactive species in the bulk of the solution appear on the curves. Three examples of this type will be given here:

Aldimine and ketimine formation follow the following overall reaction (37):

$$RNH_3^+ \overset{K_1}{\rightleftharpoons} RNH_2 + H^+ \tag{37a}$$

$$RNH_2 + >C{=}O \overset{K}{\rightleftharpoons} >C{=}NR + H_2O \tag{37b}$$

$$>C{=}NR + n_1e + m \cdot H^+ \xrightarrow{E_1} \text{products} \tag{37c}$$

$$>C{=}O + n_2e + m_2H^+ \xrightarrow{E_2} \text{products} \tag{37d}$$

At a given analytical concentration of the amine ($[RNH_3^+]$ + $[RNH_2]$), the wave of the Schiff base $>C{=}NR$ increases (36) with increasing pH in the form of a dissociation curve with an inflection point at pK_1; because the Schiff base can also be protonated; this treatment is oversimplified. The equilibrium is usually established rapidly during the preparation of the solution.

For 2-hydroxychalcones, the establishment of the equilibrium is slower. The waves at potential E of the chromanone formed (37) increase with increasing pH, corresponding to the increasing concentration of the reactive phenolate form of chalcone (38):

$$\text{C}_6\text{H}_4(\text{OH})\text{–CO–CH=CH–R} \rightleftharpoons \text{C}_6\text{H}_4(\text{O}^{(-)})\text{–CO–CH=CH–R} + \text{H}^+ \tag{38a}$$

$$\text{C}_6\text{H}_4(\text{O}^{(-)})\text{–CO–CH=CH–R} \underset{}{\overset{+\text{H}^+}{\rightleftharpoons}} \text{chromanone (O–CH(R)–CH}_2\text{–CO ring)} \tag{38b}$$

$$\text{chromanone (O–CH(R)–CH}_2\text{–CO ring)} + ne + m\,\text{H}^+ \xrightarrow{E_1} \text{products} \tag{38c}$$

$$\text{C}_6\text{H}_4(\text{O}^{(-)})\text{–CO–CH=CH–R} + ne + m\,\text{H}^+ \xrightarrow{E_2} \text{products} \tag{38d}$$

Finally, the cathodic waves of α,β-unsaturated ketones, of β-mercaptoketones $>\text{COCH}_2\overset{|}{\text{C}}\text{—SR}$, as well as the anodic waves of the mercaptans RS^- at a given analytical mercaptan concentration (38) change with pH according to the change in the acid–base equilibrium (39a) in eqs. (39):

$$\text{RSH} \rightleftharpoons \text{RS}^- + \text{H}^+ \tag{39a}$$

$$\text{RS}^- + >\text{COCH=}\underset{|}{\text{CH}} \overset{+\text{H}^+}{\rightleftharpoons} >\text{COCH}_2\underset{|}{\text{C}}\text{HSR} \tag{39b}$$

$$>\text{COCH=CH} + n_1e + m_1\text{H}^+ \xrightarrow[\text{cathodic}]{E_1} \text{products} \tag{39c}$$

$$>\text{COCH}_2\underset{|}{\text{C}}\text{HSR} + n_2e + m_2\text{H}^+ \xrightarrow[\text{cathodic}]{E_2} \text{products} \tag{39d}$$

$$n_4\text{RS}^- + \text{Hg} \underset{\text{anodic}}{\overset{E_3}{\rightleftharpoons}} (\text{RS})_{n_4}\text{Hg} + n_3e \tag{39e}$$

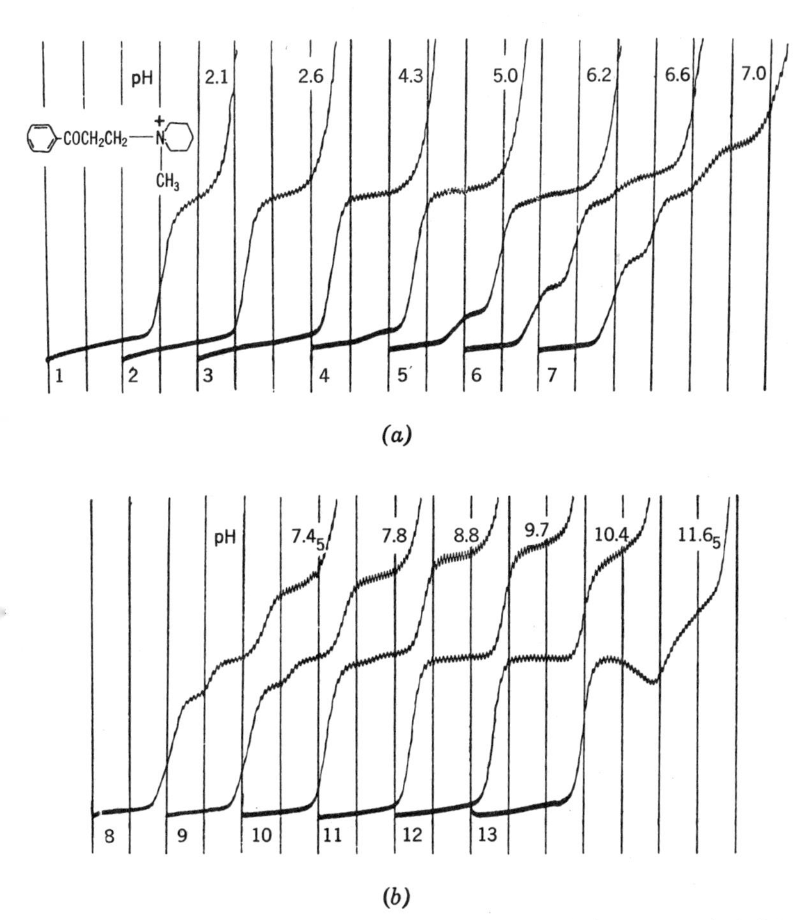

Fig. 10. Dependence of waves for β-piperidinopropiophenone methoiodide on pH. Britton-Robinson buffers, $2 \times 10^{-4}M$ depolarizer. pH values given on the polarogram. The curves were recorded 2 min. after mixing the stock solution with the deaerated buffer solution. curves (*1*)–(*3*) starting at −0.4 V., curves (*4*)–(*13*) at −0.6 V., 200 mV./absc., S.C.E., full scale sensitivity 3.2 μamp. (39).

There is still one more possibility for the observed pH dependence of polarographic waves. That occurs when a substance undergoes an irreversible chemical reaction (whereas in examples quoted so far only chemical equilibria were involved) that takes place in a time period comparable to that necessary for scanning the polarographic curve.

When recording the curves always start after the same period after the preparation of the solution, the recorded curves reflect the progress of the reaction. This case will be demonstrated on curves of β-piperidinopropiophenonemethoiodide (39) recorded at various pH values (Fig. 10). At pH < 5.0, only the wave of the Mannich base methoiodide is observed. With increasing pH, a wave at more positive potentials increases, corresponding to phenyl vinyl ketone, formed by Hoffmann degradation. At pH >9, only the wave of phenyl vinyl ketone is observed, the Hoffmann degradation taking place quantitatively before the recording of the curve is carried out.

B. REACTIONS INTERPOSED BETWEEN TWO ELECTRODE PROCESSES

The reactions taking place between two electrode processes, such as step IV in Table I, can correspond again either to acid–base equilibria or to chemical transformations, the rate of which is pH-dependent.

1. *Acid–Base Reactions*

The acid–base equilibria can be symbolized as follows (40):

$$A + n_1e \xrightarrow{E_1} B \tag{40a}$$

$$B + H^+ \rightleftharpoons BH^+ \tag{40b}$$

$$BH^+ + n_2e \xrightarrow{E_2} \text{products} \tag{40c}$$

Establishment of the acid–base equilibria can be either fast or comparable to the rate of the electrode process. An example of the former type is the reduction of aromatic and conjugated unsaturated hydrocarbons (R) (40), that in water-containing solutions or in aprotic solvents containing a sufficient amount of proton donors follows eqs. (41):

$$R + e \overset{E_1}{\rightleftharpoons} R^- \tag{41a}$$

$$R^- + HA \underset{\text{fast}}{\rightleftharpoons} \cdot RH + A^- \tag{41b}$$

$$\cdot RH + e \xrightarrow{E_2} RH^- \tag{41c}$$

$$RH^- + HA \rightleftharpoons RH_2 + A^- \tag{41d}$$

The uptake of the second electron occurs at potentials E_2 that are comparable to or even more positive than E_1. Then, only one two-electron wave is observed.

The uptake of the first electron is probably irreversible under these conditions. This would explain the observed pH-independence of half-wave potentials of such systems. A direct proof of the scheme (41) in water-containing solutions in which the first acid–base equilibrium is always shifted to the right-hand side is impossible. The reaction can moreover be complicated by the reaction $\cdot RH + R^- \rightarrow RH^- + \cdot R$.

It was possible to elucidate the mechanism by lowering the water content or the proton donor concentration in the aprotic solvent. Under these conditions two separate waves are observed corresponding to processes (41a) and (41e):

$$R + e \overset{E_1}{\rightleftharpoons} R^- \tag{41a}$$

$$R^- + e \overset{E_3}{\rightleftharpoons} R^{2-} \tag{41e}$$

The behavior in the intermediate range of proton donor concentration was not studied in such detail as to permit a decision whether the position of the equilibrium (41b) or the rate of its establishment governs the current in this region of proton donor concentration. The change from aqueous to aprotic solvents enabled distinguishing the rate of proton transfers in other two-electron processes, in particular, of carbonyl compounds and quinones.

Acid–base equilibria established with a rate comparable to that of the electrode process and hence perturbed by electrolysis, were encountered in the reduction of carbonyl compounds in alkaline media and in the reduction of phthalimide. The reduction of carbonyl compounds (41) occurs as follows (42):

$$CO + e \overset{E_1}{\rightarrow} CO^{(-)} \tag{42a}$$

$$CO^{(-)} + H^{(+)}[Na^{(+)}] \rightleftharpoons \dot{C}OH[\text{or } \dot{C}ONa] \tag{42b}$$

$$\dot{C}OH \text{ or } \dot{C}ONa + e \overset{E_2}{\rightarrow} COH^{(-)} \tag{42c}$$

$$COH^{(-)} + H^+ \rightleftharpoons CHOH \tag{42d}$$

$$CO^{(-)} + e \overset{E_3}{\rightarrow} CO^{(2-)} \tag{42e}$$

$$CO^{(2-)} + 2H^+ \rightleftharpoons CHOH \tag{42f}$$

The potentials at which the reduction steps E_1 and E_2 occur are comparable. Hence at a sufficiently low pH value at a given alkali metal ion concentration, one two-electron wave is observed. Because of the irreversibility of step E_1, the half-wave potential of this step is pH-independent. At a sufficiently high pH value the acid–base equilibrium (42b) is shifted left. Further reduction occurs at a more negative potential E_3 and two reduction steps E_1 and E_3 appear. Increase in alkali metal ion concentration can also increase the height of the more positive step.

The reduction of phthalimide (PM) (42) follows the following scheme (43):

$$\mathrm{PM} + \mathrm{H}^+ \underset{\text{fast}}{\rightleftharpoons} \mathrm{PMH}^+ \tag{43a}$$

$$\mathrm{PMH}^+ + e \xrightarrow{E_1} \cdot\mathrm{PMH} \tag{43b}$$

$$\cdot\mathrm{PMH} + \mathrm{H}^+ \overset{k}{\rightleftharpoons} \cdot\mathrm{PMH_2}^+ \tag{43c}$$

$$\cdot\mathrm{PMH_2}^+ + e \xrightarrow{E_2} \text{products} \tag{43d}$$

$$\cdot\mathrm{PMH} + e \xrightarrow{E_3} \text{products} \tag{43e}$$

In acid media, the rate of reaction (43c) with the rate constant k is sufficient so that all of the phthalimide is reduced in the form $\cdot\mathrm{PMH_2}^+$. Because potentials E_1 and E_2 are comparable, one two-electron step is observed (Fig. 11). With increasing pH, the rate of

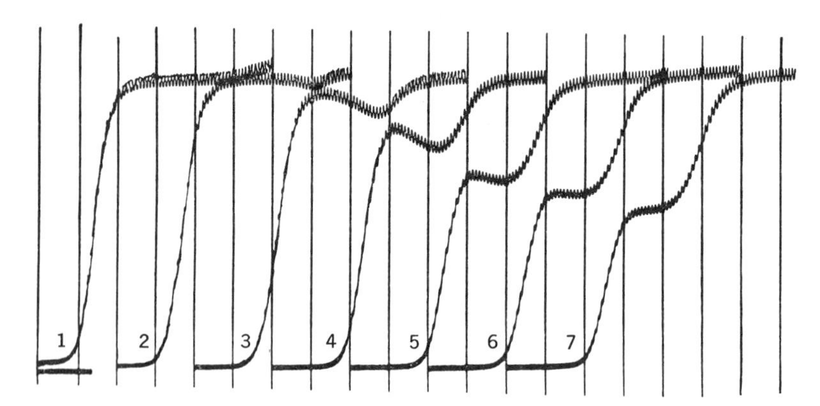

Fig. 11. Dependence of waves of phthalimide on pH. Britton-Robinson buffer, $10^{-3}M$ phthalimide, pH: (*1*) 3.95; (*2*) 4.7; (*3*) 5.65; (*4*) 6.35; (*5*) 6.87; (*6*) 7.1; (*7*) 7.45. Curves starting at −0.7 v., S.C.E., 110 mV./absc., h = 60 cm, t_1 = 3 sec., full scale sensitivity 2.8 μamp. (42).

(43c) with the rate constant k decreases. At pH > 7, the reduction occurs in two steps, the more positive E_1 and more negative E_3. pH dependence of the more positive wave $\bar{\imath}_l$ for $5.10^{-4}M$ phthalimide is given by eq. (44):

$$\frac{\bar{\imath}_l}{\bar{\imath}_d} = 1 + \frac{0.886(kt_1/K)^{1/2}[\mathrm{H}^+]}{1 + 0.886(kt_1/K)^{1/2}[\mathrm{H}^+]} \qquad (44)$$

Nonlinear dependence of the current $\bar{\imath}_l$ at pH 6–7 on the concentration of phthalimide, the shift of the $\bar{\imath}_l/\bar{\imath}_d$–pH dependence with phthalimide concentration, and pH-dependence of the half-wave potentials indicate a still more complicated mechanism. Dimerization, parallel or consecutive to protonation of the radical, seems to be a possible explanation.

2. Chemical Reactions

The systems in which chemical reactions are interposed between two electrochemical processes can be subdivided according to whether both the electrode processes take place on the same reaction center or on two electroactive groups.

An example of a chemical reaction placed between two electrochemical reactions actually occurring on the same reactive center is the reduction of *o*- and *p*-nitrophenols, *o*- and *p*-nitroanilines, and *p*-nitrosophenol.

The reduction of *o*- and *p*-nitrophenols (43) follows eqs. (45a)–(45c)(45).

$$HO{-}C_6H_4{-}NO_2 + 4e + 4H^+ \xrightarrow{E_1} HO{-}C_6H_4{-}NHOH + H_2O \qquad (45a)$$

$$HO{-}C_6H_4{-}NHOH \underset{k_{-1}}{\overset{k_1}{\rightleftharpoons}} O{=}C_6H_4{=}NH + H_2O \qquad (45b)$$

$$O{=}C_6H_4{=}NH + 2e + 2H^+ \overset{E_2}{\rightleftharpoons} HO{-}C_6H_4{-}NH_2 \qquad (45c)$$

Because the potential E_2 is more positive than E_1, the reduction occurs in one single wave. Because the lifetime of the quinoneimine intermediate is short, its hydrolysis is of no importance. On polarographic curves, the single observed wave corresponds to a six-electron reduction in acid or alkaline solutions where the rate of the dehydration reaction with the constant k_1 is acid- or base-catalyzed. In the medium pH-range where the dehydration rate is slow, the wave height corresponds to a four-electron process.*

The reduction of *p*-nitrosophenol (45–48) follows an analogous path, only the first step (45d) is reversible; that allows quantitative

$$HO\text{-}C_6H_4\text{-}NO + 2e + 2H^+ \underset{}{\overset{E_1}{\rightleftharpoons}} HO\text{-}C_6H_4\text{-}NHOH \qquad (45d)$$

treatments (48,49) to be applied.† The explanation of the more negative wave observed with *p*-nitrosophenol in acid media remains unsatisfactory.

Among systems with two electroactive groups and a chemical reaction interposed between two electron transfer processes, α-substituted ketones, α,β-unsaturated ketones, and diketones will be considered here.

The reduction of those α-substituted ketones in which the C—X bond in the grouping $COCH_2$—X is activated so that its reduction occurs at more positive potentials than that of the carbonyl group, as was observed for X = NR_2 (51,53,54), $^{(+)}NR_3$ (51), SR (50), $^{(+)}SR_2$ (15,52), OR (50), $^{(+)}PR_3$ (52), and halogens (55), follows the scheme in acid media (46):

The radical formed can dimerize, interact with solvent, be adsorbed, or be further reduced at more negative potentials.

* Electrolysis at pH 2.5 gave (44) 5% of quinoneimine and 89% of *p*-aminophenol. The suggested scheme was verified by chronopotentiometric measurements (45).

† The transfer of four electrons in acid media was confirmed coulometrically and the *p*-aminophenol formed in controlled potential electrolysis was identified spectrophotometrically (49). The scheme was verified using chronopotentiometric and potentiostatic methods, showing that $k_1 = 0.6\ sec^{-1}$ at pH 4.8 which corresponds to a half-life of 0.5 sec for the quinoneimine.

$$\mathrm{ArCOCH_2 - X + 2}e \xrightarrow{E_1} \underset{\underset{\mathrm{O}}{\vdots}}{\mathrm{ArC}}\mathrm{\cdots CH_2^{(-)} + X^-} \tag{46a}$$

$$\underset{\underset{\mathrm{O}}{\vdots}}{\mathrm{ArC}}\mathrm{\cdots CH_2^{(-)} + H^+} \underset{k_{-1}}{\overset{k_1}{\rightleftharpoons}} \mathrm{ArCO{-}CH_3} \tag{46b}$$

$$\mathrm{ArCOCH_3 + H^+} \underset{k_{-2}}{\overset{k_2}{\rightleftharpoons}} \mathrm{Ar}\underset{\underset{\mathrm{OH^{(+)}}}{\|}}{\mathrm{C}}\mathrm{CH_3} \tag{46c}$$

$$\mathrm{Ar}\underset{\underset{\mathrm{OH^{(+)}}}{\|}}{\mathrm{C}}\mathrm{CH_3} + e \overset{E_2}{\rightleftharpoons} \mathrm{Ar}\underset{\underset{\mathrm{OH}}{|}}{\dot{\mathrm{C}}}\mathrm{CH_3} \tag{46d}$$

On polarographic curves the two-electron wave at the potential E_1 is followed by another wave at a more negative potential E_2. The half-wave potential of this latter wave and its decrease with increasing pH at pH >6 due to a decrease of reaction rate with constant k_2 is analogous to that observed for acetophenone. The height of the more negative wave is nevertheless smaller than would correspond to a one-electron reduction. This is attributed to the fact that the enolate $\underset{\underset{\mathrm{O}}{\vdots}}{\mathrm{ArC}}\mathrm{\cdots CH_2^{(-)}}$ is an electroinactive reduction product that must be first transformed into the electroactive keto form. The rate of this transformation with constant k_1, which is acid–base catalyzed, limits the wave height of the more negative wave.

For some compounds of this type, and in particular for derivatives of branched ketones $COCR_2X$, a smaller wave of the keto group reduction is observed even at higher pH values where the unprotonized form of the ketone is reduced in a two-electron step. The rate of enol-keto transformation is again supposed to be the factor governing the current.

A similar picture was observed for α,β-unsaturated ketones. The first two-electron reduction step, split in acid solutions into two one-electron steps, is followed by another reduction wave, the potential of which is identical to that of the corresponding saturated ketone but the height of which is lower than predicted and reaches the theoretical height of a two-electron process only at pH >8. The following reduction scheme (47) has been suggested (56):

pH <2:

$$\left.\begin{array}{l} \mathrm{Ar{-}\underset{\underset{OH^{(+)}}{\|}}{C}{-}CH{=}CHR \rightleftharpoons Ar{-}\underset{\underset{O}{\|}}{C}{-}CH{=}CHR + H^{+}} \quad (47a) \\ \mathrm{Ar{-}\underset{\underset{OH^{(+)}}{\|}}{C}{-}CH{=}CHR + e \rightleftharpoons Ar{-}\underset{\underset{OH}{|}}{\dot{C}}{-}CH{=}CHR} \quad (47b) \end{array}\right\} \text{1st wave}$$

The dimer can dimerize, react with solvent, or be adsorbed.

$$\left.\begin{array}{l} \mathrm{Ar{-}\underset{\underset{OH}{|}}{\dot{C}}{-}CH{=}CHR + e \rightarrow Ar{-}\underset{\underset{OH}{|}}{\overset{-\delta}{C}}{\cdots}CH{\cdots}\overset{-\delta}{C}HR} \quad (47c) \\ \mathrm{Ar{-}\underset{\underset{OH}{|}}{\overset{-\delta}{C}}{\cdots}CH{\cdots}\overset{-\delta}{C}HR + H^{+} \rightleftharpoons Ar{-}\underset{\underset{OH}{|}}{C}{=}CH{-}CH_2R} \quad (47d) \end{array}\right\} \text{2nd wave}$$

$$\left.\begin{array}{l} \mathrm{Ar{-}\underset{\underset{OH}{|}}{C}{=}CH{-}CH_2R \underset{\text{slow}}{\overset{\text{acid catalysis}}{\rightleftharpoons}} Ar{-}\underset{\underset{OH^{(+)}}{|}}{C}{-}CH_2CH_2R} \quad (47e) \\ \mathrm{Ar{-}\underset{\underset{OH^{(+)}}{|}}{C}{-}CH_2CH_2R + e \rightleftharpoons Ar{-}\underset{\underset{OH}{|}}{\dot{C}}{-}CH_2CH_2R} \quad (47f) \end{array}\right\} \text{3rd wave}$$

The radical can undergo similar reactions as after the first step.

pH 4–6.

First, two waves are attributed to the same processes (47a)–(47d) as at pH <2, followed by (48):

$$\left.\begin{array}{l} \mathrm{Ar{-}\underset{\underset{OH}{|}}{C}{=}CHCH_2R \underset{\text{slow}}{\overset{\text{base catalysis}}{\rightleftharpoons}} Ar{-}\underset{\underset{O}{\|}}{C}{-}CH_2CH_2R} \quad (48a) \\ \mathrm{Ar{-}\underset{\underset{O}{\|}}{C}{-}CH_2CH_2R + H^{+} \underset{\text{slow}}{\rightleftharpoons} Ar{-}\underset{\underset{OH^{(+)}}{\|}}{C}{-}CH_2CH_2R} \quad (48b) \\ \mathrm{Ar{-}\underset{\underset{OH^{+}}{\|}}{C}{-}CH_2CH_2R + e \rightleftharpoons Ar{-}\underset{\underset{OH}{|}}{\dot{C}}{-}CH_2CH_2R} \quad (48c) \end{array}\right\} \text{third wave}$$

with subsequent reactions of the radical as above.

pH 6–9:

$$\mathrm{Ar{-}\underset{\underset{O}{\|}}{C}{-}CH{=}CHR} + 2e \rightarrow \mathrm{Ar{-}\overset{-\delta}{\underset{\underset{O^{(-)}}{|}}{C}}{\cdots}CH{\cdots}\overset{-\delta}{C}HR} \qquad (49a)$$

$$\mathrm{Ar{-}\overset{-\delta}{\underset{\underset{O^{(-)}}{|}}{C}}{\cdots}CH{\cdots}\overset{-\delta}{C}HR} + 2H^{+} \rightleftharpoons \mathrm{Ar{-}\underset{\underset{OH}{|}}{C}{=}CH{-}CH_2R} \qquad (49b)$$

first wave (49a, 49b)

$$\mathrm{Ar{-}\underset{\underset{OH}{|}}{C}{=}CH{-}CH_2R} \underset{\text{slow}}{\overset{\text{base catalysis}}{\rightleftharpoons}} \mathrm{Ar{-}\underset{\underset{O}{\|}}{C}{-}CH_2CH_2R} \qquad (49c)$$

$$\mathrm{Ar{-}\underset{\underset{O}{\|}}{C}{-}CH_2CH_2R} + 2e \rightarrow \mathrm{Ar{-}\overset{(-)}{\underset{\underset{O^{(-)}}{|}}{C}}{-}CH_2CH_2R} \qquad (49d)$$

$$\mathrm{Ar{-}\overset{(-)}{\underset{\underset{O^{(-)}}{|}}{C}}{-}CH_2CH_2R} + 2H^{+} \rightleftharpoons \mathrm{Ar{-}\underset{\underset{OH}{|}}{CH}{-}CH_2CH_2R} \qquad (49e)$$

second wave (49c–49e)

This scheme (49) explains changes of the more negative wave with pH.

A smaller wave of the reduction product has also been found (57,58) for benzil. The first two-electron reduction wave is followed by another wave that appears at potentials of the benzoin reduction, but is considerably smaller.* It was assumed that the enol form is formed first and that the height of the benzoin wave is limited by the rate of the enol–keto conversion. No details about pH-dependence have been given.

For *p*-diacetylbenzene (60), the second wave corresponding to the reduction of *p*-$CHOHCH_3$-substituted acetophenone cannot be observed at pH 3–6 but increases at lower and higher pH values. As the product of the controlled potential electrolysis prepared at pH 3–6 is electroactive, it is assumed that rearrangement of the primary electrolysis product occurs slowly.

C. REACTIONS CONSECUTIVE TO ELECTRODE PROCESSES

First-order reactions consecutive to the electrode process, i.e., step IV consecutive to III, Table I, can be detected from the study of

* On the other hand, for 2,2′-pyridil, the more negative wave, corresponding to 2,2′-pyridoin reduction, has a height corresponding to the 2,2′-pyridoin concentration (59). The enol–keto conversion is either faster, or the enol form is reducible, or the reduction in the first step follows another path than that for benzil.

polarographic curves only when electrode process III is reversible. In such cases, waves with shapes predicted by the equation for reversible processes are expected, the half-wave potential of which is shifted when compared with the equilibrium redox potential measured potentiometrically.

Among organic substances, such a system was found for ascorbic acid which is oxidized electrochemically to an unstable electroactive intermediate transformed by a fast chemical reaction into the electroinactive dehydroascorbic acid. It is assumed that the electroinactive form is hydrated. Anodic waves of ascorbic acid and other enediols (61–63) possess the ideal shape and the predicted pH-dependence of the half-wave potentials, but these are some 200 mV. more positive than the potentiometrically determined data. The relation (50) derived (64,65) for the dependence of the half-wave potentials on drop-time

$$E_{1/2} = E^\circ - (RT/2F) \ln 0.886(D_{\mathrm{Ox}}/D_{\mathrm{red}})^{1/2}kt_1 \qquad (50)$$

where k is the rate constant of the deactivation process was verified (64). The heights of the reduction waves found (66) for the reduction of dehydroascorbic acid and their half-wave potentials do not agree with this theory, indicating that the reaction scheme is incomplete.

V. Effect of Depolarizer Concentration

Deviations from the linear dependence of the limiting current indicate, apart from the effects of adsorption phenomena and catalytic processes that are detected and confirmed also by other methods, participation of a higher order reaction. At concentrations of the electroactive substance higher than about $8 \times 10^{-4}M$, such deviations are observed, e.g., for some nitrocompounds, but this phenomenon has not been studied systematically so far.

For higher order reactions following a reversible electrode process, it has been shown (67) that the half-wave potential and the wave shape depend on the concentration of the electroactive substance and drop-time. For a second-order consecutive reaction of the type (51).

$$\mathrm{Ox} + ne \rightleftharpoons \mathrm{Red} \qquad (51a)$$

$$2\,\mathrm{Red} \xrightarrow{k} \mathrm{products} \qquad (51b)$$

the half-wave potential follows eq. (52):

$$E_{1/2} = E^\circ - 0.36(RT/nF) + (RT/3nF)\ln ckt_1 \tag{52}$$

The half-wave potential depends hence on the concentration of the electroactive substance, c, the rate constant of the deactivation reaction k, and the drop-time, t_1.

The dependence on drop-time and/or concentration was found for the half-wave potentials for the reduction of carbonyl compounds (68), the tropylium ion (69) and the quasi-reversible reductions of the N-alkylpyridinium ions. In all of these cases, dimerization of radicals is considered.

The shape of polarographic curves were also computed (70) for the system (53)

$$\text{Ox} + e \overset{E_1}{\rightleftharpoons} \text{S} \tag{53a}$$

$$2\text{S} \rightleftharpoons \text{S}_2 \tag{53b}$$

$$\text{S} + e \overset{E_2}{\rightleftharpoons} \text{Red} \tag{53c}$$

but examples of organic systems showing such behavior have not been described yet.

VI. Effects of Ionic Strength, Solvent, and Other Factors

The common feature of the effects such as ionic strength, solvent kind and composition, temperature, etc., on polarographic curves is that these changes in the medium usually affect more than one step in the overall electrode process; e.g., a change in any of these factors can affect all the steps I, II, and III (Table I). Moreover, in some instances these effects result from more than one mechanism operating simultaneously. It is difficult to separate these effects into their components or to distinguish which of the modes of action dominates and which step in the electrode process is most affected.

A. EFFECTS OF IONIC STRENGTH AND NEUTRAL SALTS

Effects of ionic strength and the kind of neutral salt present in the supporting electrolyte can be exerted both on limiting currents and on half-wave potentials.

Diffusion-controlled limiting currents are usually, apart from the effect of change in viscosity, little affected by the ionic strength and nature of the added salt. Whenever an increase in the limiting current is observed with the increase of salt concentration, the presence of maxima of the second kind should be excluded by addition of low concentrations of surface-active substances and by microscopic observation; a streaming around the dropping electrode indicates maxima.

Limiting currents governed by the rate of chemical reaction are much more sensitive to changes in ionic strength. The change in wave height* that can be affected both by the volume and surface reaction can result from primary as well as secondary salt effects, changes in double layer composition, salting-out effects, etc. No general systematic study has been reported to separate these contributions. Some types of catalytic hydrogen evolution are specifically affected by the presence of certain cations, e.g., Co^{2+} and Ni^{2+} (71), As^{3+} (72), or Li^{+} (73).

The effects of the presence of neutral salts and of the nature of the cations present on half-wave potentials is often considered negligible when compared with the effects of other factors such as pH, solvent, or structure. This assumption is not always based on sufficient experimental evidence, and in many cases, the effect of neutral salts has not been studied in sufficient detail. The half-wave potentials of both reversible, e.g., quinones, and irreversible systems, e.g., carbonyl, nitro- and heterocyclic compounds, etc., can be sensitive to salt effects, and the use of the shift or half-wave potentials with the nature of the salt or its concentration as a proof of irreversibility is rather doubtful.

Reduction waves of anions of organic acids were generally shifted to more positive potentials (74) with increasing concentration of cations. This is probably a double-layer effect and hence beyond the scope of this review.

The behavior of compounds bearing no unit charge but accessible to polarization of their dipoles in the electrical field of the electrode, such as halogens and carbonyl and nitrocompounds, will be discussed here.

The effects of cations on the reduction of alkyl halides, first mentioned by the present author (75), have been later supported by ample experimental evidence (76–78). The shifts of half-wave potentials were interpreted (79,80) as due to the changes in ψ_1-potentials, i.e.,

* The effect of cation type and concentration on the height of the wave of carbonyl reduction in alkaline media was discussed earlier, p. 119.

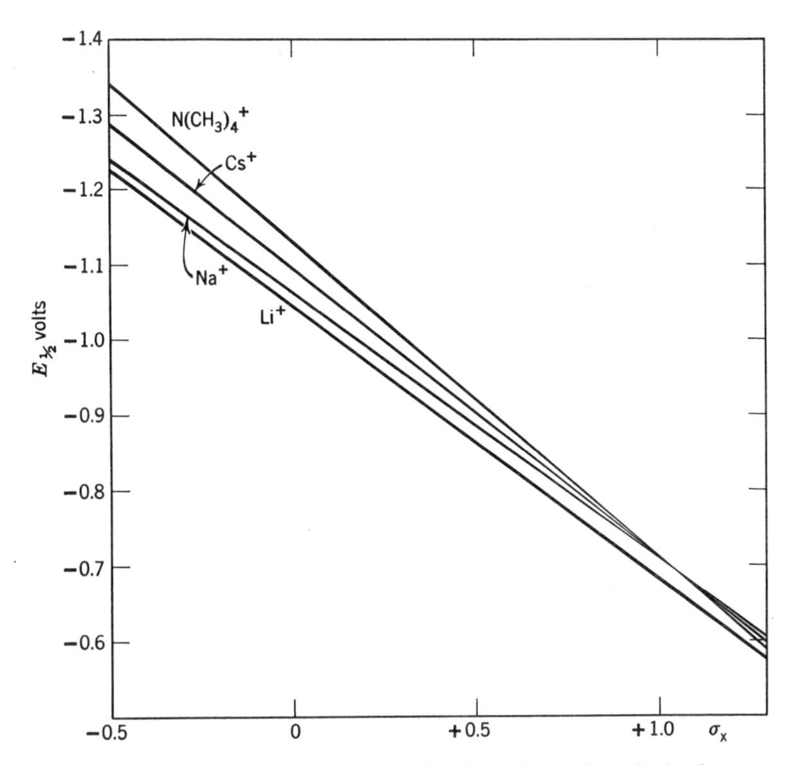

Fig. 12. Dependence of half-wave potentials $E_{1/2}$ of substituted nitrobenzenes in dimethylformamide (84) on Hammett substituent constants σ_X; plots for varying cation of the supporting electrolyte.

in the composition of the double layer. On the other hand, the salt effects on reduction waves of iodobenzenes (81) indicate participation of other factors as well.

For the half-wave potentials of the second, more negative, reduction wave of carbonyl compounds in acetate buffers (82) and for the second wave of quinones in dimethylformamide (83), a linear relation between the half-wave potential and the logarithm of the cation concentration has been observed. This has been interpreted either as due to an effect of the ψ-potential (82) or as due to ion pair formation (83). The interpretation of results in aqueous solutions is complicated by the dependence of the half-wave potential of the second wave on the buffer type and composition (27) at constant ionic strength.

The first wave of the carbonyl compounds is claimed (82) to be independent of ionic strength; the half-wave potentials of the reversible first wave of quinone reduction in dimethylformamide and acetonitrile have been found (83), on the other hand, to be dependent on cation type and concentration, even when to a lesser degree than the halfwave potentials of the second, more negative irreversible wave.

For nitro compounds, small effects have been observed (78) in aqueous and alcoholic solutions, but considerable shifts were observed in dimethylformamide in the presence of various cations (84). The

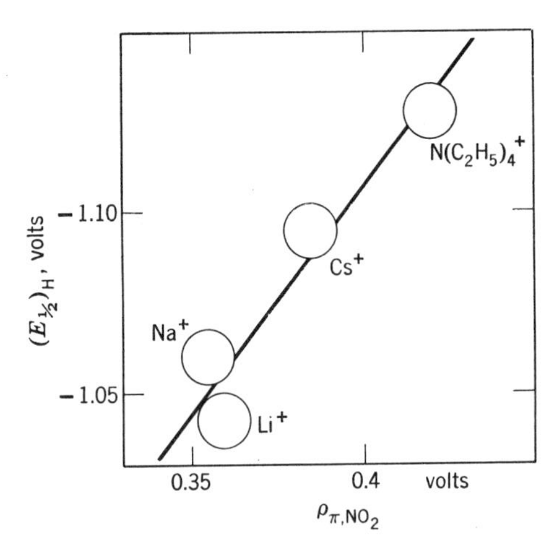

Fig. 13. Dependence of the half-wave potential of the parent unsubstituted compound $(E_{1/2})_H$ for a series of substituted nitrobenzenes in solutions containing a given cation on the value of the reaction constant ρ_π, NO_2 obtained in the same supporting electrolyte.

shifts of half-wave potentials of reversible one-electron waves were attributed (84) to ion pair formation.

Linear free energy relationships (Sec. IX-C) can be used for the characterization of the cation effects. Half-wave potentials of substituted nitrobenzenes in dimethylformamide (84) when plotted against Hammett substituent σ-constants give a linear plot, the slope of which, $\rho_{\pi,R}$ depends on the nature of the cation of the supporting electrolyte (Fig. 12). As predicted (85) the value of the reaction constant $(\rho_{\pi,R})_{Me}$ expressing the susceptibility of the electroactive group towards substituent effects depends on the composition of the supporting electrolyte in addition to the character of the electroactive group R.

Moreover, there is a further regularity in the value of the reaction constant $(\rho_{\pi,R})_{Me}$ obtained in the presence of various cations Me^{n+} and the nature of the cation. The value of the reaction constant $(\rho_{\pi,R})_{Me}$ has been found empirically to be a linear function of the half-wave potential of the unsubstituted parent compound $[(E_{1/2})_H]_{Me}$ recorded in the same supporting electrolyte (Fig. 13). In the empirical equation (54)

$$[(E_{1/2})_H]_{Me} = \mathcal{H}(\rho_{\pi,R})_{Me} + K \tag{54}$$

$\mathcal{H}$ and K seem to be universal constants for benzenoid compounds reduced in the side-chain. It is an extension of analogous equations derived (85) for the sets $[(E_{1/2})_H]_R$ and $\rho_{\pi,R}$ in which the electroactive group R was changed.

Hence the more positive the potential of the unsubstituted parent compound, the less sensitive is its half-wave potential to the effects of substituents. The sequence of the cation effects parallels that of crystallographic ionic radii. The larger the cation, the more negatively the reduction of the parent compounds occurs and the more sensitive is the electrode process towards the substituent effects.

These correlations illustrate the importance of the choice of the substance used for testing the cation effects. The dangers involved in a random choice are revealed by inspection of Figure 12. If a compound with $\sigma \doteq -0.3$ were chosen for the study, its half-wave potential would be found to be very sensitive to the nature of the cation. Oppositely for a nitrobenzene bearing a substituent with $\sigma \doteq 1.3$, the half-wave potential would be found to be practically independent of the kind of cation used. For $\sigma > 1.3$, the sequence of the cation effects would even be reversed.

Usually the half-wave potential of an unsubstituted parent compound is used for the study of the cation effects. In the present case, the sensitivity of the half-wave potentials of the unsubstituted parent compound towards cation effects is relatively good even when those of p-methoxy derivatives would be even better. But it cannot be predicted that this relation is generally valid for other electroactive groupings. Hence it seems more desirable and important to express the cation effects in terms of their influence on the value of the reaction constant $[\rho_{\pi,R}]_{Me}$ than on the value of a half-wave potential of one single substance $[E_{1/2}]_{Me}$.

The possibility of using the effects of neutral salts and type of ions on polarographic curves for the interpretation of the course of the electrode process is at present rather limited, the more so since it is impossible to distinguish between the effect on a species bearing a unit charge and on an uncharged, polarizable molecule.

B. SOLVENT EFFECTS

The possibilities that solvent effects offer for use as a diagnostic tool for elucidating the electrode process are limited just as are those of the salt effects. Few regularities in the relations between the solvent composition and polarographic behavior have been found, and, of those found, the validity was not extended to a broader scope of solvents of various types.

The nature of the solvent used can affect both the limiting currents and half-wave potentials. Apart from the effects on viscosity, the limiting diffusion currents can be affected by a change in solvation. Currents governed by the rate of a chemical reaction can be affected by changes in the rate and equilibrium constants. Those limiting currents of which the governing reaction depends on acidity can be affected indirectly by change in the acidity caused by the solvent both in the bulk of the solution and at the surface of the electrode. In some cases, the mechanism of the electrode process can change principally when two or more solvents are compared. A rather general example of that type is the comparison of reduction of various types of organic compounds in amphiprotic solvents, e.g., water or alcohols, and in aprotic solvents, such as dimethylformamide, acetonitrile, dimethylsulfoxide, etc. As mentioned in Sec. IV-B, the radicals formed in the uptake of the first electron persist in pure aprotic solvents and undergo reduction only at more negative potentials, whereas in aqueous solutions, the radical is protonized and further reduced. Hence a two- (or more) electron process is observed in the presence of water, whereas one-electron transfer takes place in the nonaqueous solvent. The polarographic curves in aprotic solvents are, in principle, simpler than those in waterlike solvents, but the limited knowledge about the chemical and electrochemical properties of these solvents, e.g., acid–base equilibria, electrolytic dissociation, ion-pair formation, solvation, etc., results in a rather limited choice of diagnostic tools for the elucidation of the electrode process, in aprotic solvents as given in Table II.

Comparison of half-wave potentials in various solvents involves certain experimental complications. Whereas the voltage drop iR is easily eliminated by proper circuitry, the problem of choosing a proper reference electrode and the diffusion potentials is a serious one. The use of pilot compounds, such as Rb^+ or naphthaquinones, proved well, but quantitative comparison can be carried out only for a series of compounds of which the half-wave potentials were measured under the same experimental conditions.

Among the substances of which the half-wave potentials are pH-independent, compounds with a carbon–halogen bond were studied in some detail. It has been shown (76,77,86–88) that with increasing concentration of ethanol or dimethylformamide, the half-wave potential is shifted towards more negative potentials. The relative effect of the change in the solvent composition is greater the more positive the reduction of the C—X bond occurs.* This effect is ascribed (89) to the desorption of depolarizer from the electrode surface rather than to a decrease in the value of the rate constant of the electron transfer k_e.

Among the half-wave potentials of pH-dependent systems, those of carbonyl derivatives in acid media will be mentioned first. In these cases in which a proton transfer precedes the electrode process proper, such as the more positive wave of benzaldehyde (86), the half-wave potentials are shifted towards more negative values with increasing concentration of organic solvents much more than they should be, if they are caused only by the change in the pH of the solution. The shift in the half-wave potential is attributed (89) to a decrease in the surface concentration of the electroactive species, in addition to a change of pH and pK_a of the protonated depolarizer.

Similarly, for the reduction of the nitro group in *p*-dinitrobenzene, a greater shift, 85 mv. with an apparent change in pH of one unit, was observed (78). The observed shifts of half-wave potentials of *p*-nitrochlorobenzene from -0.84 V. in methanol to -1.18 V. in acetonitrile and -1.41 V. in dimethylformamide in an unbuffered solution (0.1M sodium iodide) can be affected, in addition to hydrogen

* This is interesting to compare with the substituent effects, Section IX, C where the more positively the reduction occurs, the less it is sensitive to substituent effects and the smaller value of the $\rho_{\pi,R}$ constant.

bond formation between the electroactive species and the solvent and also by a change in acidity conditions at the electrode surface during the electrolysis.

Increase of organic solvent concentration in a water mixture usually decreases complications due to the adsorption of the electroactive species or the electrolysis product.

C. EFFECTS OF TEMPERATURE, IRRADIATION, AND PRESSURE

The diffusion-limiting current is known to increase by some 1.8% per degree with increasing temperature. A current increasing sharply with increasing temperature, e.g., by more than 3% per degree, can be attributed to a kinetic current governed by the rate of a chemical reaction preceding the electrode process proper. On the other hand, not all kinetic currents show this great a temperature dependence, and hence by finding the temperature coefficient of 1.8% per degree, the presence of a kinetic current component cannot be excluded. Adsorption currents show temperature coefficients that can be smaller, equal, or greater than 1.8% per degree. Hence, dependence of the limiting current on temperature is of little use for confirmation or exclusion of adsorption as the governing factor.

Effects of temperature on half-wave potentials can be easily measured but are interpreted with difficulty. It is a complex problem to separate from the observed shift the contributions corresponding to steps I, II, and III (Table I). Determination of the gross (total) activation energy from the observed shift of half-wave potentials seems to be irrelevant.

So far, no effect of the illumination of the polarographic electrode on the electrode process proper (Table I, step III) that would exert itself on the half-wave potential of the electroactive species has been found (91). On the other hand, electroactive species or electrolysis products can be formed or cleaved by photochemical reactions. These reactions are indicated by a change in the height of the polarographic limiting currents after illumination. The illumination can be carried out continuously, intermittently, or using the flash technique. Measuring the current can be done either continuously during the illumination or started after the illumination is cut off (reaction in the dark). The photochemical change can take place either at the surface of the electrode only, or in the bulk of the solution, or in both the vicinity of the electrode and the solution. The reaction can take place comparably fast with the electrode process, and a "photokinetic current" is observed. For slow reactions, the diffusion current of one of the components is measured. Among the possible types of mechanisms, the following ones were investigated (92).

Among the photochemical reactions involved in the electrode process, a reaction parallel to the electron transfer was considered (55):

$$A + n_1e \overset{E_1}{\rightleftharpoons} B \quad (55a)$$

$$A + h\nu + S \overset{k}{\rightarrow} B \quad (55b)$$

$$B + n_2e \overset{E_2}{\rightarrow} \text{products} \quad (55c)$$

The electrochemical step (55a) can be irreversible, too. Illumination in the latter case results in the decrease in current i_A and increase in current i_B. An example of that type is the behavior of ketones:

$$R_2C{=}O + H^+ \rightleftharpoons R_2COH^+ \quad (56a)$$

$$R_2COH^+ + e \overset{E_1}{\rightarrow} R_2\dot{C}{-}OH \quad (56b)$$

$$R_2CO + h\nu + (CH_3)_2CHOH \overset{k}{\rightarrow} R_2\dot{C}{-}OH \quad (56c)$$

$$R_2\dot{C}{-}OH + e \overset{E_2}{\rightarrow} \text{products} \quad (56d)$$

Experimentally, this reaction at the surface of the electrode is complicated by the reaction with the constant k taking place in the bulk of the solution and by photochemical cleavage of the ketyl $R_2\dot{C}OH$.

In the slow photochemical processes, either the oxidized electroactive form B results,

$$A + h\nu + S \overset{k}{\rightarrow} B \quad (57a)$$

$$B + ne \rightleftharpoons C \quad (57b)$$

or is decomposed:

$$B + h\nu + S \overset{k}{\rightarrow} D \quad (58a)$$

$$B + ne \rightleftharpoons C \quad (58b)$$

In both these systems, the electrochemical process can be either reversible or irreversible. Several types of photochemical reactions follow these schemes (92). A linear relationship was found (91) between the quantum yields and half-wave potentials in a series of substituted benzoquinones. For reversible systems the regeneration of the reduced form C can take place according to the scheme (59):

$$\mathrm{B} + h\nu + \mathrm{S} \xrightarrow{k} \mathrm{C} \tag{59a}$$

$$\mathrm{B} + ne \rightleftharpoons \mathrm{C} \tag{57b}$$

This process takes place with quinoid systems in which the increase of the anodic hydroquinone wave after irradiation is observed. Finally, for a reversible system, the reduced form C can be photochemically deactivated:

$$\mathrm{C} + h\nu + \mathrm{S} \xrightarrow{k} \mathrm{D} \tag{60a}$$

$$\mathrm{B} + ne \rightleftharpoons \mathrm{C} \tag{57b}$$

It is assumed that reduction of anthraquinone-2-sulfonic acid follows that scheme and that C is the semiquinone formed.

In all of these cases, the increase or decrease of the limiting diffusion current of species B or C with time is followed. The technique used depends on the half-time $\tau_{1/2}$ of the reaction with the rate constant k. For $\tau_{1/2} > 15$ sec., mean polarographic currents and for faster reactions, instantaneous currents are recorded; for $\tau_{1/2} > 0.15$ sec. with a string galvanometer and for $\tau_{1/2} > 0.0015$ sec., with an oscillograph (92).

These types of study are not restricted to optical radiations, but can be applied to any type of radiations that induce chemical reactions. The γ-radiations have already been applied in such a way, but so far no organic reactions have been studied.

Irradiation, even by diffuse daylight, can sometimes exert an unwanted effect on the polarographic behavior of organic compounds or on the reaction rates studied polarographically. It is advisable to check always the effect of light on the system studied and if necessary to prevent the solution studied from illumination.

Polarographic measurements can be carried out even at pressure of several thousands atmospheres (93). Both limiting currents and half-wave potentials change with pressure, even when the measurements of the latter are neither accurate nor comprehensive enough to allow discussion. No organic systems have been studied so far.

None of the techniques mentioned in this paragraph currently brings more general contributions to the elucidation of electrode processes.

D. SURFACE-ACTIVE SUBSTANCES

Addition of surface-active substances can affect the electrode process proper Table I (step III) and alter the half-wave potential, but these effects belong rather among the specific electrochemical problems. Among the effects of surface-active substances on limiting currents is the observation of greatest interest for the elucidation of electrode processes that in the presence of surfactants, the wave is suppressed. Only the first electrode process takes place while the successive steps are hindered and occur at a more negative potential. This influence can be used to separate this first step from others. In this way, uptake of the first electron in the four-electron reduction of nitro-compounds at higher pH values was detected (128). Also, chemical reactions occurring at the electrode surface can be suppressed by the addition of surface-active substances, as was shown for ninhydrin dehydration.

VII. Products and Intermediates

The techniques and treatments described in Secs. IV–VI enable us to detect some intermediates or to predict the products in some instances, but generally are of greatest importance for distinguishing the composition and form of the electroactive species. Before drawing any conclusions concerning the mechanism of the electrode process, it is essential to detect and identify all products and intermediates that are sufficiently stable. The methods used in achieving this goal are mentioned in this chapter. The methods can be divided according to the stability of the intermediate or the product. Detection of unstable intermediates, species of intermediate stability, and relatively stable species are discussed separately.

A. UNSTABLE INTERMEDIATES

The far most common type of unstable intermediates formed during the organic electrode processes are radicals and radical anions. The formation of a radical is often demonstrated by the presence of a one-electron step on polarographic curves. To confirm radical formation, electron spin resonance (ESR), the rotating disk method, radical scavengers, the Wall effect, and solvent effects can be used.

ESR is the most convincing method and offers the widest possibilities (94,95). In the first attempts (96) to detect electrolytically generated radicals, the electrolysis was carried out in an organic solvent, with the sample frozen and placed in an ESR spectrometer. Nevertheless, the potentialities of the combination of electrolysis and ESR were fully exploited only after the electrolysis was either carried out directly in the microwave cavity (97) or the generated radical was pumped through the sample cell placed in the cavity (98). The former system is advantageous for detecting short-lived radicals. For systems using external generation, the shorter the life-time, the faster must be the transfer. This latter technique also allows quantitative measurement of the radical concentration. Sufficiently stable radicals can be generated mainly in nonaqueous solvents, but some radicals are sufficiently stable to be followed even in water-containing media (94,99).

Among radicals of which the formation was confirmed in this way, nitrobenzene anion (97), radicals and anion radicals derived from ketones (94,96,100), and semiquinones (101,102) can be mentioned.

In some cases, more complex reaction types were elucidated. Hence, whereas all three chloronitrobenzenes yield the expected chloronitro anion similarly as do other substituted nitrobenzenes, for *o*-bromo- and all three iodonitrobenzenes, dehalogenation occurs in the course of the electrolysis of halogenonitrobenzenes in dimethylformamide (94,103). Formation of the nitrobenzene anion radical can be proved by ESR. Using mixtures of water and dimethylformamide of varying composition, the contribution of dehalogenation and that of halogenonitro anion radical formation can be varied. It has been shown (94) that the two-electron reduction resulting in substitution of halogen by hydrogen is followed by a one-electron uptake giving rise to the nitrobenzene anion radical.

Elimination of a nitro group was also reported for tertiary aliphatic nitrocompounds (104) and polynitrobenzenes (105,106). Nitrophenol anions were the predominating products in the latter case (106). Similarly, elimination of a nitro group occurs in 9-nitroanthracene (94).

In the reduction of 4-amino- and 4-fluorobenzonitriles (98) a cleavage of the radical formed in the one-electron uptake occurs. Ions NH_2^- and F^- are formed with a subsequent dimerization of the two benzonitrile radicals to 4,4-dicyanobiphenyl. This compound takes up another electron to form an anion radical.

For relatively more stable radicals, the ESR measurements make it possible to follow some of the reactions of the generated radicals,

Either a continuous generation or an interrupted generation connected with a recording of the time change in the radical concentration can be used (99,107). For evaluating concentration changes of the radical during continuous generation an analog computer modelling (108,109) would be advantageous.

The anion radical formed in the electrolysis of nitrobenzene in dimethylformamide solutions or in aqueous alkaline solutions containing 10% methanol as solvent and 0.06% triphenylphosphine oxide as surfactant to prevent the uptake of further electrons was studied in most detail (99). From measuring time changes in the reduction current of the nitro group and in the anodic currents of the anion radicals and phenylhydroxylamine at more positive potentials it was possible to determine the stoichiometric relations that correspond to overall reaction (61):

$$4C_6H_5NO_2^{(-)} + 4H_2O \rightarrow 3C_6H_5NO_2 + C_6H_5NHOH + H_2O + 4OH^- \qquad (61)$$

The time change in anion radical concentration is affected by two parallel reactions, one first-order and one second-order. The first-order rate constant is proportional to proton concentration; the second-order constant consists of two components: one is pH-independent, the other is linearly proportional to proton concentration. It is assumed that the pH-dependence is caused by equilibrium (62a):

$$C_6H_5NO_2^{(-)} + H^+ \rightleftharpoons C_6H_5NO_2H \qquad (62a)$$

which is under the conditions used shifted to the left-hand side. The first-order rate is attributed to reaction (62b):

$$C_6H_5NO_2H \rightarrow C_6H_5NO^{(+)} + OH^{(-)} \qquad (62b)$$

The cation radical can undergo further reactions with radical anions to give the products shown in the overall reaction.

The pH-dependent component of the second-order rate is ascribed to dismutation (62c):

$$C_6H_5NO_2^{(-)} + C_6H_5NO_2H \rightarrow C_6H_5NO_2 + C_6H_5NO_2H^{(-)} \qquad (62c)$$

The ionic species formed can dehydrate to nitrosobenzene and, after reaction with further radical anions, can yield products predicted by the overall reaction.

The pH-independent component of the second-order reaction rate is attributed to dismutation (62d):

$$2C_6H_5NO_2^{(-)} \rightarrow C_6H_5NO_2 + C_6H_5NO_2^{(2-)} \qquad (62d)$$

The dianion can be cleaved to nitrosobenzene which can react with anion radicals as mentioned above. In nonaqueous solutions, this equilibrium can be considered to be shifted to the left-hand side.

The reduction of camphorquinone in the presence of surface-active substances at pH >7 occurs with the consumption of one electron (107). The anion radical formed was followed using its anodic reversible wave as well as absorption maxima at 330 mμ. Because after interruption of the electrolysis, the camphorquinone concentration does not change with time, it was deduced that the stabilization reaction of the radical that in alkaline media follows second-order kinetics is not a dismutation process. At pH 9, the stabilization reaction follows first-order kinetics and corresponds to protonation of the anion radical. The rate constant determined for this reaction is in good agreement with the rate constant computed from the pH-dependence of polarographic limiting currents.

The radicals formed in the reduction of benzaldehyde and benzophenone are too unstable to be detected; products of subsequent reactions manifest themselves by anodic waves (107).

From the line broadening measured as a function of the concentration of the parent compound, it is possible to determine (94) the second-order rate constant k_{exc} of the homogeneous electron exchange reactions of the type:

$$R_1^{(-)} + R_2 \xrightarrow{k_{exc}} R_1 + R_2^{(-)} \tag{63}$$

The value of the rate constant k_{exc} for the system benzonitrile anion–benzonitrile in dimethylformamide was estimated to be 2×10^8 liter mole^{-1} sec.$^{-1}$ using the electrochemical generation of anion radicals inside the cavity. The rate of the exchange reaction k_{exc} in the benzonitrile system is little affected by the addition of water; this is explained by a similar aquo solvation of benzonitrile and its anion. On the other hand, for the pair naphthalene–naphthalene anion, the rate of the exchange reaction is slowed by addition of water, similarly as for the pair, benzophenone–benzophenone anion, in agreement with the strong solvation of the benzophenone anion radical.

Among the other methods considered for detection of unstable intermediates mentioned above, the rotating ring disk electrode (110) has so far been used only in a few cases of organic systems (111). Detection of radical formation by addition of scavengers and by an increase of the surface, e.g., by addition of glass wool or beds in a polarographic vessel, was attempted only in those cases in which radicals were either formed in the bulk of the solution or in a chemical reaction preceding the electrode process proper. The effect of nonaqueous solvents that can stabilize a radical and hence limit the electrode process to a one-electron step was discussed earlier (p. 119). Here the effect of the solvent on the stabilization of an intermediate can be only demonstrated in the example of reduction of 2,2-dinitropropane (112). In dimethylformamide solution containing a proton donor such as phenol, the observed two-electron wave is attributed to the sequence of reactions (64):

$$(CH_3)_2C(NO_2)_2 + 2e \rightarrow (CH_3)_2C{=}NO_2^{(-)} + NO_2^{(-)} \quad (64a)$$

$$HNO_2 \rightleftharpoons NO_2^{(-)} + H^+ \quad (64b)$$

$$(CH_3)_2C{=}NO_2^{(-)} + HNO_2 \rightarrow (CH_3)_2C(NO_2)NO + OH^{(-)} \quad (64c)$$

In aqueous acid media, the first reduction step corresponds to a four-electron uptake. It is assumed that $(CH_3)_2C(NO_2)NO$ is stabilized no more and is further reduced according to the scheme:

$$(CH_3)_2C(NO_2)NO + 2e + 2H^+ \rightarrow (CH_3)_2C{=}NOH + HNO_2 \quad (64d)$$

B. INTERMEDIATES AND PRODUCTS WITH LIMITED STABILITY

More stable species occurring as electrochemical intermediates or as products undergoing subsequent chemical reactions, i.e., formed in steps III, IV, or V (Table I) can be detected using rectangular voltage polarization (4,5), triangular voltage polarization by single sweeps, and oscillographic polarography with alternating current (10,113). The methods using triangular voltage polarization can be distinguished according to the rate of voltage scanning:

Intermediates with longer lifetimes can be detected using such a scanning rate that the whole studied voltage range is scanned in several minutes. Polarographic recording of the *i-E* curves is used and the hanging mercury drop proves the most suitable electrode (6,7). Those with a shorter lifetime can be studied using a scanning rate in which the time necessary for scanning the entire voltage range studied is of the order of seconds or fraction of seconds. The *i-E* curves in these single-sweep methods (9) are recorded oscillographically, using the dropping mercury electrode.

There is one condition to be fulfilled that is common to all these methods: It is assumed that the intermediate or product studied is electroactive in the potential range available in the medium used, i.e., in the same solution in which it has been generated.

Using rectangular voltage polarization, polarographic waves of reduction products are sometimes obtained even in cases where the product would be too unstable for identification using controlled potential electrolysis. Examples of that type are the waves of quinoid compounds resulting in oxidation of hydroquinones in alkaline media, in which the quinones would be unstable, or the waves of the oxidation products of dihydroxyfumaric acid.

Indentations on the oscillographic $dE/dt = f(E)$ curves show the presence of an electrolysis product. Detection of the intermediate or product is proved by comparison of the pattern of the studied system with the $dE/dt = f(E)$ curves obtained under identical experimental conditions with model substances assumed to correspond to the intermediate or product formed. Different conditions and mainly the cyclization of the electrode process sometimes give rise to electrolysis products differing from those obtained in classical polarography. The possibility of changing the potential range enables us to detect at which potential the studied species is formed.

An example of applications of that type is the study of the course of nitrobenzene electrolysis in alkaline media (114,115). By distinguishing the capacitive from the electrolytic indentations, by comparing curves obtained with dropping and streaming* electrodes, by separating indentations due to the original electroactive species from those caused by artifacts† and particularly by comparing the observed indentations with those obtained for expected intermediates and products, it was possible to consider the following reactions in 0.1M NaOH (65):

$$C_6H_5NO_2 + 4e + 4H^+ \rightarrow C_6H_5NHOH + H_2O \tag{65a}$$

$$C_6H_5NHOH \rightleftharpoons C_6H_5NO + 2e + 2H^+ \tag{65b}$$

$$C_6H_5NHOH + C_6H_5NO \rightarrow C_6H_5N{=}\underset{\displaystyle \overset{|}{O}}{N}C_6H_5 \tag{65c}$$

$$C_6H_5N{=}\underset{\displaystyle \overset{|}{O}}{N}C_6H_5 + 4e + 4H^+ \rightarrow C_6H_5NHNHC_6H_5 \tag{65d}$$

$$C_6H_5N{=}NC_6H_5 + 2e + 2H^+ \rightleftarrows C_6H_5NHNHC_6H_5 \tag{65e}$$

* Relatively slower consecutive chemical reactions cannot take place when these electrodes are employed.

† Intermediates formed at extremely negative or positive potentials that are electroactive and due to cyclization are shown on the $dE/dt = f(E)$ curves.

For hydrazobenzene, the participation of mercury salt formation cannot be excluded. In the presence of surface-active substances, the first reduction step of nitrobenzene can be split into one electron process corresponding to the formation of $C_6H_5NO_2^-$, followed by the uptake of three electrons.

Similarly, formation of a hydrazine derivative in the course of the reduction of sydnones at pH > 6 was confirmed (116) by a reversible indentation on the $dE/dt = f(E)$ curves corresponding to mercury salt formation.

Cathodic or anodic peaks on the *i-E* curves recorded in the single-sweep methods (9) can be applied for identification or verification of products and intermediates of electrode processes. Use of this technique is simpler than of $dE/dt = f(E)$ because of the absence of cyclization and changes due to capacity phenomena. On the other hand, the resolving power of the single-sweep method for identification is smaller in some cases than that of oscillographic polarography. An example of the application of the single-sweep method for identification of an electrolysis product can again be the formation of a hydrazine derivative in the reduction of sydnones (116) at higher pH values where an anodic hydrazine peak was observed.

The hanging mercury drop electrolysis (6,7) differs mainly from the oscillographic single-sweep method in the rate of voltage scanning. Moreover, polarizations with the former need not be carried out in the whole potential region but only in an appropriate potential range, and it is also possible to generate the intermediate or product first by electrolysis at a selected potential and only then to start the voltage scanning. To divide the whole curve into appropriate sections and to record the partial curves in the forward and the reverse trace in each range proved useful in cases when the curve recorded in the whole available potential range possessed a complicated form. The possibility of generation of the product at a preselected potential is of particular importance with organic systems that do not form amalgams for which the shape of the reverse trace depends on the amount of the product of the electrode process formed.

Examples of applications of *i-E* curves recorded using a hanging mercury drop electrode are the detection of the system *p*-phenylenediamine → *p*-quinonediimine (6) as the intermediate in the reduction of *p*-nitroaniline, identification of two reversible redox systems, and three irreversible reduction processes resulting in the reduction of p-dinitrobenzene (6), formation of radicals in the reduction of ketones (6), and the formation of hydrogen sulfide in the reduction of pyridoxthiol (25).

C. STABLE INTERMEDIATES AND PRODUCTS

The formation of stable intermediates and products in those cases in which the species formed is electroactive in the same solution in which it was generated, can be carried out using the methods given in Sec. VII-B. In some cases, moreover, the stable intermediates can be identified from the appearance of a separate polarographic wave. Electrolysis products that are electroinactive in the medium in which they are formed or that can be identified using other methods only, can be detected by controlled potential electrolysis.

1. *Intermediates that Give Rise to a Separate Wave*

When an electroactive species that is reduced at potential E_1 forms an intermediate that is electroactive in the same supporting electrolyte, several possibilities exist: (*a*) When the product is oxidized at potentials more positive than E_1, no change of the polarographic curve is observed.* (*b*) If the reduction product were oxidized at more negative potentials, a decrease of the limiting current would be observed. No example of this type has been described. (*c*) The reduction product of the first electron uptake is further reduced at potentials more positive than E_1 or only slightly different from E_1. In this case, only one single wave is observed on polarographic curves the height of which corresponds to the total electron uptake, i.e., the uptake corresponding to intermediate formation and that corresponding to intermediate reduction. This case is rather frequently encountered in organic polarography, but the shape of the polarographic curves does not alone allow us to distinguish the intermediate, and other methods must be adopted. (*d*) The reduction product formed at E_1 is further reduced at a more negative potential E_2. In this case, two waves appear on polarographic curves at potentials E_1 and E_2. In some cases, the identification of the process responsible for the wave at E_2 allows us to distinguish the product formed at potential E_1 and at more negative potentials.

* Methods given in Sec. VII-B are suitable for detection of an intermediate with such properties.

The two reduction processes at potentials E_1 and E_2, respectively, can correspond either to a successive reduction of the same group or to the electrolysis of two electroactive centers in the molecule.

An example belonging to the first group is the reduction of aliphatic nitro compounds in acid media. Two waves are observed the ratio of which corresponds to 2:1 and the height of which corresponds to a four- and two-electron process, respectively. It has been shown (117) that the wave height, wave shape, half-wave potentials, and dependence of the limiting current and the half-wave potentials of the more negative wave on pH are identical with those quantities observed for the protonized form of the corresponding alkyl hydroxylamine. Hence as the reduction product in the first wave, alkyl hydroxylamine can be expected and the overall scheme is as follows:

$$RNO_2 + 4e + 4H^+ \xrightarrow{E_1} RNHOH + H_2O \quad (66a)$$

$$RNHOH + H^+ \rightleftharpoons RNHOH_2^{(+)} \quad (66b)$$

$$RNHOH_2^{(+)} + 2e + 2H^+ \xrightarrow{E_2} RNH_3^{(+)} + H_2O \quad (66c)$$

Another example is the reduction of nitrones. In this case the reduction can follow two paths (67):

$$ArCH{=}\underset{\underset{O}{\downarrow}}{N}{-}R \longrightarrow ArCH_2\underset{\underset{OH}{|}}{N}{-}R \longrightarrow ArCH_2NHR; \quad ArCH{=}\underset{\underset{O}{\downarrow}}{N}{-}R \longrightarrow ArCH{=}N{-}R \longrightarrow ArCH_2NHR \quad (67)$$

For R = alkyl, one four-electron step was observed (118) and the decision between these two possibilities was excluded. For R = phenyl, on the other hand, two waves were observed corresponding to a three-electron reduction followed by a one-electron uptake. The half-wave potentials of the more negative one-electron wave of nitrone have been proved to be identical to the half-wave potentials of the second one-electron step of the Schiff base $ArCH{=}NC_6H_5$ over a broad pH range. The shapes of the pH dependence of half-wave potentials of these two waves were also the same. Moreover, the half-wave potentials of the more positive wave of $C_6H_5CH{=}NC_6H_5$ were only slightly different from the observed half-wave potentials of the first, three-electron reduction wave of nitrones. Hence it can be assumed that in the first, three-electron step of nitrone reduction, the waves of a two-electron process corresponding to a reduction of nitrone to benzalaniline coalesce with the first one-electron reduction step of benzalaniline, and the overall electrode process can be depicted as follows (68):

$$\left.\begin{array}{l} \mathrm{ArCH{=}\underset{\underset{O}{\downarrow}}{N}C_6H_5} + 2e + \mathrm{H^+} \rightarrow \mathrm{ArCH{=}NC_6H_5} + \mathrm{OH^-} \quad (68a) \\ \mathrm{ArCH{=}NC_6H_5} + e \rightarrow \mathrm{Ar\dot{C}H{-}\bar{N}C_6H_5} \quad (68b) \end{array}\right\} \text{1st wave}$$

$$\mathrm{Ar\dot{C}H{-}\bar{N}C_6H_5} + e + 2\mathrm{H^+} \rightarrow \mathrm{ArCH_2NHC_6H_5} \qquad \text{2nd wave} \qquad (68c)$$

Hence the N—O bond is reduced prior to the C=N bond in accordance with the results of some chemical reductions (118). Formation of the hydroxylamino derivative as an intermediate in the four-electron reduction step is further excluded by the observation that $ArCH_2N(OH)C_6H_5$ is reduced at considerably more negative potentials, in a potential range in which no wave has been observed in the reduction of nitrone.

In molecules bearing two electroactive centers R^1, R^2, the mutual position and possibility of interaction of the two groups is of importance. If in the molecule R^1—A—R^2 the groups are completely isolated and sufficiently separated, they behave as two molecules R^1—A and R^2—A. When a certain degree of mutual interaction is possible, the group reducible at more negative potentials R^2, affects the reduction of the group R^1 as it similarly affects other substituents (cf. Sec. IX-C). The properties of the group R^1 will be slightly modified by the group R^2 when compared with R^2 = H. When the mutual interaction is strong, as happens mainly if the two reactive centers R^1 and R^2 are in close proximity or conjugated, the properties of both groups R^1 and R^2 can be fundamentally changed and the behavior of compound R^1—A—R^2 does not resemble either that of the compound R^1—A or that of the compound R^2—A. This latter case is not suitable for treatment of the elucidation of the reduction path described below. It should be noted that the electroactive centers R^1 and R^2 can be either chemically different groups or identical groupings ($R^1 = R^2$).

To elucidate the reduction course for a substance R^1—A—R^2 bearing two, more or less separated reaction centers R^1 and R^2, comparison of the reduction waves at E_1 and E_2 of the compound R^1—A—R^2 with those of the expected product* R^1_{Red}—A—R^2 in the broadest possible pH-range is recommended. Not only the half-wave potentials, but their pH-dependence, wave shapes, wave heights, and their pH-dependence are compared. If in all these parameters the behavior of the compound R^1_{Red}—A—R^2 closely resembles that of the wave of R^1—A—R^2 at the potential E_2, it can be assumed that R^1_{Red}—A—R^2 is the reduction product formed in the wave at the potential E_1.

This type of elucidation of the electrode process can be illustrated on some examples:

β-*N*-Triethylaminoacrolein is reduced in alkaline media in two waves (51). The more negative wave possesses the same half-wave potential and shows the same pH dependence of half-wave potentials, wave heights, as well as shapes as the waves of acrolein. Hence, because the second wave of β-*N*-triethylaminoacrolein corresponds to acrolein reduction, acrolein is consequently formed in the first wave. The first wave can therefore be attributed to the reductive cleavage of the C—N bond according to the scheme:

$$OHC\text{—}CH\text{=}CH\text{—}\overset{(+)}{N}(C_2H_5)_3 + 2e + H^+ \rightarrow OHC\text{—}CH\text{=}CH_2 + N(C_2H_5)_3 \quad (69)$$

Similarly, quinoxaline dioxide is reduced principally in three waves: The third and second waves have the same height and possess the same half-wave potentials as their parent quinoxaline (119). Hence, quinoxaline must be formed in the first wave; this suggests the simultaneous reduction of the two N—O bonds in the first wave. This assumption is further supported by the fact that quinoxaline-*N*-oxide, with only one N—O bond, also shows three waves. The two more negative waves correspond to the reduction of the quinoxaline nuclei; the more positive wave appears approximately in the same potential range as that for the dioxide. The wave of the monooxide corresponding to the reduction of one N—O bond is nevertheless only half

* In the intermediate R^1_{Red}—A—R^2, R^1_{Red} is the reduced form of the electroactive group R^1. In some cases, when R^1 = Cl, Br, I $NR_3^{(+)}$, $SR_2^{(+)}$ etc. R^1_{Red} can be equal to H and R^1_{Red}—A—R^2 equal to H—A—R^2. In other cases, e.g., for R^1 = NO_2 in the medium pH range, R^1_{Red} is NHOH and R^1_{Red}—A—R^2 is HOHN—A—R^2. It is assumed here that the compound R^1_{Red}—A—R^2 can be prepared synthetically. In some cases, R^1_{Red}—A—R^2 can be generated by controlled potential electrolysis.

as high as the first wave of the dioxide, corresponding to the reduction of two N—O bonds.

Another example is the reduction of *N*,*N*′-polymethylene-bis-sydnones (120). These substances are reduced at higher pH values in two steps of equal height. Both waves show similar pH dependence of wave height and half-wave potentials as the parent sydnone. Because, moreover, the half-wave potential of the more negative wave was practically equal to that of an *N*-alkyl sydnone with a longer side chain and the first wave was shifted towards more positive potentials, the shorter the polymethylene chain and the greater the mutual interaction, it is assumed that a consecutive reduction of the sydnone rings occurs. In the more positive first wave, the first sydnone ring is reduced to the corresponding hydrazine derivative; in the more negative wave, the reduction of the second sydnone ring occurs.

The most extensive example of this type can be found in the reductions of polychlorobenzenes (121) and polyiodophthalic anhydrides (122). Halogens are reduced in consecutive steps, and by comparing the half-wave potential of the more negative wave, with half-wave potentials of corresponding model substances, it is possible to decide which of the possible isomers is formed. Hence it was shown that in 1,2,4,5-tetrachlorobenzene the chlorine atom in position 5 is most easily reductively split off, and 1,2,4-trichlorobenzene rather than 1,2,5-trichlorobenzene is formed.

In these cases in which a chemical reaction is interposed between two electrochemical steps and the reaction sequence is:

$$A \xrightarrow{E_1} B \xrightarrow[k_1]{\text{chem}} C \xrightarrow{E_2} D \tag{70}$$

it is possible that the half-wave potential and the shape of the more negative wave of substance C at potential E_2 are identical with those magnitudes observed after addition of compound C to the same supporting electrolyte, but the wave height for the second wave of substance A is smaller than for added compound C. The rate with constant k_1 limits the height of the more negative wave for such an electrode process. α-Substituted ketones of the type $COCH_2$—X and α,β-unsaturated ketones COCH=CH, discussed in Sec. V-B, belong to this category.

One example which indicates the impossibility of applying the above treatment for systems in which the mutual interaction of groups R^1

and R^2 in the molecule R^1—A—R^2 is strong remains to be mentioned briefly. The waves following the first reduction step in $C_6H_5CH=$ CH CO Py, where Py is the pyridine ring, resemble neither those of $C_6H_5CH_2CH_2COPy$ nor those of $C_6H_5CH=CH—CHOHPy$ (123). Interactions of the radicals formed offer a tentative explanation.

2. Controlled Potential Electrolysis

Controlled potential electrolysis is one of the most powerful tools used in the elucidation of organic electrode processes. One can only agree with Lund (124) that "a controlled potential investigation should be adopted as a standard procedure in the polarographic work."

There are two principal possibilities for carrying out the controlled potential work: Either the electrolysis is carried out with a dropping electrode or with a mercury pool electrode. With a dropping electrode, usually a small volume of 0.1–2.0 ml. of the electrolyzed solution is used, the concentration of which is usually only about ten times higher than the generally used polarographic concentration. Currents are hence only slightly higher than in polarography and usually do not exceed 30 μamp. The most practical way of stirring the solution is by means of the falling mercury drops. A rather thin capillary properly placed in the electrolysis vessel is essential. Both in this and the other techniques described later on, a stream of pure nitrogen is passed over the surface of the solution during electrolysis. In electrolysis with a dropping electrode, the time necessary to reach 50% of the reduced form is about 2–5 hr.

With a mercury pool electrode, larger volumes of 5.0–100 ml. can be electrolyzed; concentration of the electroactive substance can be $10^{-2}M$ or even $10^{-1}M$; the current flowing is 10–300 mamp. Intensive stirring of the mercury pool proves important; a magnetic stirrer is useful for this purpose. Electrolysis of 95% of the substance can usually be carried out in 1–5 hr.

In both of these variants when irreversible processes are studied, the second working electrode can be placed directly in the electrolyzed solution. A mercury pool electrode can be used as the second working electrode in connection with the dropping electrode and a graphite electrode in connection with the mercury pool electrode. But generally a separation of the second working electrode by a salt bridge can be recommended. A separated calomel electrode is preferably used in connection with the dropping electrode. An agar bridge and a sintered glass disk are useful to prevent diffusion to and from the electrolysis compartment.

A third auxiliary electrode, usually a stemlike calomel electrode, is immersed in the electrolyzed solution and placed in the vicinity of the working electrodes. The electrode is used for measuring the potential of the working mercury electrode and possibly for monitoring its potential. It is recommended that the course of the electrolysis be followed by recording polarographic current–voltage curves after selected time intervals. When a dropping mercury electrode is used for electrolysis, the same dropping electrode can also be applied for recording the polarographic curves. When electrolysis is carried out with a mercury pool electrode, an additional dropping electrode is immersed in the electrolyzed solution. The monitoring auxiliary electrode can be used as the reference electrode for polarography. Unless changes in the electrolyzed solution occur too rapidly the recording of the whole i-E curves at selected time intervals is preferred to the recording of limiting currents only in the region of the potential at which the electrolysis is carried out. Changes in shape of the whole polarographic curves during electrolysis usually reveal more information that is useful for interpreting the course of electrolysis.

When using the dropping electrode, the potential can often be controlled manually (125) or by using a simple transistorized potentiostat (126) which is especially useful when the electrolysis of ill-separated waves is performed. Otherwise potentiostats are applied with the following characteristics: current output of 10–300 mamp, potential range of $+1.0$ to -2.5 V., time constant of the voltage compensation of the order of seconds, and the possibility of working against an iR drop of 10–100 V.

According to the type of electrode used, the methods used for identification of electrolysis products vary. When the dropping electrode is used, the amount of product is usually too small to allow isolation. Hence, methods suitable for identifying substances in solutions are used, such as polarography and related electrochemical methods, chromatographic methods—in particular, gas, thin-layer, and paper chromatography—determination of pK_a values when the product possesses acid–base properties, and measurement of the ultraviolet absorption spectra. After addition of proper reagents, color reactions can be carried out [e.g., ninhydrine reaction for the CH_2NH_2 grouping formed in the reduction of *p*-cyanoacetophenone in acid media (127) and Nessler reagents for ammonia or nitroprusside reaction for thiol groups liberated (25)], or changes in the polarographic curves followed [e.g., the presence of carbonyl groups in the reduction products can be proved by the increase of a new, more positive, wave after condensation with primary amines (35)].

The greater amounts of the product formed when a mercury pool electrode is used enable, in addition to these methods, isolation of the product and its characterization by means of elemental analysis, determination of the melting point and preparation of derivatives, infrared and mass spectra, etc.

The yield of the product is an important proof of whether in addition to the product followed and expected, other less expected byproducts are formed in considerable ratios or not. The yield can be either determined from the amount of the isolated product or by quantitative determination of one of the products in the electrolyzed solution. Various analytical methods can be used for determining the yield. It is convenient when one of the products can be determined polarographically. This was the case, e.g., in the determination of sulfur dioxide formed in the reduction of *p*-cyanosulfonamide or saccharine in alkaline media (129), cyanide ions formed in the reduction of terephthalic dinitrile (127), or mercaptans liberated in the reduction of aza-uracil mercapto derivatives (130), pyridoxthiol (25) or a thiazolidine derivative (125). Spectrophotometric methods were used for the determination of ammonia set free in the reduction of pyridoxamine (25). Titrimetrically, the amount of protons consumed in the reduction of aryl sulfones in unbuffered solutions was determined (131) to prove the splitting off of the C—S bonds. The yield of diphenylmercury and triphenylstibine formed in the reduction of the tetraphenylstibonium ion was determined gravimetrically (132).

Mass spectrometry was used (198) in the determination of nitrogen formed in the oxidation of methylhydrazines. The dependence of the total yield of nitrogen as a function of initial concentration of the hydrazine derivative shows a linear plot for methylhydrazine, but a curved plot for dimethylhydrazines. Hence for methylhydrazine nitrogen is either formed directly in the electrode process or all consecutive or interposed reactions are rapid. Curvatures of plots for dimethylhydrazines indicate that nitrogen formation involves interposed or consecutive reactions. As nitrogen evolution stops when the electrolysis is interrupted, the latter possibility is less probable. Comparison of chronopotentiometric results (indicating transfer of 2–3 electrons) with those obtained by coulometry (6 electrons), and absence of reversal wave in chronopotentiometry indicate the role of interposed reactions in oxidation of 1,2-dimethylhydrazine. Rapidly established azo-aldimino tautomeric equilibria and a slower hydrolysis of the hydrazone were presented (198) as an explanation, together with a subsequent oxidation of the hydrazine, formed in hydrolysis. The presence of formaldehyde among electrolysis products and its yields determined by the dimedone reaction were in accordance with this scheme.

It is important to confirm that the compound to be electrolyzed is sufficiently stable in the supporting electrolyte chosen for electrolysis. Electrolysis is carried out during a time that is relatively long when compared with recording polarographic curves. Compounds considered stable during polarographic analysis can undergo slower changes, and it is necessary to check the stability over the whole time period necessary for electrolysis. If chemical transformations of the studied compounds take place, their products should be considered for detection during analyses of the mixture after electrolysis.

With controlled potential sources available, it is useful to carry out the coulometric measurements as well. For this purpose, electrolysis is best performed in a small volume using a mercury dropping electrode and stirring the solution solely by falling mercury drops. The number of coulombs consumed can be determined either graphically from the plot of log i against time where i is the diffusion current measured in proper time intervals, or continuously, during electrolysis or using a coulometer in series with the electrolysis cell.

It is necessary to eliminate traces of oxygen which is reducible and would thus consume electricity and to prevent diffusion to and from the electrolysis compartment. Because nitrogen stream is usually swept over the surface of the electrolyzed solution, care should be taken to control the volatility both of the electroactive substance and the solvent. Chemical reactions of the substance to be electrolyzed and of the products in the bulk of the solution, such as hydrolysis should be either excluded or corrected for by using a blank. At small concentrations of the electroactive substance even in the more positive potential range and at higher concentrations in the potential region near the final rise of the current, it is necessary to consider the current flowing due to electrolysis of the components of the supporting electrolyte.

An important point is the course of the log *i-t* plot which is linear for simpler types of electrode processes. An example of deviations from linearity of this plot was observed (133) in the reduction of α-furildioxime at pH 6.5–9.8. The log *i-t* plot in this case shows two linear sections with different slopes. The 6-electron polarographic waves observed were seemingly in contradiction with the product of controlled potential electrolysis, 1,2-bis(2-furyl)ethylenediamine, which would indicate an 8-electron process. These findings were explained by mechanism (71):

$$A + 6e \rightarrow B \tag{71a}$$

$$B \xrightarrow{k} C \tag{71b}$$

$$C + 2e \rightarrow \text{products} \tag{71c}$$

The first, steeper decrease of the current during electrolysis is ascribed to the 6-electron step (71a), whereas the rate of the second electrolytic step is governed by the rate of reaction (71b). The chemical transformation with constant k of the order of 10^{-4} sec.$^{-1}$ was ascribed to the cleavage of the hydrogen bond, but this type of reaction usually occurs much more rapidly. The primary reduction product B probably undergoes another type of transformation.

The log *i-t* plot can be useful in detection of the presence of regeneration (catalytic) reactions (199), antecedent equilibria (200), and some consecutive reactions (199). On the other hand these plots do not allow distinguishing of chemical reactions interposed between two electrode processes (201) and of irreversible antecedent reactions (200).

It is important to understand that the products isolated from the electrolyzed solution or detected in this solution may be different from those formed in polarography at the surface of the dropping electrode, even if they are usually identical. When the dropping electrode is used for controlled potential electrolysis, the reasons for such differences are usually the same as those that affect coulometric measurements, namely consecutive reactions of products in the bulk of the solution, including the reaction of products with the original substance and volatility of the products and/or solvent.

When a mercury pool is used for controlled potential electrolysis, the following factors can affect the results in addition to those that can play a role with the dropping electrode: The longer the solution is in contact with the surface of the electrode, the more surface reactions can take place, the rate of which may be affected by adsorption of the participating species. The higher concentration usually used in mercury pool electrolyses can increase the rate of chemical reactions of higher order. The overvoltage can be generally different at the surface of the pool electrode when compared with the dropping electrode. It is recommended that well-buffered solutions be used when the electrolysis is carried out at the pool electrode, but even then it can be questionable whether the acidity at the electrode surface during electrolysis is the same as the pH in the bulk of the solution. And finally, for electrode processes accompanied by chemical reactions, the rate of the chemical process furnishing the electroactive species can be different at the dropping and the pool electrodes. Hence the value i_d/i at a given pH measured with a dropping electrode may be different from the value i_d/i at the same pH value on a current–voltage curve obtained with a pool electrode. As electrolysis with a dropping electrode is usually carried out over a longer period than with a pool electrode, the time changes of the composition of the solution are of particular importance with the former. Some of the complicating factors can be detected and eliminated when results of the electrolysis with the pool electrode are compared with those obtained with a dropping electrode.

Some examples of the more complex cases will be given next:

The most common case is a chemical reaction subsequent to the electrochemical step that proceeds too slowly to affect the electrolysis at the dropping electrode but with a sufficient rate to affect the controlled potential electrolysis. These systems are most easily detected in cases where the product of the chemical reaction with constant k is again electroactive according to the scheme (72):

$$\mathrm{A} \xrightarrow{E_1} \mathrm{B} \xrightarrow{k} \mathrm{C} \xrightarrow{E_2} \mathrm{D} \tag{72}$$

An example of this type is the behavior of *p*-diacetylbenzene at pH between 3 and 5 (60). In this pH range a diradical is formed in a reversible reaction and on polarographic curves a single two-electron step is observed.* When electrolysis was carried out at the potential of the limiting current of the 2-electron wave, the wave of p-CH_3-$CHOHC_6H_4COCH_3$, corresponding to C in scheme (72), increased in the course of electrolysis. This compound C was formed from the diradical B by a chemical reaction B→C.

Another example of this type is the reduction of isonicotinic amide (134) at pH < 1. Polarographic curves correspond to a 2-electron process, but the controlled potential electrolysis corresponds to an overall consumption of four electrons. In the latter process, the formation of pyridine-4-aldehyde as an intermediate was proved. This compound is hydrated and gives only a small wave in this pH region† and hence contributes only little to the height of the polarographic wave. During electrolysis, the dehydration has time to take place and pyridine-4-aldehyde can be quantitatively reduced to pyridine-4-carbinol. The system follows scheme (72), in which, in this case, E_2 is more positive than E_1.

To prove the formation of intermediate B, it is sometimes useful to trap it. A chemical reagent is added transforming the primary product into an electroinactive or insoluble product. Formation of semicarbazones or 2,4-dinitrophenylhydrazones is useful for detection of less stable carbonyl compounds. Carboxy derivatives formed during the electrolysis in the presence of carbon dioxide were extensively used for detecting formation of anion and dianion radicals in dimethylformamide. Examples of systems in which this formation has been proved are naphthalene (135), phenanthrene (135), diphenylacetylene (135), benzophenone (136), and some α,β-unsaturated ketones (137). A 20% yield of the diether (138) obtained in the reduction of anthraquinone in the presence of ethyl bromide in acetonitrile is proof of dianion formation just as in the analogous experiment with the reduction of benzophenone in dimethylformamide in the presence of ethyl iodide, the formation of diphenylethylcarbinol indicates the ketyl anion as a stable intermediate (136).

When we wish to increase the rate of the consecutive chemical reaction, an increase in temperature is useful. More often, we wish to identify the less stable intermediate so as to suppress its decomposi-

* Wave of p-$CH_3CHOHC_6H_4COCH_3$, observed at lower and higher pH values, is absent.

† The height of this wave is limited by the rate of dehydration and it is observed at potentials more positive than those of the amide.

tion, and this can be achieved by lowering the temperature during electrolysis. This was done, e.g., in the reduction of carbon tetrachloride in acetonitrile, which was carried out (139) at $-20°C.$ in the presence of tetramethylethylene. As one of the products, 1,1-dichloro-2,2,3,3-tetramethylcyclopropane was identified by gas chromatography. As this compound has also been synthesized by generating dichlorocarbene from chloroform with potassium *t*-butoxide in the presence of tetramethylethylene, the above result is taken as proof of the electrolytic formation of dichlorocarbene, CCl_2. This interpretation is in accordance with the fact that whereas in water-containing solution tetrachloromethane is reduced in three successive polarographic waves of equal height, corresponding to one C—Cl bond each, in acetonitrile solution, two waves of approximately the same height were observed. This is explained (139) by scheme (73):

$$CCl_4 + 2e \xrightarrow{E_1} CCl_3^- + Cl^- \tag{73a}$$

$$CCl_3^- \overset{k}{\rightleftharpoons} CCl_2 + Cl^- \tag{73b}$$

$$CCl_2 + 2e \xrightarrow{E_2} \text{products} \tag{73c}$$

In addition to consecutive reactions, side reactions can affect the electrolysis as well. An example of this type is the reduction of *o*-benzoylbenzoic acid diethylamide (134). On polarographic curves of this compound, a 2-electron wave was obtained for the whole pH range. In contrast, during controlled potential electrolysis only 1.65 electrons per molecule were consumed. Because in the latter case a dimer was isolated, it was concluded that scheme (74) is followed and that B is the radical formed after a 1-electron transfer:

$$A \xrightarrow{E_1} B \xrightarrow{E_2} C \tag{74a}$$

$$2B \xrightarrow{k} \text{products} \tag{74b}$$

The greater importance of dimerization in large-scale electrolysis can be caused by the higher concentration used or by the adsorption reduction products.

A similar side reaction was observed (140) in the electrolysis of picric acid in acid media. Consumption of 18 electrons per molecule was reached only under a certain picric acid concentration, the value of which depends on pH. Above this concentration the apparent number of electrons consumed decreases. It is assumed that the rate of reaction (74b) increases with increasing picric acid concentration but decreases with increasing hydrochloric acid concentration. Because the value of the apparent number of electrons decreases below 17, the value of n_1 in the first step $A \overset{n_1 e}{\rightarrow} B$ must be smaller than 17.

The effect of concentration can also be observed in cases in which more easily reducible compounds are formed as products of the first step of the electrolysis. In such cases a nonlinear relation can be found between the concentration of the studied substance and the electron consumption. Here, a small, more positive wave of the intermediate appears during electrolysis, as was described (141) in the reduction of tropolone.

In general, coulometric determination of the apparent numbers of electrons transferred and their changes with initial concentration, time, and volume of the electrolyzed solutions allows the detection and study of interposed (201), regeneration (catalytic) (199) and antecedent reversible (200) reactions. On the other hand the simple consecutive reactions (199) and antecedent equilibria (200) do not affect the apparent number of the electrons transferred. For the reduction of benzyldimethylanilinium bromide (202), the number of electrons determined at 25°C. was two, whereas at −35°C. the apparent value of electrons consumed varied between 1.66 and 1.40. It was assumed that an interposed reaction is involved at −35°C. that is fast at 25°C. For the treatment of consecutive reactions, the application of reversal coulometry is suggested (203).

The effect of the rate of protonation can be demonstrated in the case of the reduction of isonicotinic acid (134). Pyridine-4-aldehyde was detected as a product of the protonated form of isonicotinic acid. But whereas the waves of the protonated form of the isonicotinic acid at the dropping electrode can be observed at $pH < 8$, the formation of pyridine-4-aldehyde was found to take place only at $pH < 6$. It seems that the rate of protonation at the surface of the mercury pool is not high enough to supply a sufficient amount of the protonated isonicotinic acid at $pH > 6$.

In some few cases it is possible to isolate different products for the same polarographic curves at different potentials. The number of electrons consumed can be either the same at all potentials or different. An example of the former type is 4,4′-dithiomorpholine which gives a 2-electron reduction wave resembling those of disulfides (142). In the electrolysis carried out at the half-wave potential, morpholine and mercuric sulfide were isolated; in that performed at the potential corresponding to the limiting current, morpholine and sulfur were formed. It is assumed that the primarily formed *N*-mercaptomorpholine attacks the metallic mercury at more positive potentials but at more negative potentials it is cleaved and forms sulfur.

Consumption of a different number of electrons at different potentials was observed in the reduction of 4-cyanopyridine (143). The amount of 4-picolylamine formed in the electrolysis decreased with increasingly negative potentials.* Competitive reductions of the protonized form, producing 4-picolylamine in a 4-electron step, and unprotonized form producing 2-electron nucleophilic substitution of CN^-, take place.

The change of the apparent number of electrons per molecule of dimethylglyoxime with the potential at which the electrolysis is carried out (144) indicates a catalytic hydrogen evolution.

When two separate waves appear on the polarographic curve, corresponding to two successive electrode processes, and when the controlled potential electrolysis is carried out at the potential corresponding to the limiting current of the first wave, the height of the more positive wave decreases in the course of electrolysis. The height of the second, more negative wave can remain unchanged, decrease, or even increase during electrolysis. Some of the schemes corresponding to each type of this behavior will be discussed next.

When the second wave remains unchanged, either scheme (75) for successive reduction of one electroactive group or scheme (76) for reduction of two independent reaction centers can be involved:

$$\begin{cases} R^1\text{—}A + n_1e \xrightarrow{E_1} R^2\text{—}A & (75a) \\ R^2\text{—}A + n_2e \xrightarrow{E_2} \text{products} & (75b) \end{cases}$$

$$\begin{cases} R^1\text{—}A\text{—}R^2 + n_1e \xrightarrow{E_1} A\text{—}R^2 & (76a) \\ A\text{—}R^2 + n_2e \xrightarrow{E_2} \text{products} & (76b) \end{cases}$$

* Single-sweep *i-E* curves show an increase of CN^- formation with increasingly negative potentials.

The unchanged height of the second wave permits exclusion of scheme (77):

$$R^1\text{—A} + n_1e \xrightarrow{E_1} \text{products}_1 \qquad (77a)$$

$$R^1\text{—A} + (n_1 + n_2)e \xrightarrow{E_2} \text{products}_2 \qquad (77b)$$

which corresponds to two independent reduction paths of the same species R^1—A.

An example following scheme (75) is the reduction of bis-pyridoxyl-disulfide (25) in which pyridoxthiol is formed in the first step and is reduced in the second step. An example of scheme (76) is the reduction of *p*-cyanoacetophenone (127) in acid media, corresponding to the successive reduction of the cyano and carbonyl groups and of 6-methyl-pyridazone-3 at pH 1.0 (145). The reduction of 2-(4-pyridyl)-thiazolidine-4-carboxylic acid derivative (125) in medium pH range can be classified as following (75) or (76), according to whether the thiazolidine ring is considered as one or two reactive centers.

For the case where the height of the second wave decreases during electrolysis, we shall restrict ourselves firstly to the simple case in which the ratio of the first and second wave i_1/i_2 remains unchanged. This can happen either in the case of two independent paths as in (77) or in reactions (75) or (76) accompanied by a deactivation of the intermediate R^2—A according to (75c):

$$R^2\text{—A} \xrightarrow[k]{\text{deactivation}} \text{products} \qquad (75c)$$

The rate with constant k is in this case too slow to affect the wave, but is fast enough to deactivate the intermediate R^2—A during the electrolysis. Such a side reaction can be complicated by the fact that the primary product R^2—A is electroinactive and must be first transformed quickly into electroactive R^2—A′ according to scheme (78):

$$R^1\text{—A} - R^2 + n_1e \xrightarrow{E_1} R^2\text{—A} \qquad (78a)$$

$$R^2\text{—A} \xrightarrow[\text{fast}]{\text{activation}} R^2\text{—A}' \qquad (78b)$$

$$R^2\text{—A}' + n_2e \xrightarrow{E_2} \text{products}_1 \qquad (78c)$$

$$R^2\text{—A}' \xrightarrow{\text{deactivation}} \text{products}_1 \qquad (78d)$$

Examples of systems in which the two waves decrease in the same ratio are thioazauracils (146) and *p*-cyanobenzophenone in acid media (127).

Among cases in which the height of the second wave decreases together with the decrease of the first wave, there are systems in which the height of the first wave decreases more than that of the second. One possible explanation is scheme (79):

$$R^1\text{—A—}R^2 + n_1e \xrightarrow{E_1} R^2\text{—A} \quad (79a)$$

$$R^1\text{—A—}R^2 \xrightarrow[k]{\text{deactivation}} \text{products}_1 \quad (79b)$$

$$R^2\text{—A} + n_2e \xrightarrow{E_2} \text{products}_2 \quad (79c)$$

When the rate of reaction (79b) is comparable with the rate of electrolysis (79a), the concentration of R^2—A decreases only due to the cleavage of R^1—A—R^2 by (79b),but that of R^1—A—R^2 by both deactivation (79b) and by electrolysis (79a). An example of such behavior is the electrolysis of the 2-(4-pyridyl)thiazolidine-4-carboxylic acid derivative (125) in acid media in which the hydrolysis can take place.

An increasing second wave can be observed, e.g., when system (80) is operating:

$$R^1\text{—A—}R^2 + n_1e \xrightarrow{E_1} R^2\text{—A} \quad (80a)$$

$$R^2\text{—A} \xrightarrow[\text{slow}]{\text{activation}} R^2\text{—A}' \quad (80b)$$

$$R^2\text{-A}' + n_2e \xrightarrow{E_2} \text{products} \quad (80c)$$

Reaction (80b) is comparatively slow so that it does not allow (during the drop-time) all of the R^2—A to be transformed into R^2—A′. When in the large scale electrolysis the time is sufficient to transform all R^2—A into R^2—A′, the wave at E_2 increases. When reaction (80b) is still slower, no wave at E_2 is observed on the polarographic curve but grows during electrolysis.

Another possibility of the increased height of the second, more negative, wave is given in scheme (81):

$$R^1\text{—A—}R^2 + n_1e \xrightarrow{E_1} R^2\text{—A} \quad (81a)$$

$$R^2\text{—A} + n_2e \xrightarrow{E_2} \text{products}_1 \quad (81b)$$

$$R^2\text{—}A \xrightarrow[\text{slow}]{\text{activation}} R^2\text{—}A' \qquad (81c)$$

$$R^2\text{—}A' + n_3e \xrightarrow{E_3} \text{products}_2 \qquad (81d)$$

To observe an increase in the height of the second wave, two conditions, in addition to those concerning reaction (81c) that are analogous to those discussed for reaction (80b), are to be fulfilled: $n_3 > n_2$ and $E_3 \approx E_1$.

In addition to the increase of the more negative acetophenone wave in the reduction of *p*-diacetylbenzene (p. 154), an increase of the second wave is observed in the reduction of pyridoxaloxime (25), where the activation step (80b) or (81c) is probably the protonation of the pyridoxamine formed in the first step and in that of 6-methyl-pyridazone-3 at pH 4.7 (145).

An important tool in the study of mechanisms of electrode processes that has been used so far to only a limited extent is the stereospecificity of some electroorganic reductions. Hence, for α,α'-dibromosuccinic acids, the *threo* form of both the free acid and its anions is reduced to fumaric acid. On the contrary, the *erythro* epimer is reduced to the fumaric acid only in the undissociated form and as a dibasic anion. The univalent anion is at least partly reduced to the maleic acid. Both *threo* and *erythro* epimers of dialkyl esters of dibromosuccinic acid are reduced, similarly to the undissociated free acids, to only the dialkyl ester of fumaric acid (147,148).

In the 1-electron reduction of $\Delta^{1,4}$-ketosteroids (149), various stereoisomers of pinacols were formed according to the pH. The protonized form of the ketosteroid, reduced in acidic solution, gives rise to a pinacol with hydroxyl groups in the α-position. In alkaline media, the unprotonized ketosteroid is reduced with the formation of the isomer with hydroxyls in the β-position. The structure of the products, prepared by controlled potential electrolysis, was supported by the rates of dehydration and periodic acid oxidations. For Δ^4-3-ketosteroids, the difference in the composition of products obtained in acidic and alkaline media was not so pronounced.

VIII. Structural Effects

The next powerful tools that can be used in elucidating the mechanisms of electrode processes are structural effects that are observed when the polarographic behavior of the individual compounds in a group of structurally related substances is compared. The usual succession of events is that after the polarogaphic activity of a compound is detected, the study is extended to a group of related compounds. First, the structure of the compounds used for comparison is chosen so that structural changes would correspond to a rather broad range. One compares, how large the change can be made without changing principally the polarographic behavior, and hence inside which group the course of the electrode process can be assumed to remain unchanged. In this way, the scope of the validity of a given mechanism is preliminarily established. Next, a more detailed change in the structural effect inside the main group is studied and interpreted. Hence the behavior of compounds bearing the same electroactive group is compared on various molecular frames such as on an aliphatic chain, a benzene or heterocyclic aromatic ring, an alicyclic ring, etc. Compounds bearing the same electroactive grouping in various positions on the given frame also belong to this group and their behavior is to be compared. Effects of an exchange of substituents, more or less separated from the electroactive group, as well as effects of steric factors, including the size of the ring on which the electroactive group is bound, are studied next. And finally, the polarographic behavior of a particular electroactive group situated on a given molecular frame in a given position is compared with that of other, related, electroactive groups on the same skeleton. The observed structural effects are then interpreted, based principally on analogies of structural effects in electrochemical reactivity and in the reactivity in homogeneous chemical reactions. Some examples of applying the interpretation of these structural effects are given in the next paragraphs.

A. SCOPE OF THE VALIDITY OF THE MECHANISM OF THE ELECTRODE PROCESS: EFFECT OF MOLECULAR FRAMES

A given atomic grouping which is assumed to undergo a change in the electrode process can hardly be expected to undergo the electrolysis by the same mechanism in all existing compounds containing this grouping. Hence the same mechanism, allowing quantitative comparison of structural effects, is restricted only to a group of compounds. This group cannot only be rather broad but also rather narrow, according to the nature of the electroactive group. Differences in properties of various groups are best demonstrated by some examples. The scope of the structural variations for which the course of the electrode process remains uchanged is rather broad for thiols or semicarbazones. All thiols studied so far give an anodic wave involving a 1-electron process corresponding to formation of a mercury compound. Similarly, all semicarbazones, e.g., derived from acetaldehyde, acetone, cyclohexanone, benzaldehyde, acetophenone, benzophenone, or pyridinealdehydes, are reduced in the protonized form in a principally 4-electron step divided in some instances into two 2-electron steps.

Another rather common type of behavior is observed when molecules with a given type of electroactive grouping are principally reduced by the same mechanism, even when structural variations are substantial, but the behavior of some inividual compounds or of a discrete group of compounds differs from that which is valid for the majority. Examples can be selected among nitro and carbonyl compounds. Nitro compounds can be principally separated into two large groups. The mechanism of nitro derivatives bearing the nitro group on a saturated carbon atom differs from that governing the reduction of aromatic nitro compounds. But even among these groups, there are exceptions, e.g., in most nitroparaffins a 4-electron reduction of the nitro group to the hydroxylamine derivative takes place, but for tertiary nitro compounds, an elimination of an NO_2^- group is reported (104). Reduction of nitrobenzenes is a 4-electron process, but for *p*- and *o*-hydroxy and amino derivatives a 6-electron reduction is reported in acid and alkaline media (43).

To the last type, which is fortunately not very common, belong substances bearing a given grouping, among which analogies of behavior are restricted to small groups only. As an example, substituted benzonitriles can be cited: *p*-cyanoacetophenone and *p*-cyanobenzaldehyde in acid solutions are reduced by 4 electrons to the corresponding aminomethylene compounds (127). Terephthalic acid dinitrile and *p*-cyanobenzoic acid and its esters show a 2-electron wave in the medium pH range corresponding to a nucleophilic substitution of the CN^- group (127). *p*-Cyanophenylmethylsulfone and *p*-cyanosulfonamide (129) give 2-electron waves corresponding to the cleavage of the C—S bond. Finally, in *p*-cyanobenzamide (134), the amide grouping is reduced with consumption of 4 electrons.

Since the second and the first groups and the transitory cases between them are most common, it is obvious that the effect of the molecular frame on the polarographic behavior of a given electroactive group should be settled first. Often the behavior of a given group bound on a saturated aliphatic chain resembles that of a group bound on an alicyclic or hydrogenated heterocyclic ring but differs from the behavior of α,β-unsaturated compounds and of those bearing the same groups on an aromatic nucleus. Substances bearing the electroactive group on a benzenoid ring resemble those bearing it on a "neutral" heterocyclic aromatic ring, such as thiophene, pyrrole, or furan, but often differ from those bearing it on a "basic" heterocyclic aromatic ring, such as pyridine or thiazole. Hence an ideal study of the scope of the mechanism should compare examples of each of the subgroups, i.e., aliphatic, alicyclic, benzenoid, various types of heterocyclics, etc., but often we restrict ourselves to comparing representatives of the main groups first i.e., saturated, benzenoid, and pyridine groups.

In some restricted subgroups, it is possible to express the effect of the skeleton on the half-wave potential quantitatively either using extrathermodynamic relationships (150–153) or quantum chemistry (154), but due to the possible changes in mechanism, it cannot be expected that the validity of such treatments can be generalized. Hence estimating the limits of the scope of a given polarographic behavior even if necessary, can give us but meager information on the mechanism of the electrode process. Only in some cases does the observed limitations of the validity of a given mechanism allow us to exclude some of the possible mechanisms.

B. POLAR AND RESONANCE SUBSTITUENT EFFECTS

It has been shown in eqs. (2–4) of Sec. II of this contribution that the half-wave potential of reversible systems is proportional to the changes in the standard free energy ΔG° of the electrode process, and those of irreversible systems are a simple function of the change in the activation free energy $\Delta G^{\pm}$. This indicates the possibility of applying to changes in the half-wave potentials, caused by structural changes, the treatments developed for handling structural effects on equilibrium and rate constants (155,156).

For this purpose we shall restrict ourselves in this discussion to groups of substances X—A—R called reaction series, in which change in the steric effects can be neglected. Benzenoid molecules bearing the substituent X in *meta* or *para* position relative to the electroactive group R, or reaction series in which the substituent X is placed at a sufficient distance* from the electroactive group R, prove to be useful models. For the shift in the half-wave potential $(\Delta E_{1/2})_X$ caused by introduction of substituent X into the parent molecule chosen as a reference compound, in which X = H or X = CH_3, the following relations can be derived (85,150) for reversible systems (82):

$$(\Delta E_{1/2})_X = \frac{2.3RT}{nF} (\Delta \log K)_X \tag{82}$$

where K is the equilibrium constant of the reaction Ox + $ne \rightleftharpoons$ $\rightleftharpoons$ Red, and $\Delta \log K = \log K_X - \log K_0$ where K_0 represents the value for the reference system for which X = H or CH_3, and for irreversible systems (83):

$$(\Delta E_{1/2})_X = \frac{2.3RT}{(\Delta \alpha n)_X F} (\Delta \log k_e{}^\circ)_X \tag{83}$$

where $k_e{}^\circ$ is the heterogeneous rate constant of the electrode process Ox + $ne \xrightarrow{k_e}$ Red at the standard potential, $(\Delta \alpha n)_X$ is the change in the product of the transition coefficient α and the number of electrons in the potential-determining step n, when comparing the compound X—A—R with the reference compound H—A—R or CH_3—A—R; $\Delta \log k_e{}^\circ$ is defined as $\Delta \log k_e{}^\circ = \log (k_e{}^\circ)_X - \log (k_e{}^\circ)_\theta$, where $(k_e{}^\circ)_\theta$ corresponds to the reference compound.

Because of the validity of the Hammett equation (155–157) for *m*- and *p*-substituted benzenoid systems (84):

$$(\Delta \log K)_X = \rho_R \sigma_X \tag{84a}$$

$$(\Delta \log k)_X = \rho_R \sigma_X \tag{84b}$$

and because of the validity of the Taft equation (155,156) for other types of the molecules X—A—R (85):

$$(\Delta \log K)_X = \rho_R{}^* \sigma_X{}^* \tag{85a}$$

$$(\Delta \log k)_X = \rho_R{}^* \sigma_X{}^* \tag{85b}$$

it is possible to combine eqs. (82) with (84a) or (83) with (84b) for *m*- and *p*-substituted benzenoid systems to get eq. (86):

$$\Delta E_{1/2} = \rho_{\pi,R} \sigma_X \tag{86}$$

* The distance should be such that the structure of the transition state for the members of the reaction series can be considered very similar.

and for other, mainly alkyl substituted compounds, eq. (82) with (85a) or (83) with (85b) to get eq. (87):

$$\Delta E_{1/2} = \rho_{\pi,R}{}^{*}\sigma_R{}^{*} \tag{87}$$

In these equations $\rho_{\pi,R}$ and $\rho_{\pi,R}{}^{*}$ are proportionality constants, called reaction constants, expressed in volts, that characterize the susceptibility of the electroactive group R on a benzenoid ring to the effects of substituents placed in *m*- or *p*-positions on this ring. Its value depends on the nature of the electroactive group R, on the type of molecular frame A on which the group R is bound, and on the com-

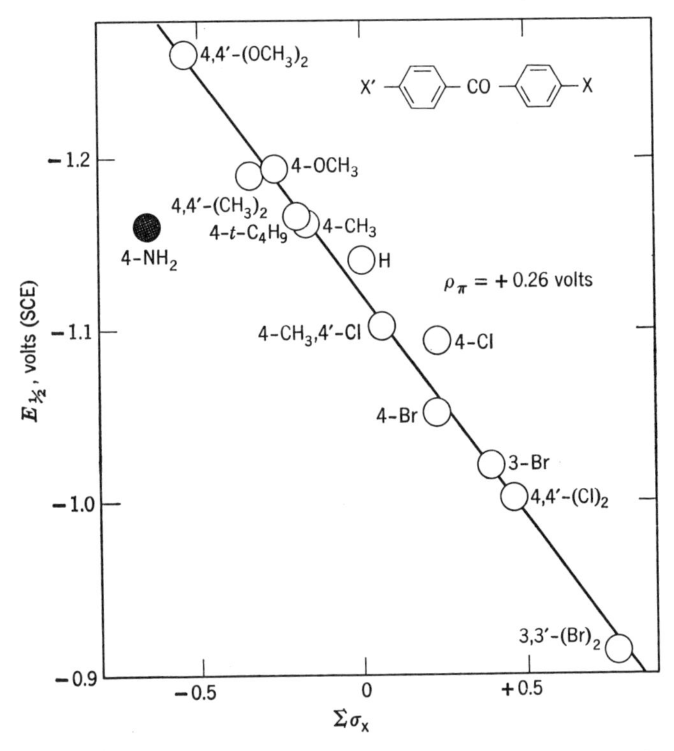

Fig. 14. Dependence of half-wave potentials $E_{1/2}$ of the reduction of substituted benzophenones on the sum of the Hammett substituent constants $\Sigma\sigma_X$. $\rho_{\pi,CHO} = +0.26$ v., point for *p*-amino derivative deviates (this wave corresponds to a two-electron process, all others to a one-electron reduction). Half-wave potentials measured in 40% methanol with an acetate buffer pH 4.7 and 0.02% gelatin.

position of the supporting electrolyte in which the polarographic curve is recorded, but it is independent of the nature of the substituent X.

σ_X is the Hammett total polar substituent constant (155–157) derived from dissociation constants of *m*- and *p*-substituted benzoic acids, σ_X^* is the polar substituent constant (155,156) based on measurement of acid- and base-catalyzed ester hydrolysis. The values of these constants that are tabulated (155–157) depend on the nature and position of the substituent, but are, in principle, independent either of the electroactive group R or of the composition of the supporting electrolyte.

In practice, the measured values of half-wave potentials are plotted against the tabulated values of substituent constants, a correlation is sought, the correlation coefficient is determined, and deviations from the linear plot are observed (Fig. 14).

Combination of eq. (82) with eqs. (84a) and (85a) for reversible systems involves no additional suppositions at all, but for the combination of eq. (83) with eqs. (84b) and (85b) for irreversible electrode processes, it is necessary to consider the role of the product αn. Originally it was assumed that the validity of eqs. (85) and (87) for irreversible systems involves the condition that the value of αn must remain practically constant throughout the reaction series studied. It was assumed that changes in the value of αn, which can be determined either from the slope of the polarographic wave, using the so called logarithmic analysis, or from the value of $dE_{1/2}/dpH$ obtained from the shifts of half-wave potentials with pH, smaller than ± 0.1 unit, are sufficiently small for the fulfillment of this condition. Recently, (150), it has been shown that at least in some reaction series in which eqs. (86) or (87) are fulfilled, the value of αn is not constant but changes regularly with changes in structure. A linear relationship between αn and the substituent constant has been found (150) corresponding to $(\Delta \alpha n)_X = \rho_{\alpha,R}\sigma_X$. The possibility cannot be excluded that such a relation even exists in other groups of compounds, where the value of αn was assumed to be constant. Either the value of the proportionality factor $\rho_{\alpha,R}$ was small and/or the error in the determination of the value of αn was too large to prove any significant correlation between αn and σ.

This observation seems to be important because it demonstrates that the value of αn can be affected by structural effects. In addition to its importance for interpreting the physical meaning of the transfer coefficient in electrochemistry, it is also important for structural correlations. The situation seems to resemble formally that situation in homogeneous kinetics where relationships have been shown to exist between $\Delta H^{\pm}$ and $\Delta S^{\pm}$, or PZ and E_a, and that these two components, separated from the experimentally obtained value of the rate constant, can both be affected by structural changes.

On the other hand, if the value of αn for one compound in the reaction series studied differs considerably from those of the rest, a deviation of the half-wave potential of this particular compound has often been observed (150).

For irreversible systems, the plot of $(\Delta E_{1/2} \cdot \Delta \alpha n)_X$ against the substituent constant [in connection with eq. (83)] rather than the value of only $(\Delta E_{1/2})_X$ has been suggested. In a few cases an improved correlation has been found when $(\Delta E_{1/2} \cdot \Delta \alpha n)_X$ was used, but generally the changes in the value of $\Delta \alpha n$ in a given reaction series are small and the error of their determination is great.* Hence the resulting relation is often much more affected by variations in the value of αn than in the value of the half-wave potential. Often a much worse or no correlation for $(\Delta E_{1/2}\ \Delta \alpha n)_X$ is found when the correlation for $(\Delta E_{1/2})_X$ is rather good.

Hence as in homogeneous kinetics in which values of rate constants often proved more useful for structural correlations than the derived values of $\Delta H^{\pm}$ and $\Delta S^{\pm}$, half-wave potentials seem to be of greater importance in such correlations than the derived value of $(\Delta E_{1/2} \cdot \Delta \alpha n)_X$.

When a deviation is observed from the linear $E_{1/2}$–σ plot, it is first determined whether the particular substances shows an extraordinarily large or small value of αn. If this is not the case, two main causes of the observed deviation can be distinguished. The deviation can be caused by the fact that the mechanism of the particular electrode process differs from those operating for substances obeying the linear relationship (Fig. 14), or alternatively, in the deviating compound, substituent effects participate that are not accounted for in the derivations of the Hammett and Taft equations (155–157) and hence are not reflected in the tabulated value of the substituent constant σ. The latter case will be discussed in some detail here.

* Data on values of αn found in the literature are rarely reliable. These data can be measured with some accuracy only using curves recorded either with a three-electrode system or using an extremely slow recording of the i-E curves.

The extraordinary substituent effects can be either specific for the deviating compound or reflect special conditions and heterogeneity during the polarographic process. This means that in the former case the substance would show deviations, if the reactivity of the reaction series were studied in chemical reactions or by other physical methods. In the latter, reactivity in the whole reaction series would behave normally in all other reactions and toward other physical methods and only in polarography would deviations be observed.

These two reasons for deviations can be distinguished when for the particular reaction series data are available on reactivity in a chemical reaction or other physical constants that follow the linear free energy relationship. Possibilities, summarized in Table III, for kinetic measurements $\log k$ can be generalized for equilibrium constants $\log K$ or for other physical quantities.

An example of this type of elucidation of sources of deviation is the comparison of half-wave potentials, nuclear magnetic resonance shifts δ_X and substituent constants σ_X (150). For substituted iodobenzenes (158), the δ–σ and $E_{1/2}$–δ correlations (Fig. 15) are split into two linear portions: one for *m*-derivatives, and another for *p*-derivatives, including 2-chloro and 2-methyl derivatives for which as a first approximation σ_{o-X} values were used and their steric effect neglected. All *o*-halogeno derivatives deviate from the $E_{1/2}$–σ plot. Comparison

TABLE III
Identification of the Reasons for Deviations from the $E_{1/2}$–σ Plot in the Reaction Series X_n—A—R

Reason	Plot		
	$\log k$–$E_{1/2}$	$E_{1/2}$–σ	$\log k$–σ
Specific polarographic	Deviation	Deviation	Correlation
Specific kinetic[a]	Deviation	Correlation	Deviation
Specific for X_1—A—R[b]	Deviation	Deviation	Deviation
Specific for X_1—A—R[c]	Correlation	Deviation	Deviation

[a] In the equilibrium or physical method involved.

[b] For the substituent X_1 in X_1—A—R, an effect is involved that is not accounted for in the change of dissociation constant of the benzoic acid with the substituent X_1 or in the ester hydrolysis rate constants; it is a nonadditive, specific substituent effect.

[c] A specific substituent effect is involved similarly as in the previous case, but influencing polarography relatively in the same degree and the rate of chemical reaction or the other quantity measured.

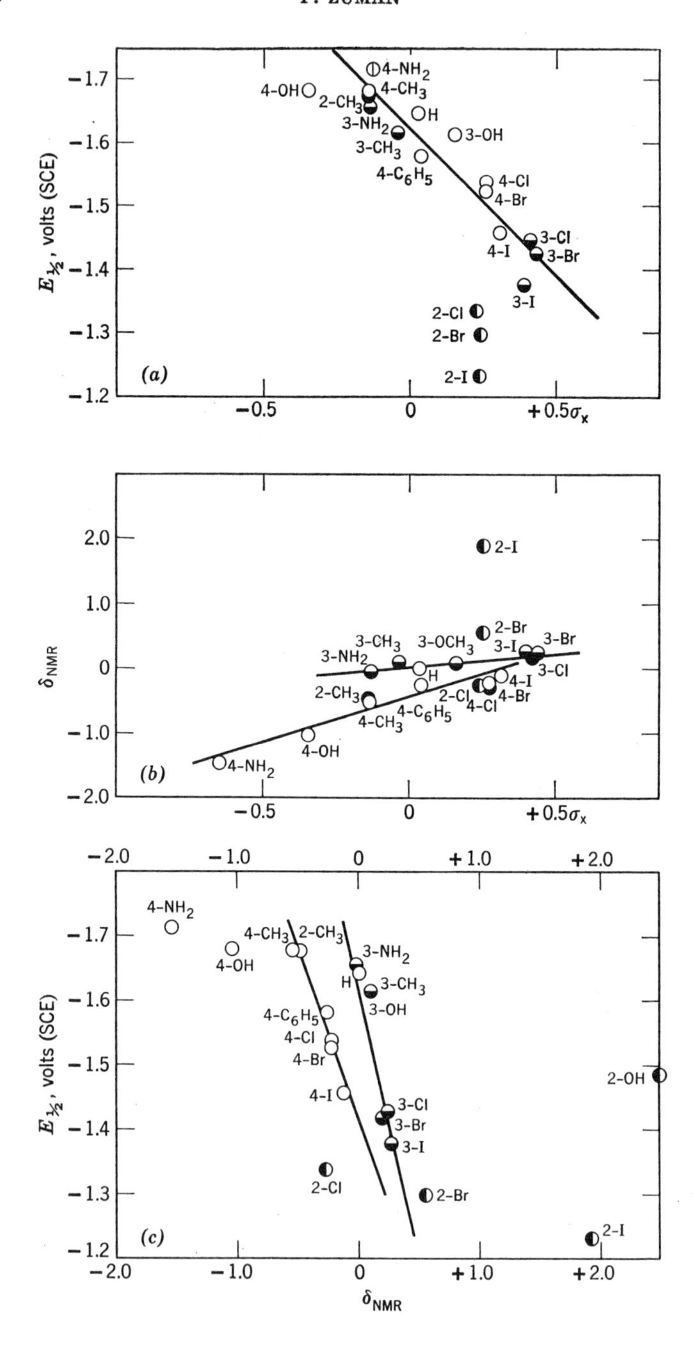

(a)
$E_{1/2}$, volts (SCE)
−1.7
−1.6
−1.5
−1.4
−1.3
−1.2
−0.5
0
+0.5σ_x
4-NH$_2$
4-OH
4-CH$_3$
2-CH$_3$
H
3-NH$_2$
3-OH
3-CH$_3$
4-C$_6$H$_5$
4-Cl
4-Br
3-Cl
4-I
3-Br
3-I
2-Cl
2-Br
2-I
(b)
δ_{NMR}
2.0
1.0
0
−1.0
−2.0
2-I
2-Br
3-CH$_3$
3-OCH$_3$
3-I
3-Br
3-NH$_2$
H
3-Cl
4-I
2-CH$_3$
2-Cl
4-Br
4-C$_6$H$_5$
4-Cl
4-CH$_3$
4-OH
4-NH$_2$
(c)
−2.0
−1.0
+1.0
+2.0
δ_{NMR}
2-OH

with δ–σ and $E_{1/2}$–δ plots indicates that this deviation is not characteristic for polarography but that it is an *ortho* effect of the bulky halogen atoms. *p*-Amino derivatives fit the δ–σ plot. Deviation from the $E_{1/2}$–σ plot shows that an effect is operating in polarography that is not expressed in $\sigma_{p\text{-}NH_2}$. *p*-Iodophenol shows deviations from $E_{1/2}$–σ and $E_{1/2}$–δ plots but not from the δ–σ plot. Thus the polarographic reduction is affected by a specific effect, probably involving the dissociation of the phenolic group that is indicated by the pH dependence of half-wave potentials of iodophenol (the half-wave potentials of other iodobenzenes are pH-independent). Similarly, the same three quantities can also be compared for the discussion of deviations (150) in the reaction series of substituted nitrobenzenes.

When for an observed deviation the possibility is excluded of the departure of the αn value from those observed for other compounds in the reaction series studied and of the specific, nonadditive effect of the substituent X_1, it is possible to conclude that a change in mechanism is the source of the deviation. Deviations from $E_{1/2}$–σ plots are usually not sufficient proof that in the substance X_i—A—R, the reduction follows another path than in other substances X_n—A—R bearing the same group R. However, an observation of the deviation from the linear $E_{1/2}$–σ plot can be a starting point for an investigation or reinvestigation of both substances X_i—A—R and X_n—A—R. Two examples of that type can be mentioned here. It was noted (85) that for the *p*-cyano group in various benzenoid compounds, deviations were observed from the $E_{1/2}$–σ plots for acid media. Examination of the reduction course of some of these derivatives leads to the discovery (127) that in *p*-cyanoacetophenone and *p*-cyanobenzaldehyde in acid media the cyano group is surprisingly reduced at more positive potentials, i.e., more easily, than the carbonyl group, and that in the first step, *p*-aminomethyl derivatives are formed. The substantially more positive (by some 0.4 V.) half-wave potential of *p*-diacetylbenzene (159) than predicted by the Hammett equation led to the reinvestigation of the polarographic

---

Fig. 15. Substituted iodobenzenes. (*a*) Dependence of the half-wave potentials of the reduction of the C—I bond on the Hammett substituent constants σ_X; (*b*) dependence of the chemical shifts δ_{NMR} for substituted fluorobenzenes on the Hammett substituents constants σ_X; (*c*) dependence of half-wave potentials of the reduction of the C—I bond on the chemical shifts δ_{NMR} for substituted fluorobenzenes. (○) *para*, (◓) *meta*, (◐) *ortho* derivatives.

behavior (60). It has been shown that the reduction of only the monoprotonized form at pH > 8 corresponds to a reduction of *p*-acetyl substituted acetophenone, whereas in the diprotonized form, a reduction to a divalent anion radical takes place. Hence a deviation from a linear $E_{1/2}-\sigma$ plot can point to interesting objects of mechanistical studies.

The sign of the reaction constant can also indicate some information concerning the mechanism of the electrode process. If with the increasing positive value of the substituent constant σ_X, i.e., with increasing electrophilic ability of the substituent, the half-wave potential is shifted to more positive values, i.e., facilitated; then, in accordance with usage in homogeneous kinetics, the sign of the reaction constant $\rho_{\pi,R}$ is taken by definition as positive. It was assumed (157) that the positive value of the reaction constant ρ gives evidence that the reaction is facilitated by a low electron density on the reaction center. Oppositely, reactions stimulated by high electron density on the reactive group are assumed to show a negative value of ρ.

In benzenoid and aromatic heterocyclic series, the overwhelming majority of reaction series corresponding to reduction processes shows a positive sign of the reaction constant $\rho_{\pi,R}$. This would, according to the definition given above, correspond to a similar type of mechanism of the potential-determining step, a nucleophilic attack. The electron or the electrode is considered the most probable nucleophilic agent. This seems to be logically expected. Among the few reaction series showing a negative sign for the reaction constant $\rho_{\pi,R}$ are electroreductions in concentrated sulfuric acid.

More complex is the situation in reaction series of aliphatic compounds. Even here, numerous reaction series show a positive value of the reaction constant $\rho_{\pi,R}{}^*$, but the number of reaction series showing a negative sign of the reaction constant $\rho_{\pi,R}{}^*$ is considerably greater than with aromatic compounds. In addition to various reaction series that show no common features, all reaction series showing the effect of alkyl groups on the reduction of organometallic compounds show a positive sign of $\rho_{\pi,R}$. If the above assumption is correct, this would indicate that the potential-determining step includes an electrophilic reaction. A chemical reaction preceding the charge transfer would perhaps be an explanation of this observation.

Even more difficult to explain is the fact that for nitroparaffins the sign of the reaction constant $\rho_{\pi,NO_2}{}^*$ changes with pH (Table IV).

TABLE IV
Effect of pH on the Values of Reaction Constants for the Reduction of Nitroparaffins

$E_{1/2}$ from ref. (163), McIlvaine buffer		$E_{1/2}$ from ref. (164), Clark-Lubs buffer	
pH	$\rho_{\pi,NO_2}{}^*$, v.	pH	$\rho_{\pi,NO_2}{}^*$, V.
2.1	−0.42	2.0	−0.45
—	—	3.0	−0.19
—	—	4.0	−0.09
5.1	−0.10	5.0	+0.13
7.0	+0.08	7.0	+0.26
8.9	+0.22	—	—

This would indicate that different mechanisms operate at different pH values.

Nevertheless, some findings raise doubts whether the explanation for the sign of the reaction constant by nucleophilic and electrophilic mechanisms is the only one possible. It is difficult to explain why for 4-alkoxy-1,2-naphthoquinones, the value $\rho_{\pi,Q}{}^* = +0.05$ V., whereas for 4-alkylamino-1,2-naphthoquinones for a similar group of alkyls the value $\rho_{\pi,Q}{}^* = -0.03$ V. has been found (162), when in both cases the *o*-quinoid system clearly represents the electroactive portion of the molecule.

The vast majority of reaction series studied so far, for which a sufficient number of half-wave potentials were measured for adequately chosen derivatives, show a good correlation for one single value of reaction constant $\rho_{\pi,R}$ when using eqs. (86) or (87). Only the half-wave potentials of *p*-substituted benzyl bromides (160) and benzenediazonium bisulfates (161) show a course of the $E_{1/2}$–σ plot that can be best approximated by two straight lines: with a positive slope for electrophilic substituents, and with a negative slope for nucleophilic substituents (Fig. 16). This behavior can be interpreted as a result of two competing potential-determining reactions: a charge transfer for electropositive substituents and an electrophilic reaction (see above) for electronegative substituents. It is interesting to note that these deviations parallel those in homogeneous kinetics of these two reaction series.

An example of information on the mechanism of the electrode process based on departures from additivity of substituent effects is the

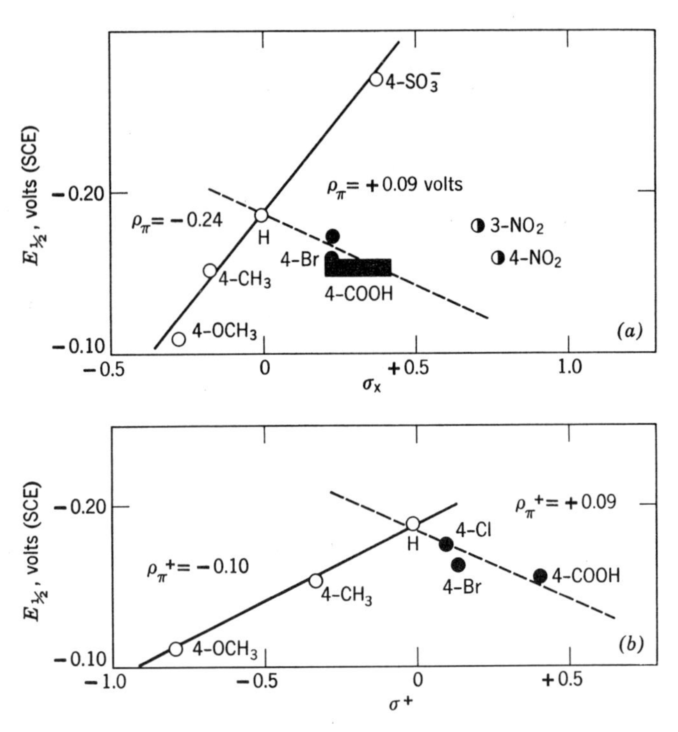

Fig. 16. Reduction of benzene diazonium salts at pH 4. pH-independent half-wave potentials: (*a*) dependence on the Hammett substituent constants σ_X; (*b*) dependence on the substituent constants for electrophilic reactions σ_X^+. Circles correspond to regression line with negative slope, full points to a line with positive slope. Halved points deviate; reduction of the nitro group is possibly involved.

treatment (150) of the half-wave potentials (165) of substituted diaryliodonium salts. The half-wave potentials of the first wave corresponding to the reduction of cation $X^1—C_6H_4—\overset{(+)}{I}—C_6H_4—X^2$ are influenced by substituents X^1 and X^2 on both phenyl rings, and their effects are additive. The second, more negative wave of diaryliodonium salts was attributed (165) to the reduction of the radical $X^1—C_6H_4—\dot{I}—C_6H_4—X^2$. But the half-wave potentials of the second waves of diaryliodonium derivatives bearing a substituent X on one ring, $X—C_6H_4—\overset{(+)}{I}—C_6H_5$, and of derivatives bearing the same substituent on both rings, X—C_6H_4—$\overset{(+)}{I}$—C_6H_4—X, are practically identical. The half-wave potentials of monosubstituted compounds fit eq. (86) and those of disubstituted compounds fit the same straight line when σ_X instead of $2\sigma_X$ is used for correlation. These observations indicate that in the second wave reduction did not involve the radical X—C_6H_4—$\dot{I}$—C_6H_4—X, but rather a species which includes only one phenyl group. It can be assumed that in the first wave one of the C—I bonds is broken either in the electrode process proper or in a consecutive chemical reaction.

In compounds, R^1R^2—A—X bearing two groupings that can be electroactive, linear free energy relationships sometimes can indicate which of the two groups present is reduced at more positive potentials, i.e., easier. To decide whether the group R^1 or R^2 is electroactive at the potentials of the more positive step, the half-wave potential of the compound R^1R^2—A—X is compared with two sets of half-wave potentials, namely those of R^1—A—X and those of R^2—A—X. The first, group R^1 is considered to be the polarographically active group and the second group as a substituent X^2. The half-wave potential is compared with the plot with a slope ρ_{π,R^1}, using R^1H—A—X as a standard and σ_{X^2} for the correlation. Then R^2 is assumed to be electroactive using ρ_{π,R^2} and R^2H—A—X as a standard, and the first group as a substituent $X^1(\sigma_{X^1})$. It is decided into which of the two relations the observed values of $\Delta E_{1/2}$ fit.

An example of this type is the behavior of monosubstituted *p*-dinitrobenzenes where the substituent is in a *m*-position relative to one, but in an *o*-position relative to the second nitro group. For the first wave of methoxy-*p*-dinitrobenzene, the shift $\Delta E_{1/2}$ relative to that of the unsubstituted *p*-dinitrobenzene fits the equation $\Delta E_{1/2} = \rho_{\pi,NO_2} \sigma_{m\text{-}OCH_3}$ but not $\rho_{\pi,NO_2}\sigma_{o\text{-}OCH_3}$. It can be deduced that the nitro group in the *m*-position relative to the methoxy group is reduced first and that *p*-hydroxyamino-*o*-methoxynitrobenzene is formed in the first reduction step.

Application of this type of treatment for the identification of intermediates can be demonstrated with the example of *m*-nitrobenzaldehyde. This substance is reduced at pH 7 in two steps. The first wave corresponds to the reduction of the nitro group to the corresponding hydroxylamino derivative. It is necessary to decide whether in the second wave the hydroxylamino or the aldehydic group is reduced. It was found that $\Delta E_{1/2}$ corresponds to $\rho_{\pi,CHO}\sigma_{3\text{-NHOH}}$ rather than to $\rho_{\pi,NHOH}\sigma_{3\text{-CHO}}$. Hence in the second wave 3-NHOH-C_6H_4CHO is reduced to 3-$NHOHC_6H_4CH_2OH$. Reduction of the latter occurs at such negative potentials that it is overlapped by reduction of the supporting electrolyte.

Three waves were observed for *p*-nitrobenzaldehyde under similar conditions. The half-wave potential of the third, most negative wave fits the $\Delta E_{1/2}$–$\sigma_{4\text{-}NH_2}$ plot (with $\rho_{\pi,CHO}$) but not $\Delta E_{1/2}$–$\sigma_{4\text{-}CH_2OH}$ (with $\rho_{\pi,NHOH}$). The third wave thus corresponds to the reduction of 4-$NH_2C_6H_4CHO$. The amino group must be formed in the second step which corresponds to the reduction of 4-$NHOHC_6H_4CHO$ to 4-$NH_2C_6H_4CHO$.

In the dibromo derivative Br^1—CH_2CH═$CHBr^2$, it was shown (150) that $\Delta E_{1/2}$ corresponds to $\Delta E_{1/2} = \rho^{I}_{\pi,Br^1}\sigma^{I}_{Br^2}$ where inductive substituent constants σ^{I} are used, and hence it was assumed that in the first step the C—Br^1 bond is reduced. This is in accord with the observation (166) that halogens bound to an olefinic carbon are usually reduced at more negative potentials than those bound on a saturated carbon and that, on the other hand, the reduction of a CH_2—X bond is facilitated by an unsaturated bond in the β-position.

The above examples illustrate the various ways in which linear free energy relationships can be used in the elucidation of the electrode process mechanism.

C. STERIC EFFECTS

Among the steric effects that affect polarographic behavior, the role of steric hindrance of coplanarity, bond length and polarization, and orientation at the electrode surface were recognized. The latter factor is actually equivalent to the effect of structure and stereochemistry of the transition state in homogeneous reactions. On the other hand, no equivalent of steric hindrance of the approach of the reactant has been proved so far. It is possible that due to the small size of the attacking electron, a much denser crowding of the neighborhood of the electroactive group by bulky groups would be necessary in polarography than in homogeneous kinetics. Data on polarographic behavior of such densely sterically hindered compounds have not been reported. There are two cases in which such an effect can be expected to operate. Whereas dialkylperoxides are generally reducible at positive potentials, di-*tert*-butylperoxide was reported to be nonreducible (166). Similarly for cyclic diketones (Table V), the nonreducibility of 3,3,8,8-tetramethyl-1,2-cyclo-octanedione can arise from a similar effect, but the steric hindrance of coplanarity of the diketo grouping cannot be excluded in this case. A more detailed examination of these and similar systems would be necessary to reach a decision as to whether this type of steric effect can affect the polarographic behavior or not.

The best understood of the steric effects in polarography is the steric hindrance of coplanarity (150,166,168) that occurs, if on the aromatic ring bulky groups are substituted in an *o*-position or positions relative to the electroactive group. Whereas in most cases the shifts of half-wave potentials of benzenoid compounds bearing the electroactive group in the side-chains amount to several tens of millivolts, for bulky groups in *o*-positions shifts of several hundred millivolts towards more negative potentials are observed. These shifts

TABLE V
Half-Wave Potentials (167) of Cyclic Diketones
(50% Isopropylalcohol, 0.1*M* Acetic Acid, 0.1*M* Sodium Acetate)

	n	$E_{1/2}$, V.
$(CH_3)_2C(COCO)(CH_2)_nC(CH_3)_2$	2	−0.75
	3	−1.16
	4	Nonreducible
	14	−1.25

were attributed to the fact that the bulky *o*-substituents prevented attainment of coplanarity of the electroactive group and the aromatic ring. Distortion from the coplanar arrangement prevents the resonance interaction between the electroactive group and the aromatic system and the observed shift towards negative potentials corresponds to the decrease in resonance interaction. The validity of this assumption was verified by comparing the contribution of the resonance effect from independent studies (150). Most experimental evidence has accumulated for *o*-mono- and di-substituted nitrobenzenes and azobenzenes; the effect of alkyl groups in positions adjacent to the carbonyl groups in *o*- and *p*-nitrobenzoic acids is interpreted as due to the hindrance of coplanarity of the carbonyl group.

To elucidate the mechanism, proof of the steric hindrance of coplanarity can be taken as proof that in the unhindered compound the electroactive group either in the ground state or in the transition state is in resonance interaction with the aromatic nuclei.

For the other recognized types of steric effects, such as those affecting the bond length, polarization, or orientation of the organic molecule during the electrode process, the available material is too limited to allow identification and separation of these factors. Hence some examples are discussed next in which these and perhaps other steric factors participate.

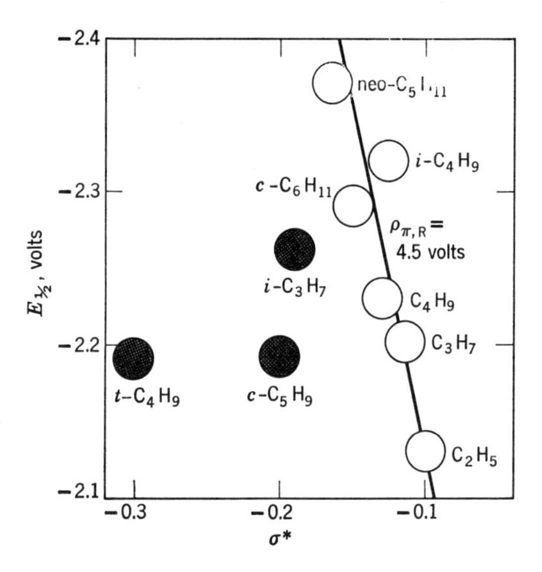

Fig. 17. Dependence of half-wave potentials for the reduction of alkyl bromides on Taft polar substituent constants σ_X^* (full points deviate).

Reduction of alkyl and cycloalkyl bromides (150,169–175) can be mentioned first. Half-wave potentials of straight-chain alkyl bromides and of those branched on the β-carbon fit eq. (87). α-Branched alkyl bromides show a shift towards more positive potentials when compared with the value predicted by eq. (87). The deviation is significant for isopropyl bromide, but is even more marked for *tert*-butyl bromide (Fig. 17). Introduction of a correction for steric effects according to Taft (155) of the type observed for acid ester hydrolysis using eq. (88) for $\delta^{\cdot}_{\pi,\mathrm{Br}} = -0.04$ V. does not markedly improve the correlation, and deviations for i-C_3H_7 and t-C_4H_9 are still observed:

$$\Delta E_{1/2} = \rho_{\pi,R}{}^*\sigma_X{}^* + \delta_{\pi,R}(E_S)_X \tag{88}$$

Because deviations increase in the same sequence, C_2H_5 < i-C_3H_7 < t-C_4H_9, as the deviations encountered in nucleophilic substitutions (176) in homogeneous solutions and that are explained by gradual change from S_N2 to S_N1 mechanism,† it was assumed that the reasons for the deviations in the values of half-wave potentials are of similar origin. Based on this analogy, it was assumed that the reduction of straight-chain alkyl bromides follows an S_N2-like mechanism. This also agrees with the positive sign of the reaction constant $\rho_{\pi,Br}{}^* = 4.6$ V. Consequently, the more positive waves of isopropyl and *tert*-butyl bromide can be ascribed to a participation† of an S_N1-like mechanism. Whereas for the S_N2-like reactions no difficulty exists in transferring the reasoning concerning the homogeneous reactions to electrode processes, for S_N1-like electrode reactions a complete analogy with homogeneous processes cannot be expected. In particular, it would be necessary to assume an increase in the dissociation rate in the electrial field of the dropping electrode. On the other hand, the other assumption tacitly involved, i.e., that the carbonium ion accepts electrons more readily than the alkyl bromide molecule, seems to be plausible‡ even when electrostatic forces only are considered.

In this connection, the behavior of cyclopentyl and cyclohexyl bromides can be mentioned. The half-wave potentials of the cyclohexyl derivatives fit eq. (87) (Fig. 17), whereas the value for the cyclopentyl derivative shows a similar deviation as the value for isopropyl bromide. It would be possible to conclude that for cyclohexyl bromide an S_N2-like mechanism is operating, whereas for cyclopentyl bromide a participation of the S_N1-like mechanism cannot be excluded. Unfortunately, this interpretation cannot be supported from the study of homogeneous kinetics, because for cycloalkyl

† These systems are better described as a participation of a continuously changed transition state. or as due to varying contributions of canonical structures with a positive charge on carbon to the resonance hydrid in the transition state, respectively. The above description is used for the sake of simplicity only.

‡ Stable carbonium ions like triphenylmethyl or tropylium ion are reduced at markedly more positive potentials than corresponding hydrocarbons.

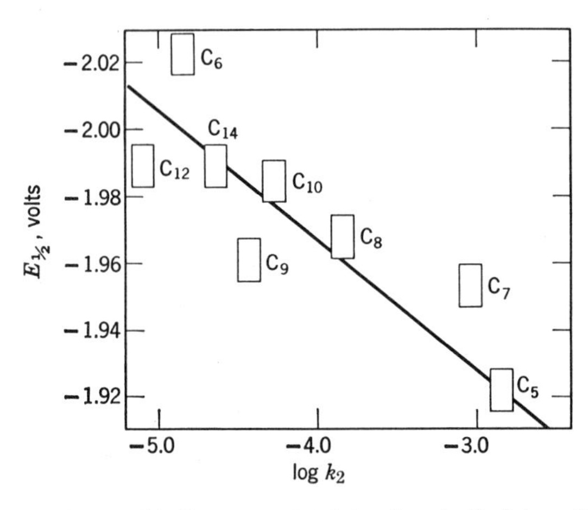

Fig. 18. Dependence of half-wave potentials of cycloalkyl bromides on logarithms of rate constants of their reaction with potassium iodide in acetone. Half-wave potentials in 99–97% dimethylformamide containing 0.1*M* $N(C_2H_5)_4Br$.

bromides, no detailed study of the effect of the ring-size on the contributions of S_N1 and S_N2 mechanisms has been reported. Reactions of cycloalkyl halides with nucleophilic reagents have been expected to follow either an S_N1 or S_N2 mechanism, according to the reagent used, for all ring sizes.

For cycloalkyl halides other than cyclopentyl and cyclohexyl, the values of σ_X^* were not reported (155) and hence another type of treatment had to be used. The half-wave potentials for these reaction series can be correlated either with rate constants of homogeneous reactions undergone by compounds of the same reaction series or with half-wave potentials of another reaction series. In both of these treatments values obtained for compounds of the same ring size are always compared.

A poor correlation has been found for the dependence of half-wave potentials of cycloalkyl bromides on the logarithms of rate constants of reactions for which an S_N1 mechanism is expected, such as solvolysis of methylcycloalkyl chlorides (177,178), cycloalkyl *p*-toluenesulfonates (179), and cycloalkyl chlorides (179). The greatest deviations were observed for the cyclobutyl derivatives. The results allow the conclusion that the S_N1-like mechanism is not the predominant one in the whole reaction series.

On the other hand, the half-wave potentials of cycloalkyl bromides have shown relatively good correlation with logarithms of rate constants according to eq. (89) of those reactions to which an S_N2 mechanism is attributed such as the exchange reaction of cycloalkyl iodides with radio-iodide (180), or reaction of cycloalkyl bromides with lithium (181) or potassium iodide (179,182) (Fig. 18):

$$(\Delta E_{1/2})_X = \rho' \Delta \log k \qquad (89)$$

Since the participation of various mechanisms for some of the ring sizes in homogeneous reactions cannot be excluded, it is not safe to deduce that the reduction of all cycloalkyl bromides follows an S_N2-like mechanism. On the contrary, as mentioned above, comparison of the behavior of cyclopentyl and cyclohexyl derivatives indicates a possibility of varying contributions of particular mechanisms. Hence it is preferable to conclude that participation of contributing mechanisms in the polarographic reduction of cycloalkyl bromides is analogous to that found in homogeneous reactions of these compounds with iodides.

Participation of various contributing mechanisms can also be deduced from the half-wave potentials of some bridgehead bromides (172–174). The shift of their half-wave potentials to more negative potentials when compared with straight-chain bromides can also be affected by steric strain, in addition to a change in mechanism. The bridgehead molecules are assumed not to be accessible to rear attack on carbon. This makes the S_N2-like mechanism less probable and frontal attack on bromine (173), or an S_N1-like mechanism is considered. The more positive reduction of the *exo*-norbornyl bromide than that of *endo*-norbornyl bromide is quoted in support of the S_N1-like mechanism. Ionization in the field of the electrode to form carbonium ions would occur more readily in *exo*-norbornyl bromide because of the anchimeric aid.

Further support for the S_N1-like reduction mechanism is sought in the fact that the reduction of *cis*-4-bromo-*tert*-butylcyclohexane occurs at more positive potentials than that of its *trans* isomer. The rear of the carbon attached to the equatorial bromine in *trans*-4-*tert* butylcyclohexyl bromide is almost as hindered as a bridgehead bromine. Therefore, an S_N2-like mechanism is less probable and the more negative wave is ascribed predominantly to an S_N1-like mechanism. The far less hindered *cis*-4-*tert*-butylcyclohexyl bromide can

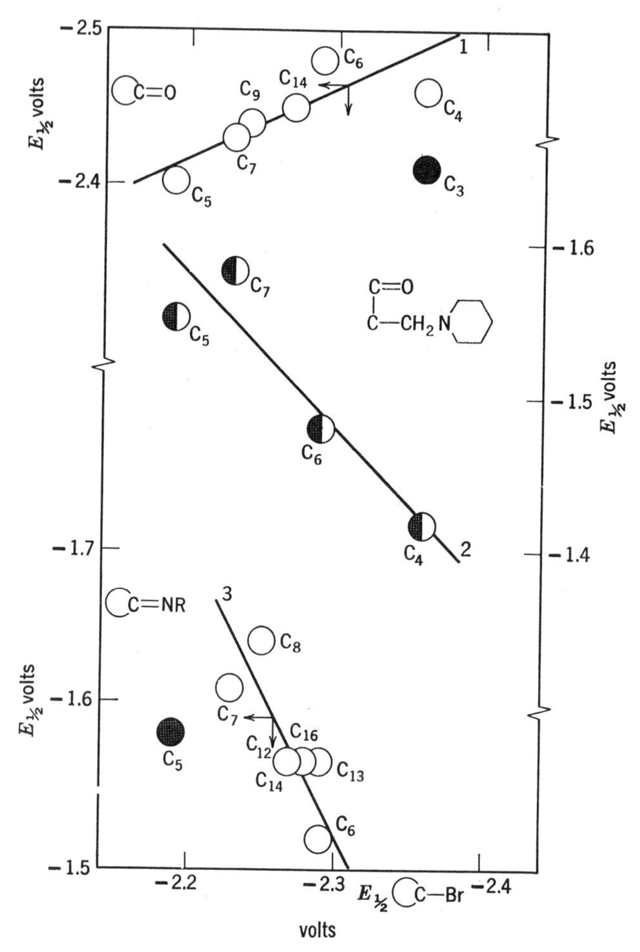

Fig. 19. Dependence of half-wave potentials of cyclanones (*1*), their betainyl hydrazones (*3*) and 1-piperidinomethyl-2-cycloalkanones (*2*) on half-wave potentials of cycloalkyl bromides. Full points deviate.

be attacked from the rear and a predominantly S_N2-like mechanism is assumed to operate in the more positive wave (174).

For some compounds, even a homolytic mechanism (173) cannot be ruled out, but it seems unlikely that this reaction mechanism dominates the whole reaction series. First, it seems improbable that at the negative potentials where the electrical double layer is negatively charged, the cycloalkyl bromide would be oriented towards the negatively charged surface of the electrode by the negative end of its dipole, as would be expected according to the halogen-bridge theory. Second, no organometallic compounds with mercury are formed (183) during the electrolysis of alkyl halides, with the exception of benzyl and allyl halides. Third, no correlation of the half-wave potentials of cycloalkyl bromides with kinetic data involving homolytic splitting of the C—Br bond has been found (171).

The second possibility for the study of cycloalkyl bromides is to correlate their half-wave potentials with half-wave potentials of other alicyclic compounds using eq. (90):

$$(E_{1/2})_{\mathrm{Br,C^1}} - (E_{1/2})_{\mathrm{Br,C^2}} = \rho_\pi'[(E_{1/2})_{\mathrm{R,C^1}} - (E_{1/2})_{\mathrm{R,C^2}}] \qquad (90)$$

where R refers to the electroactive group of the other reaction series used for comparison, C^1 to one and C^2 to another ring size.

The relatively good correlation obtained for these plots (Fig. 19) indicate that the ring size affects the half-wave potentials of cycloalkyl bromides relatively in the same way as those of cyclanones (184), their betainyl hydrazones (185), and 1-piperidinomethyl-2-cyclanones (186). Significant deviations from such relationships indicate, among other reasons, the possibility of a change in mechanism. For a larger group, such deviations are observed for 1-piperidinomethyl-2-cyclooctanone, -nonanone, and -decanone, and in fact the shape of the waves of these compounds (186) indicates a change in the mechanism of the electrode process. No explanation is known at present for the different slopes of the plots in Figure 19, that express the changed sequence. For cycloalkyl bromides and cyclanones, the cyclopentyl derivative is reduced at more positive potentials than the cyclohexyl, whereas for cyclanone betainyl hydrazones and for the Mannich bases, just the reverse is true.

The reduction of vicinal dibromides can be discussed as the next example. Principally, these compounds are reduced in one 2-electron step. Because the main product of this step has been shown to be the unsaturated hydrocarbon and because no other wave is apparent on the polarographic curve which would correspond to the reduction of the monobromo derivative, it may be assumed that an elimination process takes place with the second bromine atom participating in the formation of the transition state. Nevertheless, the increase in limiting current for some derivatives seems to indicate that in addition to the 2-electron elimination process a 4-electron substitution can participate in the electrode process.

Comparison of cyclic vicinal dibromides for various ring-size and type and for various stereoisomers (171) is made difficult by the varying value of αn. Bearing in mind the approximate character of such comparisons, the connection between half-wave potentials and the torsion angle φ between the C—Br bonds can be discussed. For molecules with a rigid structure the dependence of half-wave potentials on torsion angle, φ, resembles the Karplus curve (187). Molecules with an *anti*-periplanar arrangement ($\varphi \approx 180°$) or *syn*-periplanar arrangement ($\varphi \approx 0°$) show the most positive waves; those with the angle φ between 60 and 120°, the most negative. Similarly, those flexible dibromides which are known to be readily able to attain an *anti*-periplanar arrangement of the C—Br bond are reduced at relatively positive potentials. On the contrary, those dibromides which attain an *anti*-periplanar arrangement only with great expenditure of energy or none at all, are reduced at such negative potentials that they would correspond to rigid molecules with an angle of about 60°.

The half-wave potentials of monocyclic, flexible vicinal dibromides, as well as of some rigid dibromides, are an approximately linear function of the logarithms of the rates of homogeneous elimination reactions. Also the more positive reduction of *erythro*-5,6-dibromodecane than that of the *threo* epimer agrees with results obtained for bimolecular homogeneous reactions.

The third group of compounds in which steric effects have been studied and used the elucidation of the electrode process mechanism are α-halogenoketones.

In 2-halogeno-4-*tert*-butylcyclohexanones (188), the thermodynamically more stable equatorial halogen is reduced at more negative potentials than the axial halogen. Similarly, in the rigid system of B and C rings of steroids, the equatorial halogen is reduced at more negative potentials than the axial halogen (175). Differences in the half-wave potentials of equatorial and axial halogen vary according to the positions of both the halogen and carbonyl group in the ring system, and according to the nature of the halogen, the difference in-increases in the sequence Br < Cl < F. These differences were used for a discussion of the reduction mechanism. Whereas for alkyl halides nucleophilic substitution and perhaps bridging mechanism were considered chiefly (p. 181), for α-halogenoketones an elimination process was also considered (189). The enolate formed in the electrode process is electroinactive in the potential range studied and to be further reduced, must be transformed by an interposed chemical reaction into the electroactive ketone.

The proof of enolate formation, considered also in other reductions of carbonyl compounds (Sec. IV-B), is the difference between the shapes of the polarographic curves and products formed in controlled potential electrolysis. The latter method carried out at the potential of the limiting current of the first 2-electron wave yields Δ^4-3-ketosteroid and an equivalent amount of bromides. On the contrary, on the polarographic curves of Δ^4-2-halogeno-3-ketosteroids, no wave corresponding to Δ^4-3-ketosteroid reduction is observed apart from the 2-electron wave of the C—Br reduction. Hence, an enolate is formed at the surface of the electrode that is subsequently transformed into the Δ^4-3-ketosteroid. At the dropping electrode the time is too short for production of any significant amount of the keto form, whereas during large scale electrolysis, there is enough time to allow the transformation into Δ^4-3-ketosteroid. The formation of enolate can be realized either by an attack on the carbon of the C—Br bond or by an attack of the electron on the carbonyl carbon. In the former case, it must be assumed that the carbanion resulting from the electrode process can preferably accept hydrogen on the carbonyl oxygen rather than on the carbon from which the halogen has left. Moreover, the carbon atom bearing bromine would have to pass through a trigonal form. It is assumed that for vicinal halogenoketosteroids bearing the halogen on ring B or C, the rigidity and the shape of the molecule prevent an attack from the rear.

Hence, an attack of the electron on the carbonyl carbon is preferred. Both carbons of the C—X and C=O bonds are expected (189) to be oriented in the electrical field of the electrode with the double bond of the carbonyl group parallel to the surface of the electrode. Simultaneously, the ring attains a pseudochair form. For axial halogen this deformation is only slight, whereas for equatorial halogen the carbon-bearing halogen is separated by an axial hydrogen from the surface of the electrode. To attain the most preferred configuration relative to the electrode surface, additional energy is necessary. This explains why equatorial halogens in α-halogenoketosteroids are reduced at more negative potentials than axial halogens.

D. COMPARISON OF VARIOUS ELECTROACTIVE GROUPS

Quantitative comparison of organic molecules X—A—R bearing various electroactive groups R is limited to only those electroactive groups that are reduced or oxidized by the same mechanism. In addition to some few pairs that can be compared, there is only one more extended group of compounds that fulfill the above condition, namely halogen derivatives. In this group, the reduction occurs at more positive potentials in the sequence: F < Cl < Br < I.

Another possibility for comparing various electroactive groups is based on comparing ranges of half-wave potentials, in which occurs the reduction of benzenoid derivatives bearing one single substituent in *meta* or *para* position. These ranges (Fig. 20) were computed using eq. (86) and experimentally determined values of $\rho_{\pi,R}$ and the half-wave potential of the reference compound. The limits were calculated using the extreme values of substituent constants under the assumption that the mechanism of the electrode process of the sub-

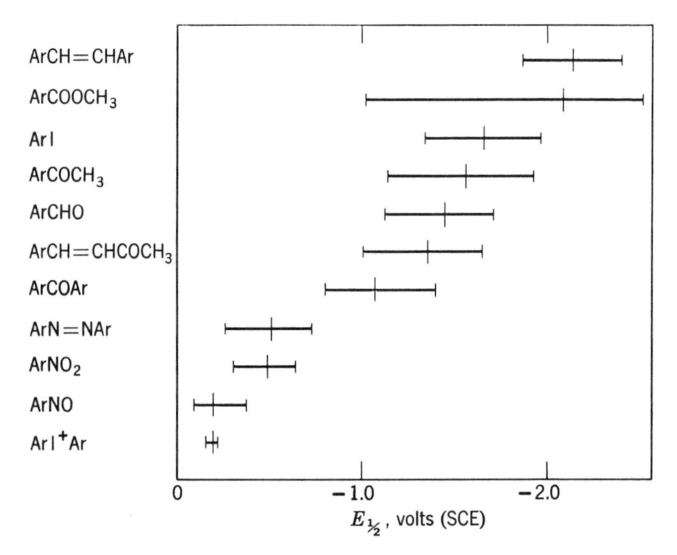

Fig. 20. The potential ranges in which the half-wave potentials of the reduction of benzene derivatives can be expected for compounds reduced by the same mechanism as the parent unsubstituted compound. For values obtained at pH 5–8 and in unbuffered media, for pH-independent systems. The middle point corresponds to the unsubstituted compound.

stituted compounds does not differ from that of the reference substance. The similarity in this group of electrode processes lies in the fact that in all cases the value of the reaction constant $\rho_{\pi,\mathrm{R}}$ is positive. It can be assumed that in all these cases the simplest nucleophilic process, i.e., electron transfer, is the potential-determining step.

Because a conjugated system of double and triple bonds should be considered as one electroactive group, comparison of systems consisting of the same building blocks but with different extension of the conjugated system should be mentioned here briefly. Early in polarography it was recognized that the more extended the conjugated system is, the more positive the reduction potential. This was often verified for systems of conjugated bonds in a chain or for polynuclear hydrocarbons with a varying number of condensed aromatic rings. The correlation between the extent of the conjugated system and the half-wave potentials has been treated successfully using quantum–chemical methods (154).

For benzenoid molecules bearing a double bond in the α position in the side chain, it was assumed that the observed shift towards more positive potentials, when the half-wave potentials of phenyl derivatives are compared with the corresponding alkyl derivatives, also corresponds predominantly to a conjugation effect. When the effect defined by the substituent constant $\sigma^*_{C_6H_5} = 0.60$ (155) is attributed to the polar effect of the substituent, it can be deduced in numerous cases (150,169) that the effect exerted by the phenyl group is polar.

Even when it is possible in some cases to compare the reactivity of various electroactive groups in the electrode process, little information of importance for elucidating mechanisms of electrode processes can be gained by such comparison. Even the sequence of reactivity of halogens, mentioned above, does not allow either for support or exclusion of any one of the considered mechanisms of the electrode processes, i.e., S_N2-, S_N1-like, or bridging mechanisms.

E. IMPORTANCE OF THE STRUCTURAL STUDIES FOR ELUCIDATION OF THE MECHANISM

Whereas the study of the effects of composition of the solution on polarographic waves supplies us mainly with information about the chemical reactions accompanying the electrode process, and whereas the study of the products and intermediates allows us often to decide which bond is broken or formed during the electrode process, structural studies allow us to gain more intimate information about the electrode process in particular about the transition state. Because with the other two approaches we can get information about the form of the electroactive species that enters the electrode process proper and the product that leaves it, the most important contributions of structural studies are perhaps those involving the stereochemistry of the transition state. We can in advantageous cases obtain information on which atom and from which side the molecule is attacked and how it is oriented and organized in space in the transition state. We can also determine whether or not nucleophilic attack on the electroactive molecule, equivalent to the electron transfer in other terminology, is the potential-determining step.

In addition to this, structural studies allow us to detect a compound that deviates in mechanism from a group of substances, for which it was previously assumed that the compound followed the same mechanism.

Because, so far as we are informed, the adsorbability changes with changing structure according to rules other than reactivity, the importance of the role of adsorption on polarographic electroreduction and electrooxidation of organic substances can be detected from structural studies. When polarographic half-wave potentials follow analogous regularities as the reactivities in homogeneous reactions, as is true in most groups of compounds in which the comparison has been carried out so far, it can be deduced that inside these groups the changes in polarographic behavior are predominantly affected by changes in chemical reactivity rather than in adsorptivity. The present picture is perhaps affected by the fact that mostly molecules of approximately the same size are compared and that the greater changes in adsorbability can be expected with compounds with longer chains. As an example in which effects of this type are indicated, the reduction of alkyl peroxides can be mentioned. In this case, the parallel between polarographic activity and chemical reactivity ceases for chains longer than about C_5 (150). The predominance of adsorption is assumed to appear for longer chains. For condensed aromatic rings, for which the role of adsorption was also observed, other structural effects can again be expected, as the π-electron systems of these rings can be in conditions resembling those in charge transfer complexes with the mercury electrode as the other partner.

Recognition of regularities in the behavior of organic substances allows us to choose rationally for further studies those substances from which maximum information can be gained. More economic research is possible according to these lines.

Last but not least, structural studies permit finding substances that can be used as models in fundamental electrochemistry for testing various theories.

IX. Polarography and Extrathermodynamic Relationships

Half-wave potentials can be treated by extrathermodynamic relationships, such as those given in eqs. (86)–(90). The reasons and conditions for such an application and the limitations and scope of applicability are given elsewhere (85,150,151,166,169,175,190). In Sec. VIII of this contribution, some examples are given as to how these relations can be used in the study of mechanisms of electrode processes. Here the question will be considered from the opposite viewpoint, namely, as a question, "How can polarography be of any help for the theory and understanding of extrathermodynamic relationships?"

Extrathermodynamic relationships were principally derived for rate and equilibrium constants of homogeneous reactions in solutions. Later, it was shown empirically that this kind of treatment can be applied to spectral and other physical data. The relation of polarography to extrathermodynamic relationships differs from most other physical methods. The half-wave potentials of reversible systems, as was shown in Sec. VIII-B, are equivalent to logarithms of equilibrium constants, whereas those of irreversible systems are proportional to logarithms of rate constants. Hence the application of half-wave potentials in extrathermodynamic relationships is not merely empirical but is a logical extension of the treatment of kinetic and equilibrium data. One specific aspect of this extension is that in most of the reactions studied in polarography, one reactant, namely the electron or the electrode, remains constant, and only the other reactant changes. That would be equivalent in homogeneous kinetics to a number of reaction series of the type X—A—R^1, X—A—R^2, etc. with one reactant B. The experimental material available in polarography [about 100 reaction series of only the *m*- and *p*-benzene substituted type corresponding to some 800 half-wave potentials with about three times as much material for other types of reaction series (150)] is considerably greater than in homogeneous kinetics for any

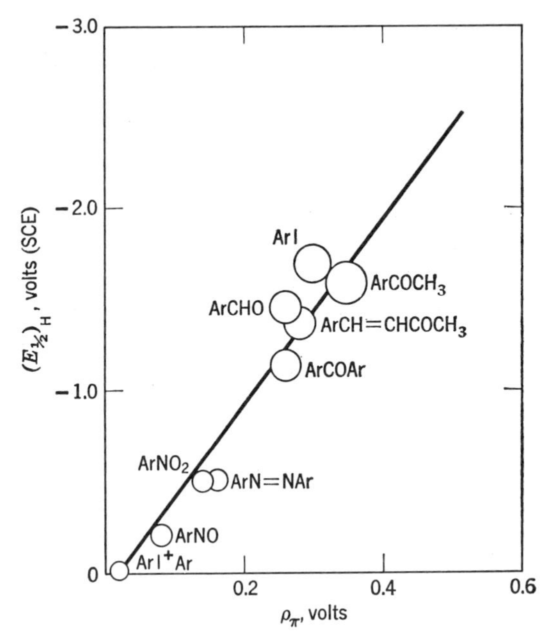

Fig. 21. Dependence of the half-wave potentials of the parent unsubstituted compound $(E_{1/2})_H$ bearing the reducible group R on the values of the reaction constant $\rho_{\pi,R}$ obtained for benzene derivatives bearing the same reducible group R at pH 5–8 and in unbuffered media for pH-independent systems.

one single reactant B and allows for several generalizations. Furthermore, all of this material has been obtained under experimental conditions that do not differ greatly. Whereas in homogeneous kinetics the choice of solvent and temperature at which the reaction is followed usually varies over wide ranges, most of the polarographic experimental work has been done in aqueous, water-containing, or water-like solvents.

This possibility of comparing a wide range of reactants X—A—R^1, X—A—R^2, etc., according to their susceptibility to substituent effects, was used in studies of connections between the properties of the electroactive group, R^1, and the values of the reaction constant ρ_{π,R^1}.

For irreversible systems, in which information about substituent constants that would characterize the effects of substituents on the reaction center in the transition state are missing, no simple relationship is found between the values of $\rho_{\pi,\mathrm{R}}$ and the substituent constants $\sigma_{p-\mathrm{R}}$, characterizing the substituent effects in the ground state of the reactant. This agrees with deductions that can be drawn from the paper by Hine (191). On the other hand, it has been possible for benzenoid reaction series to correlate the values of reaction constants $\rho_{\pi,\mathrm{R}}$ with the values of half-wave potentials of the unsubstituted reference compound bearing the same electroactive group R $(E_{1/2})_{\mathrm{H}}^{\mathrm{R}}$ and characterizing the reactivity of the group R towards nucleophilic attack by an electron (Fig. 21). The linear relation between $\rho_{\pi,\mathrm{R}}$ and $(E_{1/2})_{\mathrm{H}}^{\mathrm{R}}$ allows us to transform the three-parameter eq. (91a) into a two-parameter form (91b):

$$(E_{1/2})_{\mathrm{X}}^{\mathrm{R}} = \rho_{\pi,\mathrm{R}}\sigma_{\mathrm{X}} + (E_{1/2})_{\mathrm{H}}^{\mathrm{R}} \tag{91a}$$

$$(E_{1/2})_{\mathrm{X}}^{\mathrm{R}} = \rho_{\pi,\mathrm{R}}\sigma_{\mathrm{X}} + \mathcal{H}\rho_{\pi,\mathrm{R}} + K \tag{91b}$$

where $\mathcal{H}$ and K represent constants that are independent of the kind of benzenoid reaction series in which the nucleophilic attack occurs in the side chain but may depend on the reaction conditions.

Application of eq. (91b) is not limited to comparison of various electroactive groups studied under similar conditions, but can also be used for comparison of reaction constants $\rho_{\pi,\mathrm{R}}$ with half-wave potentials $(E_{1/2})_{\mathrm{H}}^{\mathrm{R}}$ for the reaction series with the same group R but studied polarographically under varying conditions. The correlation of values of $\rho_{\pi,\mathrm{R}}$ with $(E_{1/2})_{\mathrm{H}}^{\mathrm{R}}$ for substituted nitrobenzenes in dimethylformamide media in the presence of various cations (Fig. 13) has already been mentioned (Sec. VI-A). Similarly in those cases in which the reaction constants are a function of pH, as for instance has been found for substituted azobenzenes, there is a linear relation between $\rho_{\pi,\mathrm{R}}$ and $(E_{1/2})_{\mathrm{H}}^{\mathrm{R}}$ determined at various pH values.

When, in the first approximation, the value of the half-wave potential of an irreversible system is considered as a measure of the polarizibility of the reduced bond, the relation (91c) found empirically (85) would express a relation between the degree of polarizibility of the reduced bond and the susceptibility of the molecule to transfer the effect of the substituent on the reduced bond.

In addition to this possibility in which polarography has indicated its possibilities in expanding the number of possible approaches to the problem of the magnitude of the reaction constant ρ, polarography offers some experimental advantages. The common way of studying free energy relationships and similar problems and of determining the best values of substituent constants, σ, is to use equilibrium or, more frequently, rate constants. The extraction of comparable rate constants from experimental data is often a tedious, rather complicated, and definitely time-consuming operation, especially when the mechanism must be elucidated prior to the choice of the rate constants to be compared. When compared, the measurement of half-wave potentials is usually a simple, straight-forward procedure, after we have learned at which conditions it is preferable to carry out the measurement. Also elucidation of that part of the mechanism of the electrode process that is necessary to understand the essentials and to detect the optimum conditions for the measurement of half-wave potentials is usually simpler and less time-consuming than elucidation of the mechanism of a homogeneous chemical reaction.

Moreover, the problem, if all of the substances undergo reaction with the same mechanism in the reaction series chosen is simpler to solve in polarography than in homogeneous kinetics. The rather simple comparison of number, heights, shapes, and slopes of waves, often enable detection of the deviating compound.

Accuracy of the values of polarographic data is rather high. Most of the data given in the literature are reproducible to ± 0.01 V. or better, and under controlled conditions it is not difficult to achieve a reproducibility of ± 0.002 V. Because the average value of the reaction constant is about $\rho_{\pi,\mathrm{R}} = 0.3$ V., the difference between two half-wave potentials of compounds for which substituent constants differ by 0.033 units or 0.0066 σ units under controlled conditions is significant. As some polarographic reductions, in particular those in the aliphatic series, correspond to reaction constants of the order of units, the resolving power of polarography can be even higher. For these reasons, polarography seems to be a useful method for the determination of accurate values of substituent constants, but this possibility has been exploited only in a few cases (150) so far. For determination of substituent constants, the reaction series with medium $\rho_{\pi,\mathrm{R}}$-values seem to be the most useful, as with large values of $\rho_{\pi,\mathrm{R}}$, the uncertainty of the values of σ-constants is unnecessarily large. Polarographic measurements seem to provide a promising field of applications in such cases, in which small differences of the measured values are of importance, e.g., in σ_m–σ_p relationships, in the study of hyperconjugation, etc.

Finally, polarographic measurements can be carried out under varying conditions, and usually it is not necessary to change the techniques of measurement when changing reaction conditions, such as pH, ionic strength, temperature, dielectric constant, and the kind of solvent used. Whereas in reaction kinetics one and the same technique for determining the rate constant cannot usually be used in a temperature range broader than some tens of centigrades, polarography in a proper solvent would, in principle, provide determinations of the half-wave potentials (equivalent to rate parameters) in solutions with temperature varying over one hundred or more degrees. Even when nonpolar solvents, such as hydrocarbons, can only be used with difficulty, but mixtures of an alcohol and benzene can be used successfully, a rather broad range of solvents can be used and compared. The limiting factor can be a change in mechanism when comparing solvents of different character. Difficulties involved in the change of the potential of the reference electrode and of the diffusion potential can be circumvented by measuring the changes of differences in potentials. These differences can be measured either against one member of the reaction series studied or against a pilot ion, such as some naphthoquinone derivative, thallium, or potassium ion.

It appears advantageous to use polarographic measurements for the study of environmental effects, particularly on the value of the reaction constant.

X. Polarography and Kinetics of Organic Reactions

Polarography offers, in addition to information about reactivity in electrochemical processes that can be obtained from the measurement of half-wave potentials, data on reactivities in chemical processes. Polarographically, it is possible to determine rate constants of two types of reactions, viz., fast chemical reactions that restore the equilibria perturbed by the electrolysis and that take place in the vicinity of the electrode surface, and slower chemical reactions that occur in the bulk of the solution.

A. FAST REACTIONS AT THE ELECTRODE SURFACE

The rates of fast chemical reactions preceding, interposed, or successive to the electrode process proper, can be measured from the ratio of the kinetic current, limited by such reactions, and the diffusion-controlled limiting current. This ratio usually depends on further parameters, in particular, the drop-time and components of the supporting electrolyte. The treatment used for evaluating the rate constant depends on the system in question, but for all systems that have been studied so far, it is necessary to determine the value of the equilibrium constant (the treatment is derived only for reactions involving equilibrium) by an independent method and to measure the drop-time. Whereas the latter is rather simple, methods for the study of equilibria may be laborious or are not described at all. For substances that can undergo several successive equilibria, e.g., several successive protonations, it is sometimes not easy to decide to which equilibrium the kinetic current corresponds. In some cases the equilibrium constants can be determined by using single-sweep methods.

Demonstration of the treatment will be restricted to the simple case of acid–base equilibria of a monobasic acid of the type given in eq. (25). The value of the rate constant, k_r, can be computed from the ratio $i_{HA}/(i_{HA} + i_A)$ for any value of the hydrogen ion concentration using eq. (26). It is convenient to carry out the calculation by determining the pH value at which $i_{HA} = i_A$, i.e., at which $i_{HA} = i_d/2$. This pH value is set equal to pK' and the value of the rate constant is computed using eq. (92):

$$\log k_r = 2\, pK' - pK_a - \log t_1 - 2 \log 0.886 \tag{92}$$

In a similar way, using the appropriate equation, it is possible to compute the rate constants for other types of chemical reactions accompanying the electrode process as well.

The values of these rate constants, in addition to purely chemical complications such as the effects of general catalysis, tautomeric equilibria, solvation phenomena, etc., can be affected by the effects of the electrical field of the electrode on the diffusion, rate, and equilibrium constants of both the reactions involving the electroactive particle and the buffer, or by adsorption of one or both forms of the electroactive species of the buffer components. Hence, it is essential for the study of fast reactions to compare the computed rate constants with the results obtained by other techniques.

B. REACTIONS IN THE BULK OF THE SOLUTION

The chemical reactions occurring in homogeneous solutions can be advantageously followed using the polarographic method from the change of limiting currents with time when at least one component of the reaction mixture is electroactive and gives a measurable polarographic wave. As has been mentioned earlier (Sec. VI-C) for the special case of photochemical reactions, various techniques can be used for the measurement of polarographic currents according to the half-time $\tau_{1/2}$ of the reaction studied. For $\tau_{1/2} > 15$ sec., it is most useful to use mean polarographic limiting currents. For halftimes of the order of seconds, measurement of the peak height using the single-sweep technique proved well. For even faster reactions, instantaneous currents are recorded: for $\tau_{1/2} > 0.15$ sec. with a string galvanometer and for $\tau_{1/2} > 0.0015$ sec. with an oscillograph. For slow reactions with half-times greater than 3 hr., many of the advantages of the polarographic analysis in kinetic measurements are lost.

The limiting currents measured in kinetic studies should be, if possible, linearly proportional to the concentration of the electroactive compound. Diffusion-controlled currents are most frequently used for these purposes, but principally, the application of other types of polarographic currents e.g., kinetic or catalytic, is also possible.

When two or more components are polarographically active, it is usually necessary that their waves be well separated. Sometimes, it is, nevertheless, possible to apply the polarographic method even in cases when the half-wave potentials of waves of reactants and products (or intermediates) differ too little to secure the formation of two separated waves. Such application is possible when the wave heights of the reactant and product differ sufficiently in height under chosen conditions. Differences in wave heights can be caused either by the fact that the number of electrons consumed in the reduction of the reactants differs from that consumed in the reduction of the product [e.g., by hydrolysis of acyl derivatives of nitrophenol and nitroaniline (192)], or by the different diffusion coefficients of the reactant and product, [e.g., in 3,5-dinitrobenzoic acid and its sterol ester (193)] or in the character of the limiting current, which can be, e.g., used when reactant or product give a much smaller kinetic current than the current of the other component of the reaction mixture. An example of application of the latter type is the polarographic study of pyridoxal-5-phosphate hydrolysis (194). In the range between pH 2 and 5 pyridoxal-5-phosphate is reduced in a diffusion-controlled 2-electron step. The resulting pyridoxal gives, on the other hand, under these conditions only a very low kinetic wave, limited by the rate of dehydration of the aldehydic group or by the rate of opening of the hemiacetal ring. During hydrolysis, the limiting current decreases from the original height, limited by diffusion, to some 20–30% of the original height and is then limited by the antecedent reaction in the vicinity of the electrode surface. This decrease has been used in the computation of the rate constants of hydrolysis.

Changes of the heights of polarographic waves can be recorded in three different ways. For reactions with half-times $\tau_{1/2}$ between 15 sec. and 5 min. and sometimes slower, when the polarographic equipment used allows a slow shift of the recording paper, the applied voltage is kept constant at a chosen value usually corresponding to the potential range of the limiting current of the particular reactant or product. At this potential, change of the current with time can be recorded continuously. Markings are registered after chosen time intervals, but because the shift of the paper in the polarographic recording system is usually regular, it is often sufficient to measure the shift of the paper within a given time unit.

For slower reactions with 3 min. $< \tau_{1/2} <$ 60 min., it is usually sufficient to record the current at a chosen applied voltage only during short periods of time, near the time intervals in which the concentration is to be determined. The arrangement is principally the same as that used for faster reactions, and the only difference is that the shift of the paper is not continuous and is started only a short time before the measuring interval. When both methods which record the current at only one selected potential are used, it is recommended that a whole polarographic i-E curve be recorded after the reaction has finished to confirm on one side that the potential chosen really corresponds to the limiting current and, on the other side, that no unexpected reactions take place, as indicated by the occurrence of additional waves on polarographic curves. To choose a proper applied voltage for the recording of the current–time curves, it is usually necessary in preliminary experiments to record whole i-E curves at chosen time intervals, as described in the next polarograph.

For reactions with half-times 20 min. $< \tau_{1/2} <$ 3 hr., usually the recording of whole polarographic i-E curves proves best. These curves are usually recorded at a scanning rate of about 100–200 mV./min. after chosen time intervals. Either the time at the beginning of the recording is registered and the time necessary for the shift of paper between the starting point and that corresponding to a voltage of the region in which the limiting current is measured (determined once) is added to the starting time, or the time when the curve reaches the limiting current, or generally a predetermined potential, is registered. With these slower reactions, the precision of the measurement of time intervals that can be achieved in this way is sufficient in most instances. This technique allows us to detect all of the changes which occur on polarographic curves.

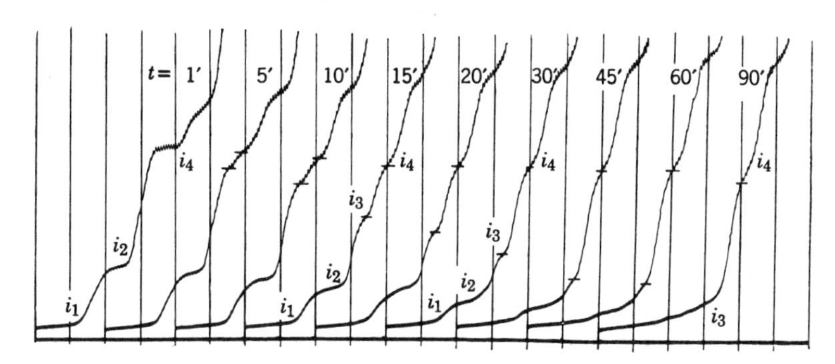

Fig. 22. Reaction of chalcone with hydroxyl ions. $2.10^{-4}M$ chalcone in $2M$ NaOH with 20% ethanol. Recording started after times given in the polarogram at −0.6 v., S.C.E., 200 mV./absc., 400 mV./min., h = 77 cm. full scale sensitivity 3.5 μamp. (195).

The advantages of the application of the polarographic method in reaction kinetics can be summarized as follows: Concentration changes of one or more components can be measured continuously and directly in the reaction mixture. Very dilute solutions can be followed, and it is sometimes possible to detect, identify, and follow the intermediates.

The dependence of current or concentration, on time during the kinetic run can be recorded automatically. The reaction mixture can often be placed directly into the polarographic cell. The presence of buffers, neutral salts, colored material, some solvents, soluble polymers, and other high molecular weight substances usually affect the measured currents only slightly and hence do not usually interfere. When it is necessary to perform the reaction outside the polarographic cell, e.g., in sealed ampoules or irradiated quartz vessels, it is usually rather simple to add the aliquot to a proper supporting electrolyte and to record the curve after irradiation. Furthermore, it is sometimes possible to measure polarographically the concentration change for more than one component of the reaction mixture at the same time. This can be achieved most simply when the last of the aforementioned techniques is used and the whole polarographic *i-E* curve can be recorded. The heights of various separated waves can be measured and their time changes evaluated. Sometimes, it is possible, nevertheless, to measure the time changes of the limiting currents at two chosen potentials, but two or more recording systems are then necessary, one for each applied voltage. A particularly advantageous example of the former type of measurement is the alkaline cleavage of chalcone (195), in which the time changes in separate waves of the reactant (chalcone), intermediate (ketol) and products (benzaldehyde and acetophenone) can be followed (Fig. 22). Even if not all applications of polarography in the study of kinetic problems are so successful by far, it, nevertheless, demonstrates well the possibilities that this method offers.

Polarographically it is also possible in some cases to detect and indicate the intermediates. Because the chemical reactivity often parallels the electrochemical activity, the intermediates often undergo, when stable enough, polarographic electrolysis and show a wave on polarographic curves. When this wave is separated from those of reactants and products, its presence can indicate formation of the intermediate. An example of this type is the detection of ketol formation (195) in the alkaline chalcone cleavage mentioned above. The character of the wave and its dependence on the composition of the supporting electrolyte can give us the first indication concerning the nature of the intermediate. Furthermore, polarographic curves recorded in reaction mixtures of varying composition can provide us with information on optimum conditions for formation of the intermediate. This will be demonstrated with the example of the behavior of tropenone-3-methoiodide (196). In buffers of pH about 9, the wave of tropenone-3-methoiodide decreases with time and a new wave of the intermediate (**1**) at more positive potentials is formed, but simultaneously the two waves of tropone, which is the final product, are formed. Hence in eq. (93), the velocities v_1 and v_2 are comparable.

$$\text{Tropenone-3 methiodide } (\overset{\oplus}{N}(CH_3)_2) \xrightarrow{v_1} \underset{(\mathbf{1})}{\text{intermediate with } N(CH_3)_2} \xrightarrow{v_2} \text{tropone} + NH(CH_3)_2 \quad (93)$$

When tropenone-3 methiodide is left to react at pH < 8, the wave of intermediate **1** is very small. The velocity $v_2 > v_1$ and the conditions are unsuitable for the formation of the intermediate. Finally at pH > 10, oppositely, the wave of the tropenone-3 methoiodide disappears almost instantaneously, and the wave of the intermediate **1** achieves its maximum height and only rather slowly decreases producing waves of tropone. Hence after transferring some tropenone-3 methoiodide into $0.02M$ sodium carbonate and acidifying after some 30 sec., the intermediate **1** can be isolated with a yield of over 80%.

Finally, the sensitivity of the polarographic method which allows kinetic measurement to be made in 10^{-3} to $10^{-5}M$ solutions and if necessary, also in small volume, offers two more advantages. First, in these dilute solutions, some reactions are not complicated by consecutive and side reactions of higher order. Hence the reaction can be simpler to study at these low concentrations than in less dilute solutions. Second, the amount of an organic substance needed for an exhaustive study, in which often hundreds of concentration–time curves ought to be measured under different conditions to make the elucidation of the mechanism possible, is substantially smaller when the polarographic method is used than when using classical analytical methods. This can be demonstrated by the oxidation of diols with periodic acid. When the course of this reaction is followed by an iodimetric titration, it is possible with 20 mg of the diol to carry out nine titrations, just enough to construct one concentration–time curve. By applying polarography (197), it is possible with the same quantity of sample to record 100 complete concentration–time curves when working with 10 ml. of the solution or even 500 curves when the volume is reduced to 2 ml.

Hence the only serious limitation of polarography as an analytical tool in reaction kinetics is usually the condition of polarographic activity of one of the components.

XI. Conclusions

By adopting the idea that organic electrode processes, when properly treated and cleared from complications implied by their heterogeneity, are only a special case of organic reactions, both polarography and physical organic chemistry can gain. Polarography gains considerable support in the elucidation of mechanisms of electrode processes; physical organic chemistry gains information about a large group of reactions of various reactants with one single reagent the electron or electrode. A more systematic comparison of the oxidation–reduction processes at the electrode with oxidation–reduction processes and other types of reactions, such as substitutions or eliminations in solutions, could bring broader insight into relationships of these reactions. It is to be hoped that polarography, both as a tool for evaluating the reactivity of organic compounds and as an analytical tool for studying reaction rates, will become more widely adopted in physical organic chemistry.

References

1. Perrin, C. L., in *Progress in Physical Organic Chemistry*, Vol. 3, S. G. Cohen, A. Streitwieser, and R. W. Taft, Eds., Interscience, New York, 1965, p. 165.
2. Zuman, P., in *Advances in Analytical Chemistry and Instrumentation*, Vol. 2, C. N. Reilly, Ed., Interscience, New York, 1963, p. 219.
3. Zuman, P., *Organic Polarographic Analysis*, Pergamon, London, 1964.
4. Kalousek, M., *Collection Czech. Chem. Commun.*, *13*, 105 (1948).
5. Rálek, M., and L. Novák, *Collection Czech. Chem. Commun.*, *21*, 248 (1956).
6. Kemula, W., *Advances in Polarography*, Vol. I, Pergamon, London, 1960, p. 135.
7. Říha, J., in *Progress in Polarography*, Vol. 2, P. Zuman and I. M. Kolthoff, Eds., Interscience, 1962, p. 383.
8. Kemula, W., and Z. Kublik, *Advances in Analytical Chemistry and Instrumentation*, Vol. 2, C. N. Reilly, Ed., Interscience, New York, 1963, p. 123.
9. Vogel, J., in ref. 7, p. 429.
10. Kalvoda, R., ref. 7, p. 449.
11. Heyrovský, J., and D. Ilkovič, *Collection Czech. Chem. Commun.*, *7*, 198 (1935).
12. Breyer, B., and H. H. Bauer, *Alternating Current Polarography and Tensammetry*, Interscience, New York, 1963.
13. G. C. Barker, ref. 7, p. 411.
14. Kůta, J., and I. Smoler, ref. 7, Vol. 1, p. 43.
15. Zuman, P., and S. Tang, *Collection Czech. Chem. Commun.*, *28*, 829 (1963).
16. Saveant, J. M., *Compt. Rend.*, *257*, 447 (1963).
17. Koutecký, J., *Collection Czech. Chem. Commun.*, *18*, 597 (1953).
18. Brdička, R., *Collection Czech. Chem. Commun.*, *12*, 212 (1947).
19. Volke, J., and V. Volková, *Collection Czech. Chem. Commun.*, *20*, 1332 (1955).
20. Volke, J., *Collection Czech. Chem. Commun.*, *22*, 1777 (1957).

21. Majranovskij, S. G., and V. N. Pavlov, *Z. Fizh. Khim.*, *38*, 1804 (1964).
22. Hanuš, V., and R. Brdička, *Chem. Listy*, *44*, 291 (1950).
23. Koutecký, J., *Collection Czech. Chem. Commun.*, *19*, 1093 (1954).
24. Ryvolová, A., and V. Hanuš, *Collection Czech. Chem. Commun.*, *21*, 853 (1956).
25. Manoušek, O., and P. Zuman, *Collection Czech. Chem. Commun.*, *29*, 1432 (1964).
26. Rosenthal, I., C. H. Albright, and P. J. Elving, *J. Electrochem. Soc.*, *99*, 227 (1952).
27. Zuman, P., and L. Turcznányi, unpublished results.
28. Majranovskij, S. G., V. M. Belikov, C. B. Korčemnaja, V. A. Klimova, and S. S. Novikov, *Izv. Akad. Nauk SSSR, Otd. Khim. Nauk.*, *1960*, 1675, 1787; *1962*, 605.
29. Turyan, Ya. I., Yu. M. Tyushin, and P. M. Zajcev, *Dokl. Akad. Nauk. SSSR*, *134*, 850 (1960); *Kinetika Kataliz.*, *4*, 534 (1963).
30. Kucharzyk, N., M. Adamovský, V. Horák, and P. Zuman, *J. Electroanal. Chem.*, *10*, 503 (1965).
31. Tirouflet, J., and E. Laviron, *Ric. Sci. 29: Contrib. Teor. Sper. Polarografia*, *4*, 189 (1959).
32. Volke, J., *Z. Physik. Chem.*, Leipzig, Sonderheft, 1958, p. 268.
33. Veselý, K., and R. Brdička, *Collection Czech. Chem. Commun.*, *12*, *313* 1947.
34. Jencks, W. P., *Progress in Physical Organic Chemistry*, Vol. 2, S. G. Cohen, A. Streitweiser, and R. W. Taft, Eds., Interscience, New York, 1964, p. 63.
35. Kůta, J., *Collection Czech. Chem. Commun.*, *24*, 2532 (1959).
36. Zuman, P., *Collection Czech. Chem. Commun.*, *15*, 839 (1950).
37. Tirouflet, J., and A. Corvaisier, *Bull. Soc. Chim. France*, *1962*, 540.
38. Zuman, P., and I. Šestáková, unpublished results.
39. Zuman, P., and V. Horák, *Collection Czech. Chem. Commun.*, *27*, 187 (1962).
40. Hoijtink, G. J., *Ric. Sci. Suppl. 30, Contrib. Teor. Sper. Polarografia*, *5*, 217 (1960).
41. Zuman, P., unpublished results.
42. Ryvolová, A., *Collection Czech. Chem. Commun.*, *25*, 420 (1960).
43. Stočesová, D., *Collection Czech. Chem. Commun.*, *14*, 615 (1949).
44. Vertyulina, L. N., and N. I. Malyugina, *Žh. Obshch. Khim.*, *28*, 304 (1958).
45. Testa, A. C., and W. H. Reinmuth, *J. Am. Chem. Soc.*, *83*, 784 (1961).
46. Holleck, L., and R. Schindler, *Z. Elektrochem.*, *60*, 1138 (1960).
47. Suzuki, M., *J. Electrochem. Soc. Japan*, *22*, 112 (1954).
48. Tachi, I., and M. Senda, *Advances in Polarography*, Vol. II, Pergamon, London, 1960, p. 454.
49. Alberts, G. S., and I. Shain, *Anal. Chem.*, *35*, 1859 (1963).
50. Lund, H., *Acta Chem. Scand.*, *14*, 1927 (1960).
51. Zuman, P., and V. Horák, *Collection Czech. Chem. Commun.*, *26*, 176 (1961).
52. Saveant, J. M., *Compt. Rend.*, *257*, 448 (1963); *258*, 585 (1964).
53. Micheel, F., and E. Heiskel, *Chem. Ber.*, *94*, 143 (1961).
54. Coombs, D. M., and L. L. Leveson, *Anal. Chim. Acta*, *30*, 209 (1964).
55. Delaroff, V., M. Bolla, and M. Legrand, *Bull. Soc. Chim. France*, *1961*, 1912.

56. Zuman, P., and J. Michl, *Nature, 192*, 655 (1961).
57. Pasternak, R., *Helv. Chim. Acta, 31*, 573 (1948).
58. Prévost, C., P. Souchay, and C. Malen, *Bull. Soc. Chim. France, 20*, 78 (1953).
59. Holubek, J., and J. Volke, *Collection Czech. Chem. Commun., 25*, 3293 (1960).
60. Kargin, J., and P. Zuman, unpublished results.
61. Cattaneo, C., and G. Sartori, *Gazz. Chim. Ital., 72*, 351 (1942).
62. Vavřín, Z., *Collection Czech. Chem. Commun., 14*, 367 (1949).
63. Brdička, R., and P. Zuman, *Collection Czech. Chem. Commun., 15*, 776 (1950).
64. Kern, D. M. H., *J. Am. Chem. Soc., 75*, 2473 (1953); *76*, 1011 (1954).
65. Koutecký, J., *Collection Czech. Chem. Commun., 20*, 116 (1955).
66. Ono, S., M. Takagi, and T. Wasa, *J. Am. Chem. Soc., 75*, 4369 (1953).
67. Hanuš, V., *Chem. Zvesti, 8*, 702 (1954).
68. Majranovskij, S. G., *Izv. Akad. Nauk SSSR, Otdel. Khim. Nauk*, 1961, 2140.
69. Zuman, P., J. Chodkowski, and F. Šantavý, *Collection Czech. Chem. Commun., 26*, 380 (1961).
70. Brdička, R., *Z. Elektrochem., 47*, 314 (1941).
71. Brdička, R., M. Březina, and V. Kalous, *Talanta, 12*, 1149 (1965).
72. Zuman, P., and R. Zumanová, *Collection Czech. Chem. Commun., 21*, 123 (1956).
73. Ždanov, S. I., and P. Zuman, *Collection Czech. Chem. Commun., 29*, 960 (1964).
74. Vopička, E., *Collection Czech. Chem. Commun., 8*, 349 (1936).
75. Zuman, P., *Proc. First Intern. Polarog. Congr. Prague, 1951*, Part III, p. 145 (1952).
76. Lothe, J. J., and L. B. Rogers, *J. Electrochem. Soc., 101*, 258 (1954).
77. Levin, E. S., and Z. I. Fodiman, *Zh. Fiz. Khim., 28*, 601 (1954).
78. Hummelstedt, L. E. I., and L. B. Rogers, *J. Electrochem. Soc., 106*, 248 (1959).
79. Reinmuth, W. H., L. B. Rogers, and L. E. I. Hummestedt, *J. Am. Chem. Soc., 81*, 2947 (1959).
80. Majranovskij, S. G., V. A. Ponomarenko, N. V. Barashkova, and A. D. Sneygova, *Dokl. Akad. Nauk. SSSR, 134*, 387 (1960).
81. Ershler, A. B., unpublished results, according to a private communication by A. N. Frumkin.
82. Majranovskij, S. G., *J. Electroanal. Chem., 4*, 166 (1962).
83. Peover, M. E., and J. D. Davies, *J. Electroanal. Chem., 6*, 46 (1963).
84. Holleck, L., and D. Becher, *J. Electroanal. Chem., 4*, 321 (1962).
85. Zuman, P., *Collection Czech. Chem. Commun., 25*, 3225 (1960).
86. Schwabe, K., ref. 7, Vol. 1, p. 333.
87. Cisak, A., *Roczniki Chem., 36*, 1895 (1962).
88. Majranovskij, S. G., N. V. Barashkova, and Yu. B. Volkestein, *Electrokhim, 1*, 72 (1965).

89. Majranovskij, S. G., *Talanta, 12,* 1299 (1965).
90. Holleck, L., and B. Kastening, *Rev. Polarog., Japan, 11,* 129 (1963).
91. Berg, H., and F. A. Gollmick, *Collection Czech. Chem. Commun., 30,* 4192 (1965).
92. Berg, H., *Abhandl. Deut. Akad. Wiss., Kl. Chem. Geol., Biol., 1,* 128 (1964).
93. Hills, G. J., *Talanta, 12,* 1317 (1965).
94. Adams, R. N., *J. Electroanal. Chem., 8,* 151 (1964).
95. Geske, D. H., in *Progress in Physical Organic Chemistry,* Vol. 4, A. Streitwieser, Jr. and R. W. Taft, Eds., Interscience, New York, 1967, p. 125.
96. Austen, D. E. G., P. H. Given, D. J. E. Ingram, and M. E. Peover, *Nature, 182,* 1784 (1958).
97. Maki, A. H., and D. H. Geske, *J. Chem. Phys., 30,* 1356 (1959).
98. Rieger, P. H., I. Bernal, W. H. Reinmuth, and G. K. Fraenkel, *J. Am. Chem. Soc., 85,* 683 (1963).
99. Kastening, B., *Electrochim. Acta, 9,* 241 (1964).
100. Stradinš, J. P., *Electrochim. Acta, 9,* 711 (1964).
101. Broadbent, A. D., and H. C. Zollinger, *Helv. Chim. Acta, 47,* 2140 (1964).
102. Rinkel, H., and W. Windisch, *Z. Physikal. Chem., 227,* 281 (1964).
103. Fujinaga, T., Y. Deguchi, and K. Umemoto, *Bull. Chem. Soc. Japan, 37,* 822 (1964).
104. Hoffmann, A. K., W. G. Hodgson, D. L. Maricle, and W. H. Jura, *J. Am. Chem. Soc., 86,* 631 (1964).
105. Bernal, I., and G. K. Fraenkel, *J. Am. Chem. Soc., 86,* 1671 (1964).
106. Chambers, J. Q., III, and R. N. Adams, *J. Electroanal. Chem., 9,* 400 (1965).
107. Kastening, B., *Collection Czech. Chem. Commun., 30,* 4033 (1965).
108. Janata, J., and O. Schmidt, *J. Electroanal. Chem., 11,* 224 (1966).
109. Janata, J., O. Schmidt, and P. Zuman, *Collection Czech. Chem. Commun., 31,* 2344 (1966).
110. Levich, V. G., *Acta Physicochim. USSR, 17,* 257 (1942); *19,* 117, 133 (1944).
111. Frumkin, A. N., L. N. Nekrasov, V. G. Levich, and Yu. B. Ivanov, *J. Electroanal. Chem., 1,* 84 (1959–60).
112. Wasserman, R. A., and W. C. Purdy, *J. Electroanal. Chem., 9,* 51 (1965).
113. (a) Heyrovský, J., and R. Kalvoda, *Oszillographische Polarographie mit Wechselstrom,* Akademie Verlag, Berlin, 1960.
(b) R. Kalvoda, *Techniques of Oscillographic Polarography,* Elsevier, Amsterdam, 1965.
114. Kemula, W., and Z. Kublik, *Roczniki Chem., 32,* 941 (1958).
115. Volke, J., *Chem. Zvesti, 14,* 807 (1960).
116. Zuman, P., *Collection Czech. Chem. Commun., 25,* 3245 (1960).
117. Petrů, F., *Collection Czech. Chem. Commun., 12,* 620 (1947).
118. Zuman, P., and O. Exner, *Collection Czech. Chem. Commun., 30,* 1832 (1965).
119. Horn, G., *Acta Chim. Acad. Sci. Hung., 27,* 123 (1961).
120. Zuman, P., *Collection Czech. Chem. Commun., 25,* 3252 (1960).
121. Levin, E. S., and Z. I. Fodiman, *Zh. Fiz. Khim., 28,* 701 (1954).
122. Elving, P. J., and C. L. Hilton, *J. Am. Chem. Soc., 74,* 3368 (1952).

123. Laviron, E., and P. Zuman, unpublished results.
124. Lund, H., paper delivered at the 19th International Congress of Pure and Applied Chemistry, London, 1963.
125. Manoušek, O., and P. Zuman, *Collection Czech. Chem. Commun.*, *29*, 1718 (1964).
126. Peizker, J., and O. Manoušek, personal communication.
127. Manoušek, O., and P. Zuman, *Chem. Commun.*, *8*, 158 (1965).
128. Kastening, B., and L. Holleck, *Talanta, 12*, 1259 (1965).
129. Manoušek, O., O. Exner, and P. Zuman, *Collection Czech. Chem. Commun.*, in press.
130. Manoušek, O., J. Krupička, J. Gut, and P. Zuman, unpublished results.
131. Bowers, R. C., and H. D. Russel, *Anal. Chem.*, *32*, 405 (1960).
132. Morris, M. D., P. S. McKinney, and E. C. Woodbury, *J. Electroanal. Chem.*, *10*, 85 (1965).
133. Gelb, R. I., and L. Meites, *J. Phys. Chem.*, *68*, 2599 (1964).
134. Lund, H., *Abhandl. Deutsch. Akad. Wiss.*, *Kl. Chem.*, *Geol. Biol.*, *1*, 434 (1964).
135. Wawzonek, S., and D. Wearring, *J. Am. Chem. Soc.*, *81*, 2067 (1959).
136. Wawzonek, S., and A. Gundersen, *J. Electrochem. Soc.*, *107*, 537 (1960).
137. Wawzonek, S., and A. Gundersen, *J. Electrochem. Soc.*, *111*, 324 (1964).
138. Wawzonek, S., R. Berkey, E. W. Blaha, and M. E. Runner, *J. Electrochem. Soc.*, *103*, 456 (1956).
139. Wawzonek, S., and R. C. Duty, *J. Electrochem. Soc.*, *108*, 1135 (1961).
140. Meites, L., and T. Meites, *Anal. Chem.*, *28*, 103 (1956).
141. Tanaka, N., T. Nozoe, T. Takamura, and S. Kitahara, *Bull. Chem. Soc. Japan*, *31*, 827 (1958).
142. Lund, H., paper read before the 19th International Congress of Pure and Applied Chemistry, London, 1963.
143. Kardos, A. M., P. Valenta, and J. Volke, *J. Electroanal. Chem.*, *12*, 84 (1966).
144. Spritzer, M., and L. Meites, *Anal. Chim. Acta*, *26*, 58 (1962).
145. Pflegel, P., O. Manoušek, and G. Wagner, private communication.
146. Manoušek, O., J. Krupička, J. Gut, and P. Zuman, unpublished results.
147. Elving, P. J., A. J. Martin, and I. Rosenthal, *J. Am. Chem. Soc.*, *77*, 5218 (1955).
148. Rosenthal, I., and P. J. Elving, *J. Am. Chem. Soc.*, *73*, 1880 (1951).
149. H. Lund, *Acta Chem. Scand.*, *11*, 283 (1957).
150. Zuman, P., *Substituents in Organic Polarography*, Plenum Press, New York, 1967.
151. Zuman, P., *Ric. Sci. 30*, *Contrib. Teor. Sper. Polarog.*, *5*, 229S (1960).
152. Zuman, P., *Z. Chem.*, *5*, 161 (1963).
153. Zuman, P., *Collection Czech. Chem. Commun.*, *27*, 630 (1962).
154. Zahradník, R., and C. Parkányi, *Talanta*, *12*, 1289 (1965), where earlier papers are quoted.
155. Taft, R. W. Jr., *Separation of Polar, Steric, and Resonance Effects in Reactivity in Steric Effects in Organic Chemistry*, M. S. Newman, Ed., Wiley, New York, 1956, pp. 556–675.

156. Leffler, J. E., and E. Grunwald, *Rates and Equilibria of Organic Reactions*, Wiley, New York, 1963.
157. Hammett, L. P., *Physical Organic Chemistry*, McGraw-Hill, New York, 1940.
158. Bennett, C. E., and P. J. Elving, *Collection Czech. Chem. Commun.*, *25*, 3213 (1960).
159. Philp, R. H., R. L. Flurry, and R. A. Day, Jr., *J. Electrochem. Soc.*, *111*, 328 (1964).
160. Klopman, G., *Helv. Chim. Acta*, *44*, 1908 (1961).
161. Kochi, J. K., *J. Am. Chem. Soc.*, *77*, 3208 (1955).
162. Zuman, P., *Collection Czech. Chem. Commun.*, *27*, 2035 (1962).
163. Miller, E. W., A. P. Arnold, and M. J. Astle, *J. Am. Chem. Soc.*, *70*, 3971 (1949).
164. Stewart, P. E., and W. A. Bonner, *Anal. Chem.*, *22*, 793 (1950).
165. Bachofner, H. E., F. M. Beringer, and L. Meites, *J. Am. Chem. Soc.*, *80*, 4274 (1958).
166. Zuman, P., *Chem. Listy*, *48*, 94 (1954).
167. Leonard, N. J., H. A. Laitinen, and E. H. Mottus, *J. Am. Chem. Soc.*, *75*, 3300 (1953).
168. Prevost, C., P. Souchay, and C. Malen, *Bull. Soc. Chim. France*, *1953*, 78.
169. Zuman, P., *Advances in Polarography*. Vol. 3, Pergamon, London, 1960, p. 812.
170. Lambert, F. L., and K. Kobayashi, *J. Am. Chem. Soc.*, *82*, 5326 (1960).
171. Závada, J., J. Krupička, and J. Sicher, *Collection Czech. Chem. Commun.*, *28*, 1664 (1963).
172. Krupička, J., J. Závada, and J. Sicher, *Collection Czech. Chem. Commun.*, *30*, 3570 (1965).
173. Sease, J. W., P. Chang, and J. L. Groth, *J. Am. Chem. Soc.*, *86*, 3154 (1964).
174. Lambert, F. L., A. H. Albert, and J. P. Hardy, *J. Am. Chem. Soc.*, *86*, 3154 (1964).
175. Zuman, P., *Talanta*, *12*, 1337 (1965).
176. Ingold, C. K., *Structure and Mechanism in Organic Chemistry*, Cornell University Press, Ithaca, 1953, p. 306.
177. Brown, H. C., R. S. Fletcher, and R. B. Johannsen, *J. Am. Chem. Soc.*, *73*, 212 (1951).
178. Brown, H. C., and M. Borkowski, *J. Am. Chem. Soc.*, *74*, 1894 (1952).
179. Roberts, J. D., and V. C. Chambers, *J. Am. Chem. Soc.*, *73*, 5034 (1951).
180. Van Straken, S. F., R. V. V. Nichols, and C. A. Winckler, *Can. J. Res.*, *29*, 372 (1951).
181. Fierens, P. J. C., and P. Verschelden, *Bull. Soc. Chim. Belges*, *61*, 427, 609 (1952).
182. Schotsman, L., P. J. C. Fierens, and T. Verlie, *Bull. Soc. Chim. Belges*, *68*, 580 (1959).
183. Hush, N. S., and K. B. Oldham, *J. Electroanal. Chem.*, *6*, 34 (1963).
184. Kabasakalian, P., and J. McGlotten, *Anal. Chem.*, *31*, 1091 (1959).

185. Prelog, V., and O. Haefliger, *Helv. Chim. Acta*, *32*, 2028 (1949).
186. Mühlstädt, M., and R. Herzschuh, *J. Prakt. Chem.*, *20*, 20 (1963).
187. Karplus, M., *J. Chem. Phys.*, *30*, 11 (1959).
188. Wilson, A. M., and N. L. Allinger, *J. Am. Chem. Soc.*, *83*, 1999 (1961).
189. Delaroff, V., M. Bolla, and M. Legrand, *Bull. Soc. Chim. France*, *1961*, 1912.
190. Zuman, P., *Rev. Polarography, Japan*, *11*, 102 (1963).
191. Hine, J., *J. Am. Chem. Soc.*, *81*, 1126 (1959).
192. Holleck, L., and G. Melkonian, *Naturwiss.*, *41*, 304 (1954); *Z. Elektrochem.*, *58*, 867 (1954).
193. Berg, H., and H. Venner, *Ric. Sci.* *29*, *Suppl.*, *4*, 181 (1957).
194. Zuman, P., and O. Manoušek, *Collection Czech. Chem. Commun.*, *26*, 2314 (1961).
195. Čárský, P., P. Zuman, and V. Horák, *Collection Czech. Chem. Commun.*, *30*, 4316 (1965).
196. Zuman, P., G. Fodor, and V. Horák, unpublished results.
197. Zuman, P., and J. Krupička, *Collection Czech. Chem. Commun.*, *23*, 598 (1958).
198. King, D. M., and A. J. Bard, *J. Am. Chem. Soc.*, *87*, 419 (1965).
199. Geske, D. H., and A. J. Bard, *J. Phys. Chem.*, *63*, 1057 (1959).
200. Bard, A. J., and E. Solon, *J. Phys. Chem.*, *67*, 2326 (1963).
201. Bard, A. J., and J. S. Mayel, *J. Phys. Chem.*, *66*, 2173 (1962).
202. Mayel, J. S., and A. J. Bard, *J. Am. Chem. Soc.*, *85*, 421 (1963).
203. Bard, A. J., and S. V. Tatwawadi, *J. Phys. Chem.*, *68*, 2676 (1964).